Hawks

Long-tailed Hawk • Dark Chanting-goshaw goshawk • Eastern Chanting-goshawk • Li Goshawk • Ovambo Sparrowhawk • Rufous- • Madagascar Sparrowhawk • Eurasian Sp Frances's Sparrowhawk • Levant Sparrowh • Chestnut-flanked Sparrowhawk • African Little Sparrowhawk • Red-thighed Sparrowhawk • Black Sparrowhawk • Henst's Goshawk • Northern Goshawk

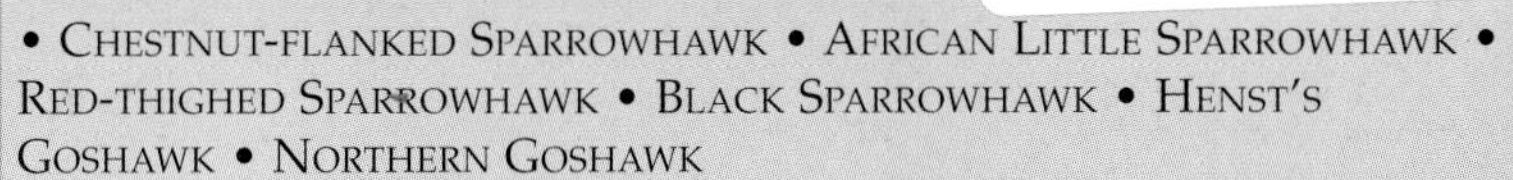

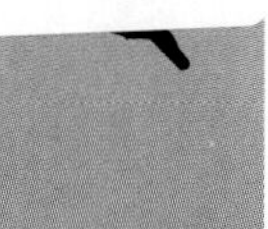

Kites

Black Kite • Red Kite • African Cuckoo-hawk • Madagascar Cuckoo-hawk • Black-shouldered Kite • African Swallow-tailed Kite • Bat Hawk

Falcons

Lanner Falcon • Peregrine Falcon • Saker Falcon • Eleonora's Falcon • Taita Falcon • Red-necked Falcon • Merlin • African Hobby • Eurasian Hobby • Amur Falcon • Red-footed Falcon • Grey Kestrel • Dickinson's Kestrel • Sooty Falcon • Banded Kestrel • Common Kestrel • Greater Kestrel • Fox Kestrel • Lesser Kestrel • Madagascar Kestrel • Mauritius Kestrel • Seychelles Kestrel • African Pygmy Falcon

Owls

Milky Eagle-owl • Cape Eagle-owl • Pharaoh Eagle-owl • Eurasian Eagle-owl • Spotted Eagle-owl • Akun Eagle-owl • Fraser's Eagle-owl • Pel's Fishing-owl • Vermiculated Fishing-owl • Rufous Fishing-owl • Marsh Owl • Short-eared Owl • Abyssinian Long-eared Owl • Madagascar Long-eared Owl • Eurasian Long-eared Owl • African Wood Owl • Madagascar Red Owl • Common Barn Owl • African Grass Owl • Maned Owl • Madagascar Hawk-owl • Hume's Owl • Tawny Owl • Red-chested Owlet • African Barred Owlet • Pearl-spotted Owlet • Chestnut-backed Owlet • Little Owl • White-faced Scops Owl • African Scops Owl • Sokoke Scops Owl • Eurasian Scops Owl • Pallid Scops Owl • Sandy Scops Owl • Madagascar Scops Owl • Pemba Scops Owl • Anjouan Scops Owl • Grande Comore Scops Owl • Seychelles Scops Owl • African Bay Owl • Shelley's Eagle-owl • Albertine Owlet • São Tomé Scops Owl

SASOL

BIRDS OF PREY

of Africa and its Islands

ALAN & MEG KEMP

ILLUSTRATED BY PETER HAYMAN

To Lucy and Justin, our brood, and to
Michael, Valeska, Jean, Harry and Ethel, our parents.

New Holland Publishers (UK) Ltd
London • Cape Town • Sydney • Singapore

24 Nutford Place
London W1H 6DQ
United Kingdom

80 McKenzie Street
Cape Town 8001
South Africa

14 Aquatic Drive
Frenchs Forest, NSW 2086
Australia

First published in 1998

Illustrations by Peter Hayman, with the exception of pages 275 (main), 281 (top two), 283 (top left & right), 289 (main), 317 (top two perched), 321 (perched), 323 (top), 325 (bottom left) which were illustrated by Ian Lewington (© D. & P. Chamberlain).

Project manager: Pippa Parker
Editor: Jenny Barrett
Cover design: Dominic Robson
Designers: Dean Pollard and Dominic Robson
Cartographers: Eloise Moss and Mark Seabrook

Reproduction by Hirt & Carter (Pty) Ltd, Cape Town
Printed and bound by Tien Wah Press (Pte.) Ltd, Singapore

2 4 6 8 10 9 7 5 3 1

ISBN 1 85974 100 2

Front cover: African Marsh Harrier – N. Dennis
Spine: Scops Owl – R. du Toit
Back cover: Verreaux's Eagle – A. Froneman
Title page: Eurasian Black Vulture – B. Lundberg/BIOS

SPONSOR'S FOREWORD

We live on a continent blessed not only with great natural beauty but also an unsurpassed diversity of animals and plants. Moreover, in among the region's fauna is a grouping of birds that captures the human imagination in a special way, for it embodies the essence of Africa. These are the birds of prey. The enchantment, the very spirit of the continent's wild places resides in the haunting call of the fish eagle, in the majesty of vultures painting lazy circles in the sky, in the acrobatic virtuosity of the bateleur, in the secretive ways of the nocturnal owl.

Sasol has embarked on a programme to sponsor natural history publications for a number of reasons, but most especially to encourage active and prospective birders to become better acquainted with our precious natural heritage. The funding helps make the books more affordable, exposing them to a wider readership, and to that extent Sasol is also promoting a more general awareness of the environment and the need to conserve it.

We are delighted with the role that *Sasol Birds of Southern Africa* (published in 1993) has played in environmental education and in popularizing birding as a pastime. We believe that *Sasol Birds of Prey of Africa and its Islands*, will prove a worthy companion volume, and that it will enhance even further the pleasures of birdwatching.

Pieter Cox
Managing Director, Sasol Limited

Sasol and the Environment

'Protecting the environment is an obligation, not a choice'

Sasol believes that the quality of the air, water and soil should be protected for the continued benefit of all ecosystems. In this way, the needs of present and future generations will be met, enabling them to live in an environment of acceptable quality.

Sasol's management is committed to acting responsibly towards the environment, and to considering the effects of Sasol's operations on the environment when taking decisions.

AUTHORS' ACKNOWLEDGEMENTS

Our knowledge of birds of prey and owls and their field identification has expanded with the help of many people over many years. Ian Sinclair, who initiated this book, has always encouraged us, allowed us to use plates from his own books, and has expertly checked our texts for those species which we have never actually seen. We would also like to thank, for the value we have found in their published works, and in our correspondence and informed discussions with them, the following: David Allan, George and Keith Begg, the late Richard Brooke and Leslie Brown, Tom Cade, Bill Clark, Peter Colston, Rob Davies, Bob Dowsett, Francoise Dowsett-Lemaire, Rudi Erasmus, Carl Jones, Clem Haagner, Rich Harland, Tony Harris, Ron Hartley, Kotie Herholdt, Josep del Hoyo, Andrew Jenkins, Tim Liversedge, Richard Liversidge, Gordon Maclean, Gerhard Malan, the late Peter Mendelsohn, John Mendelsohn, Bernd-Ulrich Meyburg, Petiros Mundhlovu, Peter Mundy, Ken Newman, Ian Newton, Chris Olwagen, Des Prout-Jones, the late David Reid-Henry, Rob Simmons, John Snelling, the late Walter Spofford, Peter Steyn, Warwick Tarboton, and Anthony van Zyl. We also thank all the photographers that submitted their pictures for consideration, and especially those authors and photographers whose published works are not referenced in a book of this kind but whose contributions were essential.

On the production side, Pippa Parker steered the project with a skilled hand, Jenny Barrett was a wonderfully critical editor, Dominic Robson and Dean Pollard designed the attractive and functional layout and made magical adjustments to the plates, Peter Hayman accurately painted the new artwork, Carmen Swanepoel found all the photographs and Laura Stratten sorted out the Afrikaans names. We are indebted to Sasol Limited for their generous sponsorship.

Alan and Meg Kemp

ARTIST'S ACKNOWLEDGEMENTS

I must express my sincere thanks to Dr Robert Prys-Jones and his staff at the British Museum, Tring, United Kingdom, for allowing me access to, and for their help in, examining the museum's skin and spirit collection; to Ian Dawson and Robert Hume for their help in sourcing references; and to Pippa Parker for guidance and encouragement.

Peter Hayman

CONTENTS

J. Knobel

Tawny Eagle

INTRODUCTION

The identification of birds of prey is one of the most exciting wildlife challenges on the African continent and its surrounding islands. The African savanna supports the highest diversity of birds of prey in the world, while other species are abundant on its steppes and in its rainforests. The islands adjacent to Africa have their own special fauna, much of it unique, though some species are related to mainland species, and others to more distant cousins in Asia.

Throughout the whole African region, birds of prey are among the most dominant birds, top of the food chain and often the first to signal problems in ecosystems under their domain. They are an obvious barometer of environmental health and for this reason alone they deserve our understanding. Many species remain poorly known, so learning to identify African birds of prey must be the first step in our awareness and further study.

Most people can instantly recognise birds of prey and owls by their hooked bill, strong feet and rapacious demeanour. Many birds of prey are striking in appearance, large and powerful, colourful and handsomely marked, or fast and agile in flight. They look as though they should be easy to identify, but this can be deceptive. Often small details of colour or nuances of proportion are necessary to confirm their identity, rather than the impressive initial appearance.

This book is intended to make the identification of African birds of prey easier, for both the beginner and the experienced observer. It does not gloss over the problems, but it does present the necessary facts in a condensed, accessible form and with the support of numerous illustrations. It guides the reader in how to break this wonderful diversity down into manageable groups. It then helps one to identify those details most important for the accurate identification of each species. Our hope is that out of recognition, and the knowledge that this conveys, will come appreciation and conservation.

THE MAIN GROUPS OF BIRDS OF PREY

The birds of prey of Africa can be divided into ten main groups for the purposes of field identification. These ten groups, described below, form the basis upon which the book has been arranged into chapters. The groupings are not strictly by taxonomy; in fact the placement of some species within a group is sometimes somewhat arbitrary in that regard. Rather, the birds are arranged within groups into approximately five orders of body size, from very large, through large, medium and small to very small; within each group, they are arranged primarily in groups of similar species, with larger species featured near the beginning and smaller species near the end of each group wherever possible. Some groups are easier to define than others, and each group differs in the number of species that it includes. When in the field, allocating any bird of prey you see to one of the ten groups will help narrow down the number of species you have to consider, and so help in the identification of the species.

Secretarybird (1 species) An unmistakeable, very large eagle-like bird with long legs, long centre tail feathers, a bare, orange face and long, floppy crest feathers. It walks on the ground for much of the time but also has a distinctive shape in flight. Endemic to Africa.

Vultures (11 species) are very large to large birds of prey, often with colourful areas on the head and neck. Most adopt the much-lampooned hunched silhouette when perched. They are often seen soaring in flight and most have long, broad wings with long primary 'fingers' at the end. They appear small-headed on the wing, compared to eagles, and flight identification relies mainly on wing shape and underwing pattern. All – even long-necked species – fly with the neck retracted. Since most feed on carrion, they have heavy bills which they use to tear up carcasses, and short, bare legs with long toes and stubby claws to hold down food. They often gather in numbers to feed, bathe and roost. Some of the species nest in loose colonies. A few are commensal with humans.

Bare-legged eagles (12 species) are an assortment of several groups of unrelated very large to large birds of prey. All have stout, bare legs and feet with long, recurved claws. They are usually seen alone or in pairs. Most have long, broad wings and often soar in flight with long primary 'fingers' at the wing tips. Several species are boldly marked and/or brightly coloured. The Osprey, a large, aberrant fishing-hawk is included here.

Booted eagles (17 species) are very large to medium-sized birds of prey. All have the legs feathered down to the feet, a feature not found in any other group except owls. The nape and crown feathers are often elongated. Most occur alone or in pairs, but migrant species sometimes congregate in loose flocks. They often soar in flight and have long, broad wings with obvious primary 'fingers' at the wing tips. Most are agile fliers and rapacious hunters. Some occur in dark and light phases.

Buzzards (10 species) are all medium-sized birds of prey, stockily built and with broad, rounded heads. They have short, stout, bare legs and feet. Their wings are broad and rounded, and most have short, often reddish, tails. They often soar in flight but hunt mostly from perches or by hanging in the wind. Most have a broad, dark trailing edge and wing tip. Several species have drab brown plumages and all show much individual variation in hue and pattern.

Harriers (9 species) are medium-sized, long-winged and long-tailed birds of prey. They usually fly low over the ground with alternating bouts of flapping and gliding. They are buoyant and light on the wing, making agile twisting dives at prey. They have long, thin, bare legs and feet for reaching out at prey, and an obvious owl-like facial disc, indicating good hearing ability. In most species, adult males and females are differently coloured. The larger harrier-hawks or gymnogenes, similar in overall design, are also included in this group.

Goshawks and sparrowhawks (20 species) range from large to very small short-winged hawks, most adults of which have fine barring on their underparts. At rest, the wing tips usually reach only a short way down the long tail. All have long, slender, bare legs and feet and a small, rounded head with a delicate bill. Their short, rounded wings and long tails enable them to manœuvre through the densely wooded habitat that most species prefer. Softpart colours, rump and tail patterns, and details of breast markings are especially important in identification. They are often secretive and difficult to observe. The Lizard Hawk, though sometimes known as the Lizard Buzzard, is placed among the goshawks and sparrowhawks, where it belongs.

Kites (7 species) are a varied collection of medium-sized to small birds of prey. Most have short, stocky, bare legs and toes, so often perch rather horizontally. In most species the eyebrow ridge is poorly developed, making them look less rapacious than other birds of prey. Most fly with the wings slightly bent, forming an angled carpal or 'wrist' joint, and most have long, rather pointed wings. Their tail dimensions are varied, being short, medium or long in size and forked or square in shape. They are usually solitary, but some species assemble in flocks, several roost communally and a few are commensal in numbers with humans.

Falcons and kestrels (23 species) are medium-sized to very small, long-winged birds of prey. They have a large, rounded head with a robust, parrot-like bill. The notches on the cutting edge of the bill that are characteristic of this group are difficult to detect in the field. Many have dark 'moustache' stripes on the cheeks. When perched, their pointed wings usually reach the tip of the tail. In flight, most are fast and agile with long, narrow wings and a long tail. Several kestrel species regularly hang and hover in flight. In some smaller species the sexes are differently coloured. Most falcons are found alone or in pairs, but some congregate at food sources, and migrant kestrels hunt and roost in large flocks. Like owls, they do not build their own nests but use natural ledges, cavities or old stick nests of other birds.

Owls (43 species) are large to very small birds of prey. Their large, forward-facing eyes are set in a well-defined facial disc. Many species also have ear-like tufts of feathers on the large, rounded head. These features, together with the nose-like bill, give owls the characteristic 'face' known to many people. Most owls have dumpy bodies, broad, rounded wings and a short tail. They are usually strictly nocturnal and fly silently on muffled wing feathers. All are vocal and readily identified by their calls. All lay white, rounded eggs but none build their own nest.

Taxonomic classification

The formal classification of birds of prey recognises two separate and unrelated orders. These are the diurnal birds of prey, the Falconiformes, and the nocturnal birds of prey, the Strigiformes. Each order includes a number of separate families, within each family are a number of genera, and each genus is made up of a number of species. The pair of scientific names given for each species comprises the generic and specific names.

Order Falconiformes

This is a disparate collection of diurnal birds of prey, grouped into four distinct families but with few overall traits in common. All have the typical short, hooked bill of birds of prey and most have strong, grasping feet with long, curved claws. They appear to have a common ancestor with waterbirds, in particular storks, as indicated by the anatomy and behaviour of America's turkey vultures and condors, and the Secretarybird of Africa.

Family Sagittariidae There is only one species in this distinctive African family, this being the Secretarybird which has the body and head of an eagle but long legs, nesting habits and behaviour that resemble those of storks.

African genus: *Sagittarius*.

Family Accipitridae This is the largest family of diurnal birds of prey. It includes species which range widely in body size and biology, from tiny sparrowhawks to huge eagles and vultures. Other groups in the family are goshawks, harriers, buzzards, kites, cuckoo-hawks and bathawks.

African genera: *Aviceda, Pernis, Macheiramphus, Elanus, Chelictinia, Milvus, Haliaeetus, Gypohierax, Gypaetus, Neophron, Gyps, Aegypius, Necrosyrtes, Torgos, Trigonoceps, Terathopius, Circaetus, Dyrotriorchis, Eutriorchis, Polyboroides, Circus, Micronisus, Kaupifalco, Melierax, Accipiter,*

Urotriorchis, Butastur, Buteo, Aquila, Hieraaetus, Lophaetus, Spizaetus, Stephanoaetus and *Polemaetus*.

Family Pandionidae There is also only one species, the Osprey, in this distinctive family. This medium-sized raptor occurs on all ice-free continents; in Africa it is mainly a migrant, entering the region from some of its breeding areas in Europe and around the northeast African coast. Ospreys are large hawks with various anatomical specialities for feeding on fish. Their closest relatives are among the Accipitridae.

African genus: *Pandion*.

Family Falconidae This is a clearly defined family with its own special moult pattern, eggshell colour and behaviour. In Africa, it comprises medium to small falcons, kestrels and one pygmy falcon. Although it shares several aspects of behaviour with owls, such as not building a nest, the two families are not usually considered to be related.

African genera: *Falco* and *Polihierax*.

Order Strigiformes

The members of this order, the owls, are easily recognised by their large, forward-facing eyes, well-developed facial disc and large ear openings. They share with diurnal birds of prey the hooked bill and strong grasping feet with curved claws. Other aspects of anatomy, such as the lack of a crop and an arch on the forearm bone, suggest a more distant relationship. None are nest-builders, and all lay white, rounded eggs. They fall into two families, both of which are represented in Africa. Their closest relatives are probably nightjars, another nocturnal order of birds.

Family Tytonidae A small group of barn, grass and bay owls. The heart-shaped facial disc, the claw on the centre toe with a comb-like edge, and the solid sheet of bone between the eyes are characteristic features. The bill projects downwards rather than forwards.

African genera: *Tyto* and *Phodilus*.

Family Strigidae The main family of 'typical' owls, many with ear-like tufts of feathers projecting from the head. The inner toe is shorter than the middle toe, the breast bone has two notches, and there is a thin or perforated bone between the eyes.

African genera: *Otus, Bubo, Scotopelia, Glaucidium, Athene, Strix* and *Asio*.

IDENTIFICATION CHALLENGES

Raptors and owls present some specific challenges for those wishing to identify or study them in the field. These problems, discussed below, need not be insurmountable, and some practical ways of addressing them are presented here.

Birds of prey are few and far between. Most birds of prey are solitary creatures which invariably occupy large home ranges. One may see a particular species only at long intervals; only rarely are similar species seen in close proximity, and this makes it difficult to develop and practise one's skills in field identification. To overcome this problem, *familiarise yourself with common or distinctive species in your area*. Make a list of the commonest species for each area that you visit and get to know their distinguishing features. Then, when a species you do not know makes its appearance, you will immediately recognise it as such and be able to compare it to the most similar of the species you already know. If you manage to identify a species, watch it for a while so that you gain a good impression of what it looks like in different light, as it flies away, soars overhead or performs some distinctive behaviour.

Good sightings are hard to come by. Birds of prey are frequently seen in flight, and owls by night; all are alert to and often shy of humans. This makes it difficult or impossible to see all the important identification features at each sighting. *Look carefully for those small details of colour or form that constitute key features.* Narrow down the number of species you think the bird in question could be and then remind yourself of their key identification features. *Look out for unusual or distinctive behaviour*, such as calls, courtship or hunting techniques. Make it easier for yourself with the best aids you can afford – binoculars, telescope, camera, video. Don't be too hard on yourself. Remember, anyone putting a name to more than 80 percent of the birds of prey they see is probably using some guesswork!

Size, shape and movement of the birds are difficult to assess. These subjective criteria are important in the identification of birds of prey, but they are difficult to judge in the field and to describe on paper. African birds of prey range in size from pygmy falcons and owlets at 60 g to eagles and vultures weighing in at 8 kg. They differ widely in the shape and relative size of their bills, necks, wings, legs and tails. Each has its own distinctive way of perching, walking and flying. In flight, the wings and tail can be held in a variety of positions. The only solution to this problem is to *practise!* Spend as much time as you can spare observing the birds in the field; develop the subjective skills to judge size, shape, and rate of movement. Practise ways of comparing species at different distances and of retaining a mental picture of them in your head. Make full use of opportunities when two or more species are together, so that you can compare and learn each one's shape and behaviour.

Most bird of prey species have at least two different plumages. Many birds of prey have different juvenile and adult plumages, in some the sexes are also differently coloured, others have different colour forms or morphs, and in a few there are even different intermediate plumages. All of these variations need to be considered when making an identification, including what they might look like when moulting from one plumage to another, or in rare cases of albinism. *Learn the basic differences* between juveniles and adults, and between males and females. List and learn those species with colour forms and variants, and match this knowledge with the size and shape of each species, which is often your initial point of departure.

There are so many species of birds of prey in Africa. There are roughly 110 species of diurnal birds of prey and 45 species of owl on Africa and its islands to distinguish from each other. Add to this the above-mentioned plumage variations within species, and the number of forms rises to well over 300. *Get to know which species should be in a particular area, in that habitat, at any given time of year.* When you see a bird of prey, first decide to which of the ten main groups given in this book the bird belongs and then exclude any species that is unlikely to be there. Once you have isolated the group to which the bird belongs, check the key identification features for the most similar species that remain on your list. Only if you are still unsure after that should you begin to think that you may have seen something new or unusual. Birds of prey are great fliers and many migrate over habitats they would not normally occupy, so they might turn up anywhere. Generally speaking, however, they are usually seen in the areas where we think they belong.

Tips to Identifying Birds of Prey

1. **Learn to distinguish juveniles from adults.** Juveniles usually have light buff tips to the feathers on the upperparts, giving a scaled or scalloped effect. They are usually browner than adults and their softpart colours are often different. Juvenile owls of several species leave the nest with the head and body covered in down but later, when they moult, the juvenile plumage is very like that of the adult.
2. **Be aware of subadults.** When birds are moulting from juvenile to adult plumage they have, for a time, a mixture of old and new feathers. This can produce some odd-looking birds that are not usually illustrated in any book. Try to imagine what a mixture of plumages might look like, but concentrate initially on features of size, shape and behaviour.
3. **Note softpart colours.** If you can, note the colours of the bill, cere, eyes, bare facial skin, legs, feet and claws. In most birds of prey the bill is black, the cere yellow, the eyes brown, the bare facial skin yellow, the legs and feet yellow, and the claws black – look out for and make special note of blue, green, red, orange, pink or white.
4. **Study the head and face.** Usually the head is one part that is clearly visible, because the bird wants to keep an eye on you! The proportions and details of the bill, cere, eyebrows, lores, eyes, bare facial skin, crown and nape are distinctive for each species. The face is especially useful for distinguishing species that are easily confused, such as the brown eagles, sparrowhawks or buzzards. Each species of bird of prey has a temperament of its own. Try in your mind to match the face with the temperament and habits. In some species, the sexes can also be distinguished by fine facial details of colour or proportions.
5. **Look out for larger females.** Female birds of prey are always larger than their male counterparts, only slightly so in some groups (vultures, kites, kestrels, owls) but markedly so in others (goshawks and sparrowhawks, eagles, falcons). In several groups, such as sparrowhawks and eagles, large females of one species may overlap in size with small males of another species.
6. **List and learn colour forms.** Take note of which species have dark and light forms (e.g. Booted and Wahlberg's eagles), melanistic forms (e.g. Gabar Goshawk and Ovambo Sparrowhawk), grey and rufous forms (several owls, including African Scops and Spotted Eagle owls), or marked individual variation in plumage colour and pattern (e.g. buzzards and brown eagles). Then, when you see an odd bird you will know which species to suspect.
7. **Bear in mind similar species.** Take note of those species, often unrelated to one another, which occupy the same habitat and range and which share similar plumage colours and patterns. Cuckoo-hawks and sparrowhawks, goshawks and hawk-eagles, Bat-Hawks and falcons are some examples.
8. **Note special behaviour.** Record any calls you hear, jot down details of hunting behaviour or prey, describe any courtship flights, and note the structure and placement of nests. Birds of prey lead full and active lives, many aspects of which are as useful for identification as are details of plumage colour and anatomy.
9. **Make note of habitat types.** Each species is found most commonly in a distinct habitat type. Forest, woodland, savanna, grassland, steppe and desert are the primary divisions. Within each habitat, each species has further preferences, depending on its special requirements for feeding and nesting sites. Some are generalists and widespread (e.g. Black-shouldered Kite, Black Kite, Barn Owl), while others are specialists and restricted in range (e.g. Taita Falcon, Bearded Vulture, various island species).
10. **Take your location into account.** Many species have limited or fragmented ranges, or are replaced by similar species in different parts of Africa. Verreaux's Eagle is absent from the hills of western Africa, despite the presence of its principal prey of hyrax. The long-eared owls of northern Africa, eastern Africa and Madagascar are different species. Knowing what to expect at each location can reduce your options for identification to about half the total number of species found on Africa and its islands.

IN THE FIELD

Birds of prey are usually difficult to find, therefore, you need to consider several techniques that might improve your chances. In whatever habitat you search for birds of prey, it is worth looking for good vantage points on a hilltop or cliff edge, and spending time there. Scan the surrounding area for birds of prey on obvious perches, and let your eye sweep across the skyline for birds on the wing. During the heat of the day, lie on your back and search the heavens for soaring birds high overhead. Don't be in too much of a hurry – birds of prey move around as they conduct their lives, and if you are patient (and lucky) they will come to you.

Always be alert to signs that a bird of prey is in the area. Keep an eye open for 'whitewash', the characteristic splash of droppings that accumulates below favourite perching, roosting and nesting sites. Also listen for calls, especially from forest birds of prey and from nestlings. Few diurnal birds of prey are very vocal, but the study of their calls as an aid to identification has been neglected and deserves further attention. Most are not as quiet as we think and many species have characteristic calls that can confirm their identification, especially species living in dense forest. Of course, calls are a primary method for locating and identifying owls in the dark of night.

Listen to animals that might warn you when a bird of prey is on the move – after all, they have a special interest in knowing the whereabouts of their main predators! Be alert to animals suddenly taking fright. Listen for cries of alarm from birds and mammals, several of which have special 'bird-of-prey' warning calls. Drongos, in particular, make a habit of chasing birds of prey, and even mimic the calls of their opponents, often giving away the presence of, say, a secretive sparrowhawk. Be aware of the mobbing behaviour of small birds, when a number of species gather round and scold a predator. This is often directed at a resting hawk or sleeping owl which you might otherwise overlook. Do take care as you approach, however, since snakes and even leopards can attract the same attention!

Try to be out early, as many birds of prey start hunting from first light. Stay out until at least mid-morning, when thermals of rising air become established, encouraging these birds to take to the wing and soar. Stay out the whole day if you really want to study an area. Midday is good for soaring birds, except over continuous forest canopy where thermals only rarely form. Then, in the late evening, many species hunt actively to fill their stomachs for the night ahead. At nightfall and during the following hours is also a good time to see owls emerge from their roosts and take up open perches. Thereafter, the night watch takes over from the day brigade.

Be conscious of the weather. Cold, cloudy days often keep the larger birds of prey grounded, making them more accessible for you to find. Feel the heat on sunny days and judge when thermals will start forming and birds will begin to soar. Listen to weather reports – the arrival of a weather front, with driving winds and rising air, may carry a group of birds of prey on the move, especially at the beginning and end of the migration season. Remember also that a mountain range will deflect strong winds to form 'standing waves', ridges of rising air that stretch many kilometres in the lee of the mountain, and are often used by soaring birds of prey. Learn to understand conditions in your area and head for the hills to see birds that might make use of the weather fronts.

Code of Ethics for Birders

Keep disturbance to a minimum. Birds of prey lead busy and demanding lives, and will not benefit from unnecessary disturbance. Approach a perched bird slowly and quietly, and try not to stand or park in the open and stare at it. Watch its movements for signs of restlessness, especially if it bobs its head, mutes, lowers a resting foot or rouses its feathers. Be especially careful if it is feeding – it has probably worked hard for its meal and needs those nutrients.

Stay well away from nesting areas. Be especially careful if you come across the nest of a bird of prey. The bird may continue to brood the eggs or chicks, but it is more likely to be frozen in fear at your approach than accepting of your presence. Some birds, such as the Bateleur, desert the nest for the least reason. The trail of your scent leading to ground, tree or cliff nests also immediately raises the chances of its discovery by a predator. Even if a bird dives and calls at you, it is probably more scared than aggressive, and that energy could be better spent on raising its offspring. Be especially careful of colonial breeders, such as some vultures and falcons, where the effect of your disturbance is multiplied by the number of breeding pairs present at the colony.

Respect the hunting needs of birds of prey. Predators need to kill prey to survive. This is the only lifestyle they know. Try not to come in front of a hunting bird or disturb its prey. Stay to one side and keep your distance. Do not interfere at the kill, which may be a tough struggle, and during which the bird of prey itself may be vulnerable to injury. Keep your feelings for the prey to yourself.

Be sensitive to territorial requirements. Birds of prey declare the boundaries of their territories by calls or by attack. Repeated visits to nest areas often lead to heightened aggression. Playing back of calls, especially for owls, causes these birds to call back and come up in defence. If you don't have the patience to listen for and find an owl, use call-up methods as sparingly as possible – a few phrases, at one locality, preferably early in the evening. Some owls spend days revisiting and reproclaiming their ownership, to the detriment of all their other commitments. Try using other call-up techniques or predator-callers, such as those that make sounds of prey, of other predators or of animals in distress. These are more natural in the daily lives of birds of prey and less threatening to their territorial ownership.

Use study techniques sparingly. Any interference with a bird of prey lowers its chances of survival to some degree. Use rings, wingtags, radio-transmitters, visits to nests, handling of chicks, even your presence as an observer, as sparingly as possible. Design your observations and studies with care. Be able to justify to yourself and to anyone else any disturbance you may impose on a bird of prey.

Travel as much as you can. An alternative to spending long periods at a lookout is to travel long distances in search of birds. This is especially good in open habitats, but of little use in dense woodland or forest. Direct your search to interfaces between habitat types, where hills meet flats or where dry bush adjoins riverine forest (as these ecotones often support a higher total number of species than do the surrounding habitats alone). Visit patches of special habitat: clumps of forest, exotic plantations, rocky outcrops, gorges or cliffs. Use roads lined by tall trees or utility poles that might be used as perches. Make enquiries about concentrations of potential food, such as the emergence of termite alates, swarms of locusts, mass roosts of queleas or swallows, nesting colonies of queleas or weavers, or plagues of rodents. Spend time around waterholes and rivers; birds of prey are often attracted to the prey animals which come to drink. If you do pass through forest, travel slowly and look out for birds of prey perched above the road, along paths or in clearings, on the lookout for potential prey items which may be crossing these open spaces.

Select particular habitats to survey, if you can. African birds of prey are most obvious, most abundant and most diverse in extensive areas of wooded savanna. In the best national parks and game reserves, on a good day, one might encounter 25–30 species. The species total rises if the habitats are diversified by the presence of cliffs, riverine forest and wetlands. On the other hand, forests and woodlands remain the most difficult habitats in which to search. But often they offer special rewards of rare and interesting species.

Do not despair if you cannot reach pristine habitats. Birds of prey are also mobile and adaptable. They will make use of plantations of exotic trees, quarries, man-made lakes, even cliff-like buildings within cities. Sparrowhawks regularly sneak about in suburbia, while owls and occasionally even Bat Hawks hunt under street lights. Many species can be seen hunting along road verges or using utility poles as perches. A growing number of species build their nests on electricity pylons, and a number of species nest within city limits.

Aim to discover nest sites, but when you do, ensure that you keep disturbance in such areas to a minimum. Birds of prey often nest year after year in the same general area. Nests of many species are easy to spot and activity is often concentrated in the nest area. To find nests, search below cliffs and large trees for tell-tale signs, such as remains of prey or regurgitated pellets of indigestible bones, fur, feathers and scales; keep an eye out for moulted feathers, or for pluckings of fur and feathers torn off from prey; note moulted downy breast feathers, especially for owls, which are dropped to expose the brood patch, and indicate a nearby nest. Also be on the lookout for birds of prey carrying food in flight back to the female or chicks at the nest. Often this shows as a lump in the feet just under the tail, but some species, such as snake-eagles, carry more obviously with a snake's tail dangling from the bill. Again, be cautious and careful around the nest – don't rush in and disturb a meal.

AFRICAN HABITATS

There are eight major habitat zones which are described and mapped here. On the ground, most of these zones merge gradually into one another, rather than changing abruptly, so that the dividing lines between them are somewhat arbitrary. Furthermore, each zone includes small areas of different habitat or topography that cannot be illustrated at this scale, but which are important to the distribution of several raptor species. These include such features as swamps, marshes, rivers, lakes, rocky hills, cliffs, gorges or patches of forest, woodland or palms. In the same way, it is also not possible to describe the habitats for the separate small oceanic islands that are occupied by raptors. The extent of man-made modification in all habitats, but especially in evergreen forest, woodland, savanna and on Madagascar, is also not illustrated. Some raptor species have particular habitat requirements that do not conform to these major zones, while others extend across several adjacent zones, especially across the woodland-savanna-grassland-steppe continua that support the greatest diversity of species.

DESERT

There are four main desert zones in Africa, of which the Sahara is by far the largest, followed by the Nubian, the Namib and the Horn of Africa. Each includes extensive areas of sand dunes and gravel plains. Most are interrupted by rocky hills and mountains, some are crossed by dry watercourses with scattered trees and a few support wooded oases: each of these

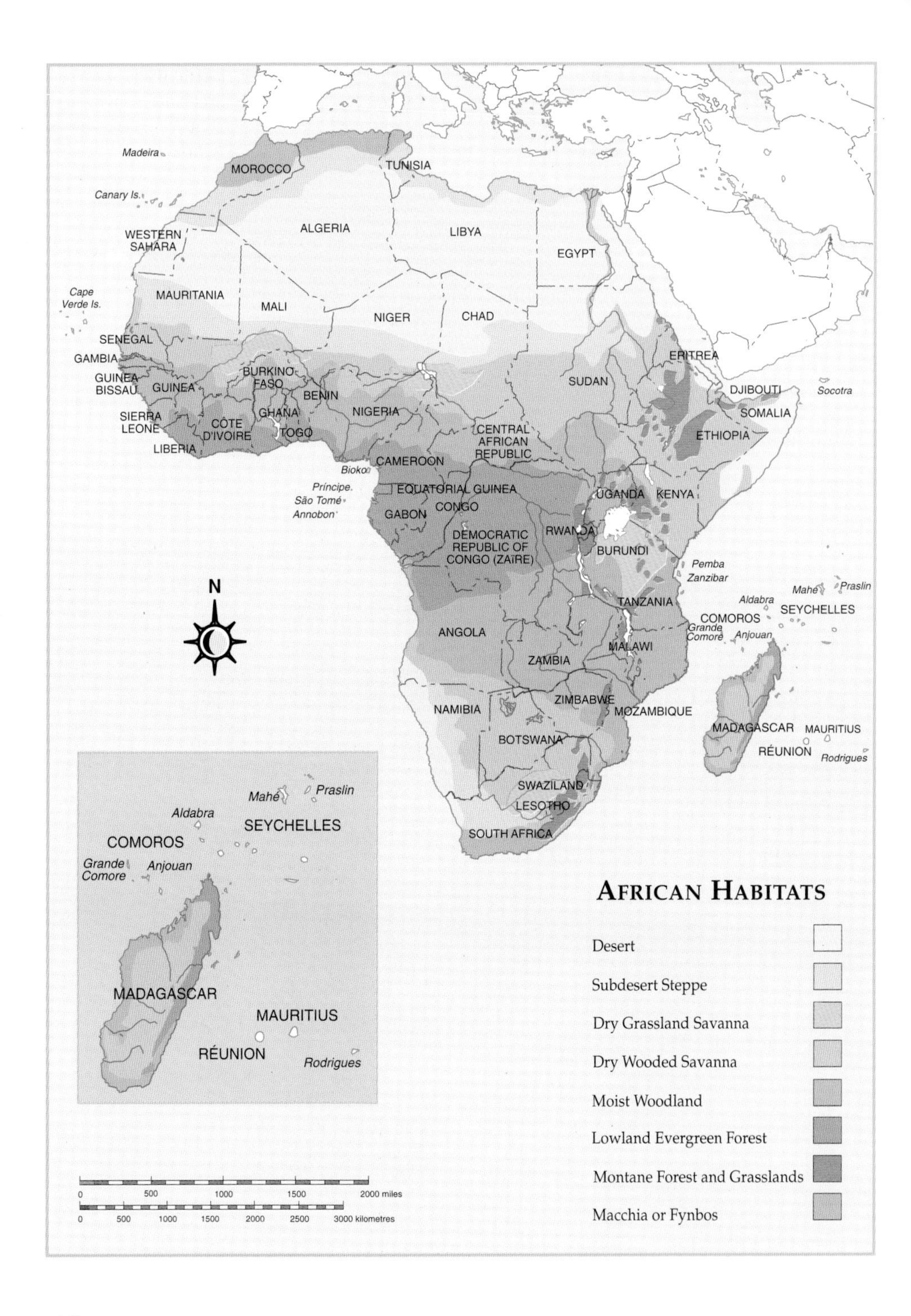
Madeira
Canary Is.
MOROCCO
TUNISIA
ALGERIA
LIBYA
EGYPT
WESTERN SAHARA
Cape Verde Is.
MAURITANIA
MALI
NIGER
CHAD
SENEGAL
GAMBIA
GUINEA BISSAU
GUINEA
BURKINO FASO
BENIN
NIGERIA
SUDAN
ERITREA
DJIBOUTI
Socotra
SOMALIA
ETHIOPIA
SIERRA LEONE
CÔTE D'IVOIRE
GHANA
TOGO
LIBERIA
CENTRAL AFRICAN REPUBLIC
CAMEROON
Bioko
Príncipe
São Tomé
Annobon
EQUATORIAL GUINEA
GABON
CONGO
UGANDA
KENYA
RWANDA
BURUNDI
DEMOCRATIC REPUBLIC OF CONGO (ZAÏRE)
Pemba
Zanzibar
TANZANIA
Aldabra
Mahé
Praslin
SEYCHELLES
COMOROS
Grande Comore
Anjouan
N
ANGOLA
ZAMBIA
MALAWI
ZIMBABWE
MOZAMBIQUE
NAMIBIA
BOTSWANA
MADAGASCAR
MAURITIUS
RÉUNION
Rodrigues
SWAZILAND
LESOTHO
SOUTH AFRICA
Mahé
Praslin
Aldabra
SEYCHELLES
COMOROS
Grande Comore
Anjouan
MADAGASCAR
MAURITIUS
RÉUNION
Rodrigues
0 500 1000 1500 2000 miles
0 500 1000 1500 2000 2500 3000 kilometres
AFRICAN HABITATS
Desert
Subdesert Steppe
Dry Grassland Savanna
Dry Wooded Savanna
Moist Woodland
Lowland Evergreen Forest
Montane Forest and Grasslands
Macchia or Fynbos

intrusions is especially attractive to birds of prey. All grade around their edges into arid subdesert steppe and open dry savanna. Only the Sooty Falcon and Pharaoh Eagle-owl have their main breeding range within this habitat.

Subdesert steppe

Desert margins and extensive areas of arid flats, especially in the Karoo biome of southern Africa and on the perimeter of the Ethiopian highlands, support sparse low bushes and, after infrequent rains, sparse grass. These areas can be productive for reptiles and in summer for insects, which attract some birds of prey, especially near cliff faces and along drainage lines. This habitat is often visited in summer by species from adjacent dry grassland savanna and dry wooded savanna.

Dry grassland savanna

Only the central highlands of South Africa, and to a lesser extent the plains of eastern and western Africa, support open grassland, which in the former are now scattered with exotic trees. These areas, and the herds of large mammals which they support, are especially important to Rüppell's and Cape vultures, Greater Kestrels, and some migrant kestrels and harriers, but all grade imperceptibly into dry wooded savanna with scattered bushes and small trees.

Dry wooded savanna

This is one of the two most extensive habitat zones in Africa and comprises mainly grassland covered with scattered low fine-leafed deciduous trees and bushes at varying densities. It grades at one extreme into moist woodland and at the other into dry grassland savanna or subdesert steppe. The main tracts are in southern Africa, including the so-called Kalahari Desert, East Africa and sub-Saharan western and northeastern Africa. There are no species restricted to this habitat, but several, such as the Secretarybird, African White-backed Griffon, Lappet-faced Vulture, Black-shouldered Kite, Common Barn Owl, African Scops Owl, Milky Eagle-owl, and Pearl-spotted Owlet are at their most abundant in this zone, as are several migrants, such as the Black Kite and Steppe and Lesser Spotted eagles.

Moist woodland

This is the other of the two most extensive habitat zones in Africa and comprises mainly tracts of broad-leafed deciduous trees and bushes divided by patches of grassland. It extends from the borders of lowland evergreen forest across much of western and central Africa, to grade imperceptibly into the dry wooded savanna to the west, northeast and south. A few species, such as Western and Southern Banded snake-eagles and the Grey Kestrel, are virtually confined to this habitat while others, such as the White-headed Vulture, African Hawk-eagle, African Cuckoo Hawk and African Barred Owlet, are at their most common there. Many other species, such as the Hooded Vulture, Brown Snake-eagle, Bateleur, Tawny Eagle, Martial Eagle, African Harrier-hawk, Gabar Goshawk, Shikra and Spotted Eagle-owl extend across both this habitat and dry wooded savanna, while a few species, such as the Swallow-tailed Kite, Grasshopper Buzzard and Ayres's Hawk-eagle, shift between these zones as the seasonal rainfall and deciduous trees alter the vegetation density.

Lowland evergreen forest

Evergreen rainforest occurs in two major blocks on the lowlands of the African continent, along the west coast and in the basin of the Congo River, and also along the east coast of Madagascar. It has been reduced in many areas to patches of primary forest linked by secondary forest and cultivation, especially in West Africa and Madagascar. These forests are the sole habitat of Cassin's Hawk-eagle, Congo and Madagascar serpent-eagles, Long-tailed Hawk, Chestnut-flanked and Red-thighed sparrowhawks, and various owls including Madagascar Red, Sandy Scops and Maned owls, Akun and Shelley's eagle-owls, and Red-chested and Chestnut-backed owlets.

MONTANE FOREST AND GRASSLAND

Montane habitats in sub-Saharan Africa often extend to high altitudes, enjoy a relatively high rainfall and support restricted habitat patches of forest, moist grassland, alpine vegetation and rocky slopes. The main areas are the Ethiopian, Cameroon and East African highlands, the western and eastern Rift Valley mountains, and the eastern highlands and Drakensberg of southern Africa. Some forest species, such as the Crowned Eagle, African Goshawk and Black Sparrowhawk, are shared with lowland forests. Others such as the Bearded Vulture, Mountain, Augur and Jackal buzzards, Rufous-breasted Sparrowhawk, Cape Eagle-owl, and Congo Bay Owl, Albertine owlet and Long-eared owls are most common within some of these areas.

MACCHIA OR FYNBOS

This low, dense Mediterranean-type scrub is confined to areas of winter rainfall along the northwest African coast and around the southern tip of Africa. It is associated in both areas with high, rugged mountains and these factors combine to support a variety of birds of prey. Several European breeding species extend only onto northern Africa within this habitat, and in the south several species, such as the Black Harrier and Booted Eagle, have their core breeding range within this habitat.

HOW TO USE THIS BOOK

A number of design features make this book easy to use. Each of the ten groups of species has its own colour tag and icon to facilitate location in the book. Most species cover a double-page spread, and both pages are headed by the common English name of the species. Only a few, rare owl species share a page.

Use the right-hand page for quick reference. At the top is a broad coloured band in which the **key features** for identification are described (including an indication of the bird's size), the **most similar species** are listed, and a **distribution map** shows at a glance where the species is found. Below this are the main illustrations, labelled to assist identification of the different ages, sexes, forms and poses.

In most cases, the left-hand page also includes at least one photograph of the species, set within the descriptive text. The species text is presented in two parts. The first part is aimed at the most useful identification features, depending on whether the bird was seen **at perch**, **in flight** or engaged in some **distinctive behaviour**. The second part provides a more detailed description of the **adult**, **juvenile** and, where appropriate, of the **subadult**, including differences in plumage and size between the sexes. Juveniles and subadults are only described to the extent that they differ from adults. Many of the terms used in these sections can be found marked on the **illustrated glossary** (pp. 24, 25 and inside back cover), which shows the body parts of a bird of prey, or in the **glossary of terms** (pp. 331 and 332). There is also a description of the **distribution, habitat and status** and a comparison between the species under discussion and **similar species**.

Size is an important component of identification for birds of prey. In the **key features**, size is allocated to one of five categories: very large (over 70 cm tall), large (50–70 cm tall), medium-sized (30–50 cm tall), small (20–30 cm tall) and very small (less than 20 cm tall). Size is indicated here by an estimate of the bird's average height when perched in an upright position, rather than its length as a stretched-out museum skin (as given in most books on the subject). Its estimated wingspan in flight is also given for assessing size on the wing, but this measure varies widely within height categories due to different wing shapes. In most birds of prey, females are larger than males, so under the **adult** description the size of the female (where known) is indicated as the percentage by which she is larger than the male (based on the ratio of their average wing lengths). The range of male and female body masses is also presented, as an indication of how bulky they are and to allow comparison between species. Where a single figure is given, this is either an average or the weight of a single specimen. For a few species, no weight information was available at the time of going to press.

Distribution maps are an important aid to identification. Birds of prey are more mobile than many other groups, many are migrants and some are so difficult to locate that their exact range is poorly known. The maps presented here attempt to summarise our present knowledge. Areas in which the birds breed, or where they are resident throughout the year, are shaded **orange**. Areas where non-breeding birds are commonly seen or are regular visitors are shaded in **dark green**, while those areas where the birds are more sparsely distributed or where vagrants have occurred are shaded **light green**. Species found only on the Indian Ocean islands have different distribution maps; this serves not only to magnify the sometimes very limited areas where the birds are found but also helps distinguish these species from mainland species. The maps may not cover every record of a species; some birds wander far from their normal range or habitat. Various aspects of the range and abundance are also covered in the text.

In the text that deals with *distribution* of each species, different parts of Africa have been described in a very general way as northern, southern, central, eastern and western Africa, to avoid confusion with the politically specific terms of North Africa, South Africa, Central Africa, East Africa and West Africa. So, for example, 'western Africa' refers to the countries which generally lie on the western side of the continent, and not exclusively to those on the Gulf of Guinea; similarly 'eastern Africa' extends as far north as Somalia and as far south as Mozambique, rather than only Kenya, Uganda and Tanzania.

At the end of the book there is a **glossary of terms**, a list of selected titles for **further reading**, **indexes** of scientific, and English common names, and a **language table** that shows the equivalent French, German, Spanish and Afrikaans names Most names follow the *Handbook of Birds of the World* (Volume 2 for diurnal birds of prey and in press for owls), with only a few exceptions. The latter are mainly owls.

A few additional species have been reported for the African continent and its islands, historically and only as unconfirmed or very rare vagrants. These species, mentioned for completion, are White-tailed Sea-eagle *Haliaeetus albicilla* (NE Egypt), Rough-Legged Buzzard *Buteo lagopus* (N Africa) and Brown Fishing-owl *Ketupa zeylonensis* (Seychelles, NE Egypt).

*first three sections summarize species' **main** physical and behavioural characteristics for field identification*

English common name

scientific name (comprising genus and species names)

Falco columbarius

AT PERCH
Appears stocky, and often stands rather horizontally. Dark streaking all over evident on grey male and brown female. Rather short wings only reach dark end-bar on tail.

IN FLIGHT
Very fast with quick, direct, driving wingbeats. Soars, often with tail fanned to show barring. Never hovers. Note dark, barred wings, underwing coverts and tail, from above and below, with pale tips to wing and tail feathers.

DISTINCTIVE BEHAVIOUR
Hunts on the wing, mainly for small birds, with fast, level flight interspersed with glides. Chases down prey, often after agile, swerving flight. Stops to land on the ground or on low, open perches. Call a fast 'kik kik kik'.

name of photographer

A. Rouse/NHPA

Adult male

S. Dalton/NHPA

Juvenile taking off

caption for photograph

ADULT Male dark grey above with fine, black streaks. Whitish eyebrow. Below pale rufous with fine, dark brown streaks, including neck and cheeks. Faint dark malar stripe. **Female** browner with heavier streaking, grey or brown rump and more prominent pale forehead and eyebrow. Flight feathers and tail dark grey (male) or brown (female) above and below with narrow, pale grey bars, and with broad, dark end (male) to white-tipped tail. Bill black with grey base; eyes dark brown; cere, eye-ring and slender, bare legs and long toes dull yellow. Female about 9% larger (male 150–210 g, female 189–255 g).
JUVENILE Both sexes similar to adult female, but with browner rump. Facial skin blue.

DISTRIBUTION, HABITAT AND STATUS Rare, non-breeding migrant to coastal lowlands of northwestern Africa and Nile Delta during northern winter. Old specimen record from Durban in South Africa. Rare vagrant to Senegal and Sudan.
SIMILAR SPECIES Only overlaps, and female might be confused, with larger, more slender **Common Kestrel** (p. 232) (deep rufous back spotted with black; below buff streaked with black; grey tail unbarred (male) or narrowly barred with rufous (female); broad, dark subterminal band) and female or juvenile **Lesser Kestrel** (p. 238) (heavily streaked buff underparts; rufous, evenly barred tail; dark 'moustache'; weak feet with white claws), both of which hover regularly.

214 ALTERNATIVE NAME: Pigeon Hawk

more detailed description of bird in different plumages, and of its distribution and status

information in parenthesis pertains to bird mentioned in bold type, highlighting points of difference between it and the main species under discussion

alternative English common name(s)

symbol indicates key features, that is, the species' key, or essential, identification features; the first section to turn to in the effort to establish identity

distribution map shows the approximate range of the species; the colours are significant: ***orange*** *indicates breeding range;* ***dark green*** *indicates where the bird is commonly seen or a regular visitor;* ***light green*** *indicates sparse distribution or where it is a vagrant*

stylized silhouette or icon: an aid to quick identification, indicates the group within which the species falls; see key on inside front cover

common name provides the running head

Merlin

Small, stocky falcon, sexes different. Male dark grey above and pale rufous below, streaked with black. Female browner, especially above, with heavier streaks below. In both sexes, note whitish eyebrow and faint 'moustache' stripe. In flight, dark barred underwing and tail with pale tips. Juvenile similar to adult female. Very small (about 20 cm tall, 50–67 cm wingspan).

Common Kestrel, female Lesser Kestrel

symbol indicates most similar species (those with which the bird could most easily be confused)

colour and tint supplement the icon as an identification aid, indicating group within which the species falls; see key on inside front cover

215

caption gives basic information on the specimen portrayed:

ad. – adult bird
subad. – subadult bird
juv. – juvenile bird
imm. – immature bird
♂ – male
♀ – female

name of photographer (of photograph featured among illustrations)

ILLUSTRATED GLOSSARY

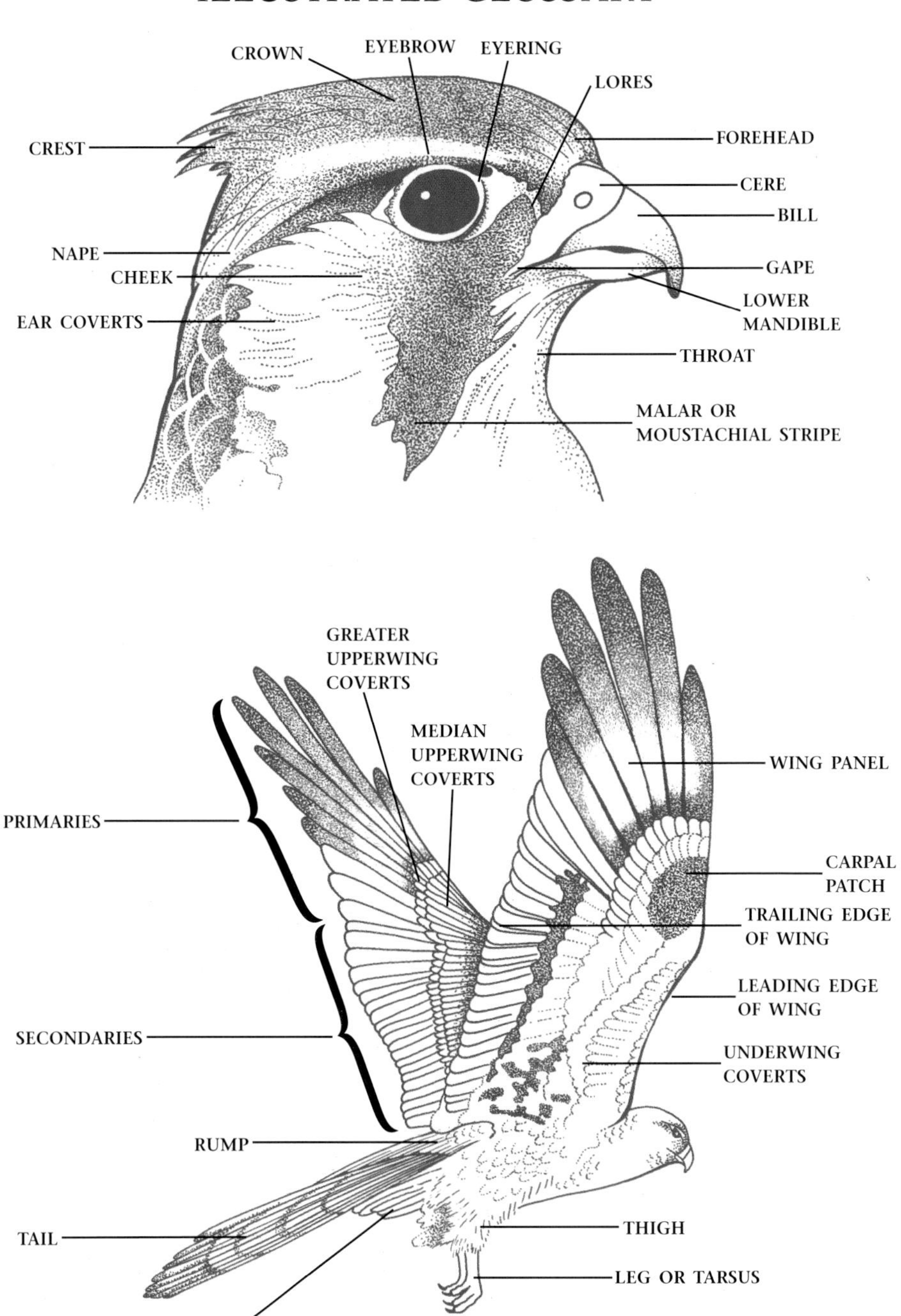

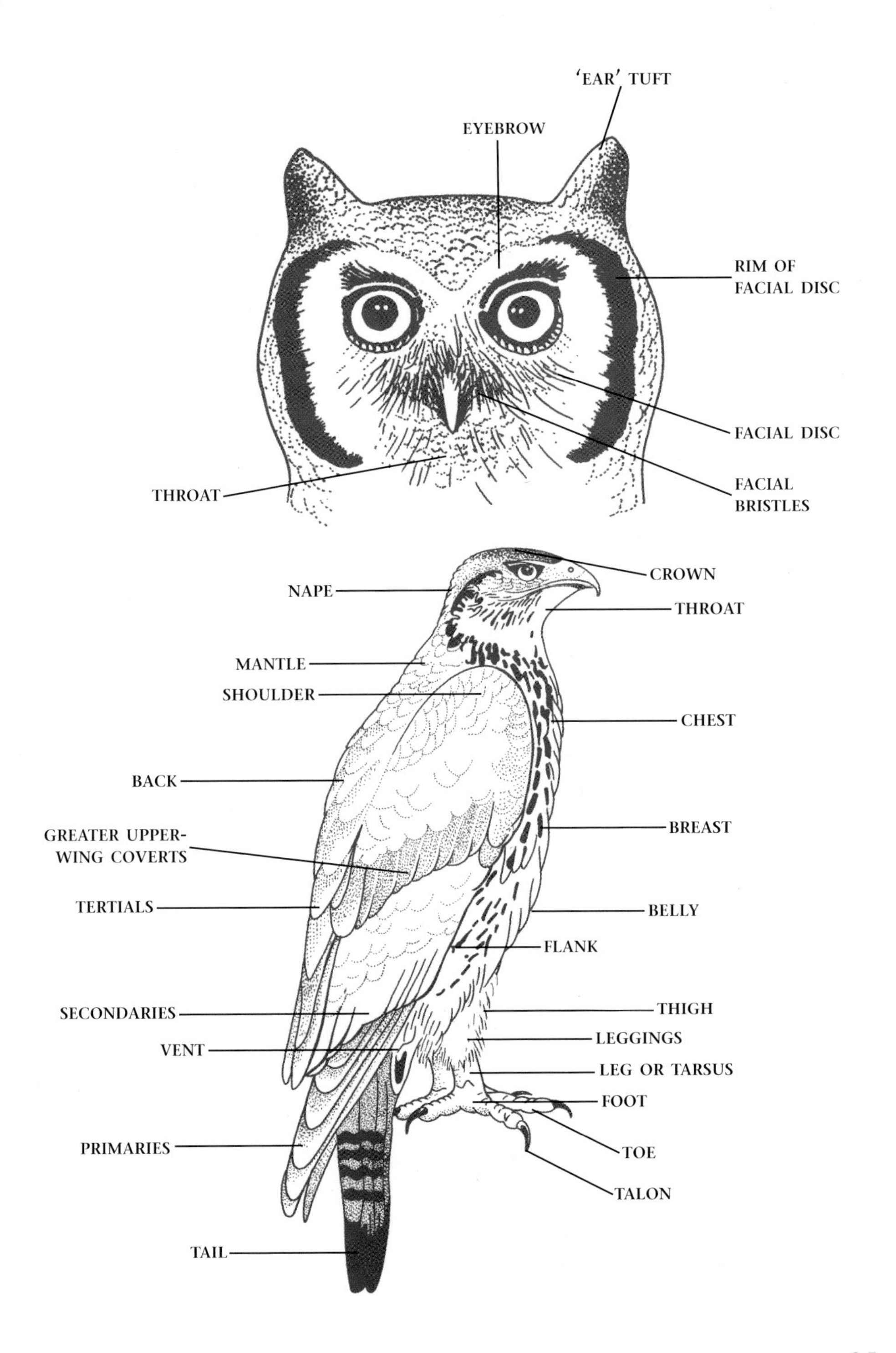
'EAR' TUFT
EYEBROW
RIM OF
FACIAL DISC
FACIAL DISC
FACIAL
BRISTLES
THROAT
CROWN
NAPE
THROAT
MANTLE
SHOULDER
CHEST
BACK
BREAST
GREATER UPPER-
WING COVERTS
TERTIALS
BELLY
FLANK
SECONDARIES
THIGH
LEGGINGS
VENT
LEG OR TARSUS
FOOT
PRIMARIES
TOE
TALON
TAIL

Secretarybird

Sagittarius serpentarius

AT PERCH

Spends most of the day walking on the ground. Usually seen perched only at dusk and dawn, when it stands at the top of a small tree, often on an old nest, or lies down to roost. The long, bare legs, long tail, loose crest, patterned plumage and colourful softparts are all notable.

IN FLIGHT

Runs or leaps into flight if disturbed, making off with deep, slow wingbeats. Flies mainly during the heat of the day, taking off to soar high in thermals on outstretched wings, held at a slight dihedral and with upturned tips, the head and long neck protruding in front and the long legs and tail behind. The black flight feathers, black belly and thighs and darkly banded, graduated tail contrast with the pale body and white underwing and vent.

DISTINCTIVE BEHAVIOUR

Walks purposefully across the veld in search of small animals which it usually stamps on and kicks to death before stooping down to swallow them. Often takes off to soar when moving to a nesting, roosting or watering site. Prior to breeding, performs ponderous pendulum-like flights high overhead with raucous croaking calls. Nests and roosts on top of small, flat-topped trees.

J. Knobel

Adult hunting on foot

I. Davidson

Adult

ADULT Above grey, paler towards rump and breast. Nape has long, black, club-shaped, erectile crest feathers. Thighs and belly black, vent white and fluffy. Flight feathers black, underwing coverts white. Tail grey with dark grey base, broad, black, subterminal band and white tip. Bill blue-grey; eyes dark brown or yellow; cere yellow; bare facial skin around eyes deep orange; very long, bare legs with heavy scaling and short, pink, stubby toes. **Female** often slightly darker grey and larger, but otherwise sexes similar in plumage (both sexes 3 405–4 270 g).

JUVENILE Similar to adult, but with brown wash over grey plumage areas, paler orange face, black bill, grey eye, and fine, grey barring on the white underwing coverts and vent. Moults directly into adult plumage.

DISTRIBUTION, HABITAT AND STATUS Widespread in non-forested areas of sub-Saharan Africa. Occurs from semi-desert to grassy openings in forest and woodland. Nomadic, as drought, grazing and fire alter its habitats. Most common in open savanna and grassland with scattered, small thorn trees.

SIMILAR SPECIES Unlike any other bird of prey in Africa. Confusion, if any, likely only with long-legged cranes, storks and herons, all of which have long, straight bills. Resembles only much smaller **African Harrier-hawk** (p. 128) in basic colour and bare face.

Large, long-legged, eagle-like bird. Grey, black and white. Note long central tail feathers. Long, loose crest feathers also conspicuous. Bare, red face. Short, hooked, blue-grey bill. Very large (about 120 cm tall, 212 cm wingspan).

Unique, unlike any other African bird of prey

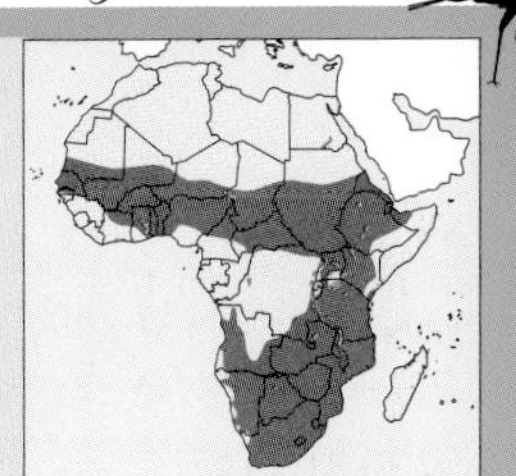

P. Steyn/Photo Access

Bearded Vulture

Gypaetus barbatus

AT PERCH

Normally perches on rocks or ledges, standing horizontally on the short legs with the very long wings hunched over the back, the long wing tips and tail projecting behind. The contrast between black back and rufous underparts is striking, the pale head with long, spiky nape, black mask and projecting 'beard' is obvious, and the red and yellow eye is notable at close range. Juvenile shares the characteristic shape, subadults often with a black head to highlight the red-rimmed eye.

IN FLIGHT

Very long, pointed wings, like a giant falcon, and broad, long, diamond-shaped tail are characteristic. Glides with wings level or slightly downcurved. Pale head of adult with black mask and 'beard' and pale bill often obvious; from below, the rufous body with spotted collar also notable. Juvenile has slightly broader wings but similar tail shape, pale body, and often black head, also easily recognised.

DISTINCTIVE BEHAVIOUR

Soars over and glides across mountainous terrain for long periods without flapping, even in strong winds, with continual adjustment of wing and tail profiles. Lands at its main food of carrion bones, walking with a waddling gait. Then takes off clutching a large bone in its feet which it drops on rocks below, before spiralling down to eat the fragments. Perches, roosts and builds a large, untidy nest on high rock ledges or in potholes.

R. du Toit

Juvenile hanging or kiting in the wind

A. Weaving

Adult

ADULT Above slaty-black; wing coverts have thin, buff shaft streaks that end in small, white spots. Head, neck and underparts white, but to a varying extent stained rufous with iron oxide as a result of dust-bathing. Head often almost white, with broad, black patch from base of bill to around eye and black 'beard' tuft projecting below base of bill. Black-tipped feathers form partial collar across breast. Flight feathers, underwing coverts and tail black, with dark slate centres and cream feather shafts. Bill pale horn-colour; eyes pale yellow with broad, red rim; short legs (bare in sub-Saharan subspecies *meridionalis*, feathered in northwestern African *barbatus*) and long toes dark grey. Sexes similar in plumage, female about 2% larger (both sexes 5 200–6 250 g).

JUVENILE Dark brown above, almost black on head and neck but pale grey-brown on the rump. Below paler brown, mottled with white feather bases. Eyes and eye-rim brown, incomplete 'beard', legs pale brown. Acquires adult plumage over 7 years, becoming more plain and rufous below; above initially more mottled, with dark and light feathers, but with a black head that later becomes pale from the bill backwards.

DISTRIBUTION, HABITAT AND STATUS Sedentary breeder with a patchy distribution in mountainous habitat. The nominate Eurasian population extends into northwest Africa. The African subspecies extends from the Arabian Peninsula to the highlands and Rift Valley of northern and eastern Africa, with a population now isolated on the central Drakensberg mountains of South Africa. Common on the Ethiopian highlands but uncommon to rare and declining elsewhere.

SIMILAR SPECIES Diamond-shaped tail, especially in brown juveniles, similar to that of **Egyptian Vulture** (p. 46) (much smaller; rounded wings; unfeathered face; long, thin bill; all-brown body and wings).

ALTERNATIVE NAME: Lammergeier

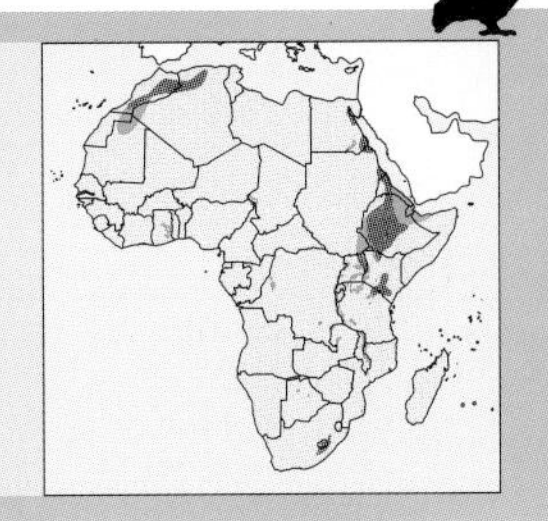

Heavy, short-legged, long-winged, eagle-like bird. Back and wings slaty-black. Head, neck and underparts rufous. Head palest, with black mask and 'beard' tuft. Eyes yellow with red rims. Bill horn-coloured. In flight, note long, broad, diamond-shaped tail. Juvenile brown with mottled upperparts, plain dark head and brown eyes. Subadult often has black head and is pale rufous below. Very large (about 80 cm tall, 250–282 cm wingspan).

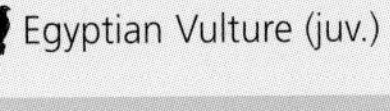

Egyptian Vulture (juv.)

R. du Toit

African White-backed Griffon

Gyps africanus

AT PERCH

Typical vulturine form, with large, hunched wings and small, bare head withdrawn or extended on long neck. Dark flight feathers contrast with pale brown coverts and dark tail with white back. Head and neck down white when fresh, against black skin, but becomes stained brown or covered with blood. Note dark crop-patch when full. Juvenile appears darker, streaky, with white fluffy head and neck, at least at fledging.

IN FLIGHT

Long, broad wings held in a shallow V-shape when gliding, secondaries form bulging trailing edge, ends with long, upturned primary 'fingers'. The pale coverts and body contrast with the dark flight feathers, and short, wedge-shaped tail with white back. Juveniles darker with white forearm-patches.

DISTINCTIVE BEHAVIOUR

Roosts or waits near carrion, on open perches. Usually seen in flight, gliding overhead or circling high in thermals. Descends to carcasses *en masse*, rushing down from the sky or nearby perches and feeding frantically. Often fights with harsh, grating screeches, after bounding attacks or after standing on a carcass with wings outstretched. Usually resorts to waterholes or rivers after feeding to sun- and water-bathe. Builds a large stick-nest in a large tree, usually on the crown.

ADULT Overall cream, pale or dark brown, plain or with fine, dark brown streaks. Back and neck ruff white. Becomes paler and less streaked with age. Flight feathers, including greater upperwing coverts, and tail brownish-black, older birds with slightly darker trailing edge to secondaries. Underwing coverts white, a few birds with dark blobs on the greater coverts. Small head and long, slender neck with black skin and sparse covering of white down, becoming stained brown. Crop down brown, bare shoulder-patches green. Bill, cere, eye and legs black. Sexes similar in plumage and size (both sexes 4 150–7 200 g).

JUVENILE Slaty-brown overall, including the back, with fine, white shaft streaks, especially on long, pointed ruff feathers. Wears to matt brown. Note pointed black flight and tail feathers. Forearm skin with white down. Head and neck skin pale green with a dense covering of white down. Moults into adult plumage over 6 years, becoming paler brown, with rounded flight feathers, then with white-tipped back feathers.

DISTRIBUTION, HABITAT AND STATUS Found through much of sub-Saharan Africa, apart from extreme desert and dense forest. Vagrant to Bioko Island. Most common in flat, open, wooded savanna with scattered, large trees for roosting and nesting. Usually the commonest vulture by far. Less common in dense woodland, on open grassland and among mountains.

SIMILAR SPECIES Similar to other larger, heavier griffon (*Gyps*) vultures with their thick, muscular, long necks. Overlaps in southern Africa with **Cape Griffon** (p. 34) (usually paler; adult pale

T. Carew/Photo Access

Juvenile in threat posture

cream; line of dark blobs on greater wing coverts; mottled back; pale secondaries below with dark trailing edge; yellow eye; blue neck skin; juvenile pale brown with pink neck skin), and in eastern and western Africa with **Rüppell's Griffon** (p. 32) (adult dark brown, spotted with cream; dark back; yellow eye; orange bill; juvenile brown with broad, white streaks), and on the Ethiopian highlands with **Eurasian Griffon** (p. 36) (usually paler; adult pale brown; dark centres to greater wing coverts; white ruff; eye and bill yellow; neck skin blue; juvenile has reddish brown ruff and back and amber-brown eyes).

ALTERNATIVE NAMES: White-backed Griffon, White-backed Vulture

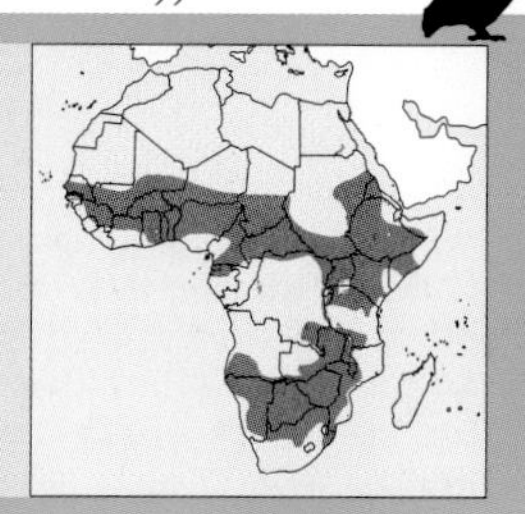

Large, pale brown vulture. Long, bare neck with black skin and small head covered in sparse, white down. White back. Black flight and tail feathers contrast with cream or brown body and wing coverts. Juvenile darker brown with white streaks, head and neck with white down and pale green skin. Very large (about 85 cm tall, 212–228 cm wingspan).

Cape Griffon, Rüppell's Griffon, Eurasian Griffon

Rüppell's Griffon

Gyps rueppellii

AT PERCH

Long, muscular neck and small head usually obvious when perched or standing on the ground, often in rather upright stance. Appears dark overall, adult speckled and juvenile streaked with cream, with all-dark flight feathers. Yellow bill, purple-blue neck and shoulder-patches, and white ruff and thigh 'chaps' render adult distinguishable at close range. Juvenile has dark face and eye against white head and neck down, and deep pink neck skin.

IN FLIGHT

Long, broad, rectangular wings, held at slight V-shape when soaring, with bulge of secondaries at rear edge, long, upturned primary 'fingers' and rather long tail spread to form wedge-shape. Flight feathers and tail all-dark, against pale, speckled body and wing coverts, white forearm-patch and covert tips forming pale lines across underwing. Juvenile more uniformly dark brown with streaking obvious only at close range.

DISTINCTIVE BEHAVIOUR

Roosts and nests in colonies on tall cliffs or in river gorges, the rocks splashed with white droppings. Overnights on trees when feeding far from colony; in Cameroon, rarely nests in trees. Soars in vicinity of colony or glides out fast and direct to search for carrion in surrounding habitat. Gathers in numbers at a carcass, often with other vulture species. Fights for food, kicking with the feet or standing with wings extended, often making harsh, grating calls.

M. Denis-Huot/BIOS

Juvenile sunbathing

D. Richards

Adult

ADULT Sooty-brown overall, each feather with broad, cream tip to give scaled effect above and paler, speckled effect below. White, downy inside to thighs. Head covered in white down, sparse on neck. Neck ruff white. Flight feathers plain brownish black. White down on forearm patagial area. Crop down black, bare shoulder-patches blue-grey. Bill orange; cere black; head and neck skin blue-grey flushing purple when excited; eye pale yellow; bare legs and long toes dark grey. Sexes similar in plumage, female about 2% larger (both sexes 6 800–9 000 g).

JUVENILE Different to adult. Dark, glossy brown overall, upperwing feathers edged cream and streaked with cream below. All feathers sharply pointed, especially long ruff feathers. Flight feathers and tail brownish-black. Head and neck densely covered in white down at fledging, bare skin deep pink. Bill black, eye dark brown. Moults to adult plumage over 6 years, spotted with darker feathers in intermediate stages.

DISTRIBUTION, HABITAT AND STATUS Widespread across the Sahel region of sub-Saharan Africa, extending south along the Rift Valley of eastern Africa, a very rare vagrant further south to northern South Africa. Favours mountainous areas and deep gorges, where these are adjacent to open semi-desert or grassland habitats with numerous game or livestock.

SIMILAR SPECIES Adult much darker than any other long-necked griffon (*Gyps*) vulture, with unique combination of feather scaling and softpart colours. Juvenile overlaps in range mainly with similar-sized **Eurasian Griffon** (p. 36) (underwing pattern on white patagium and pale line of greater covert tips; pale blue neck skin; white neck ruff and pale brown back; amber-yellow eyes) and smaller **African White-backed Griffon** (p. 30) (dark brown with fine, white streaks), but most like similar-sized **Cape Griffon** (p. 34) (pink neck skin; white neck ruff), with which it rarely overlaps in southern Africa.

ALTERNATIVE NAME: Rüppell's Vulture

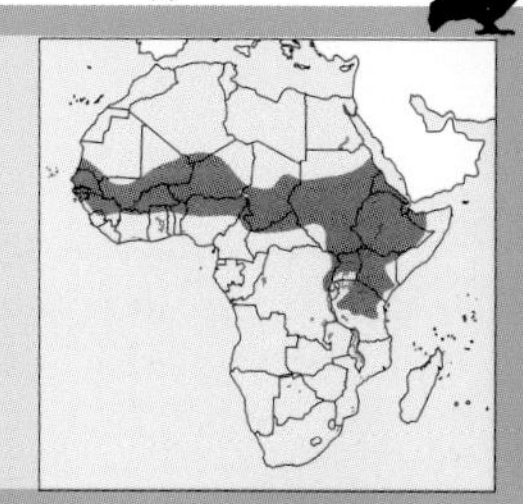

Large, dark, long-necked vulture. Sooty-brown with cream scales. Distinct patterns on wing coverts and underparts. White neck ruff and inner thighs. Orange bill; yellow eye. Blue-grey neck skin. Very large (about 90 cm tall, 241 cm wingspan).

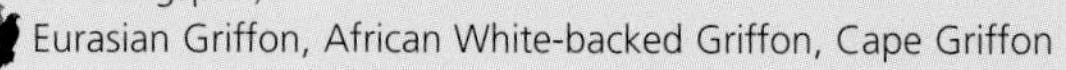

Eurasian Griffon, African White-backed Griffon, Cape Griffon

M. Denis-Huot/BIOS

Cape Griffon
Gyps coprotheres

AT PERCH
Usually sits with head tucked into shoulders and with large wings covering body and short tail. Pale cream body and wing coverts contrast with dark flight feathers, apart from line of dark blobs across centre of wing. Yellow eye obvious against dark head, neck and crop. Neck appears thick, muscular and pale blue when extended. Juvenile darker with pale streaks, dark eye and pink neck.

IN FLIGHT
Long, broad, rectangular wings held in shallow V-shape when soaring, with secondaries bulging at trailing edge, long primary 'fingers' upturned and short, square tail spread. Pale body and wing coverts contrast with dark flight feathers, with dark trailing edge to pale secondaries and, at close range, line of dark spots along greater wing coverts. Dark head and legs obvious. Juvenile has streaked body and plain, dark secondaries.

DISTINCTIVE BEHAVIOUR
Roosts and builds its stick-nests in colonies of several hundred birds on high cliffs, which become streaked with white droppings. Sunbathes and soars around the nest cliffs, gliding out over the surrounding country to search for carrion. Descends in numbers to feed at a carcass, often with other vulture species. Fights over food with harsh, grating calls, or stands with wings outstretched. Visits regular bathing pools.

R. du Toit

Subadult sunbathing

G. McIleron

Head of adult, showing coloration details

ADULT Plain pale cream overall, often stained pale rufous. Greater wing coverts, scapulars and back have dark brown centre blobs, sometimes absent on underwing. Ruff, sparse neck down and hairy head feathers white. Flight feathers and tail brownish-black above, secondaries white below with dark tips. Crop down dark brown, bare shoulder-patches blue with red edges. Bill and cere black; eyes yellow; head and neck skin blue, or face flushed pink; bare legs with long toes black. Sexes similar in plumage, female about 3% larger (both sexes 7 070–10 900 g).

JUVENILE Body and wing coverts pale brown streaked with white, especially along greater upperwing coverts, forming a pale line across wing. All feathers sharply pointed, especially long, buff neck ruff. Flight feathers brownish-black. Eye black, shoulder-patch and neck skin purple-pink, head with dense, white down.

DISTRIBUTION, HABITAT AND STATUS Restricted to southern Africa. Main colonies in South Africa, with small, outlying colonies in northern Namibia, southwestern Botswana and southern Zimbabwe. A vagrant to Zambia and southern Mozambique. Most common among mountains and over adjacent open Karoo steppe and grassland. Less common in wooded savanna.

SIMILAR SPECIES Overlaps widely with only one other long-necked *Gyps* vulture, the smaller, lighter **African White-backed Griffon** (p. 30) (usually darker brown; unspotted greater wing coverts; dark eye; black skin on head and neck; plain dark secondaries; juvenile dark brown streaked with white). **Rüppell's Griffon** (p. 32), though similar in size, is only a rare vagrant to southern Africa (adult dark brown spotted with cream; dark back; yellow eye; orange bill; juvenile brown with broad, white streaks).

ALTERNATIVE NAMES: Kolbe's Vulture, Cape Vulture

Heavy, pale, long-necked vulture. Pale cream with black flight feathers. Line of dark blobs along greater wing coverts. Bill and cere black; eyes yellow; neck skin blue. Juvenile darker brown, with pink neck skin and dark eye. Very large (about 95 cm tall, 255 cm wingspan). Endemic to southern Africa.

African White-backed Griffon, Rüppell's Griffon

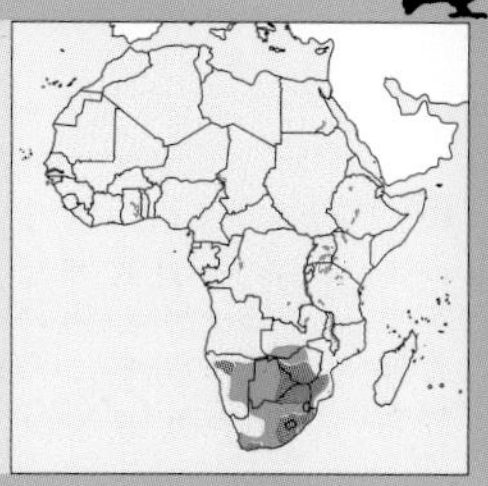

G. Mc Illeron

Eurasian Griffon

Gyps fulvus

AT PERCH

Long, muscular neck often obvious as it stands erect with neck ruff raised and extensive wings hunched. Overall sandy appearance, often stained deep rufous, with dark line of blobs along wing coverts. Note white head, neck and ruff, dark eyes and yellow bill in adult. Juvenile has long, brown or sandy ruff and black bill; moults into adult plumage over 6 years.

IN FLIGHT

Long, broad wings with sinuous trailing edge, narrowest at body, with bulging secondaries, indented at short inner primaries, and long outer primary 'fingers'. Wings held in shallow V-shape when soaring. Black flight feathers, alula-patch and short, rounded tail contrast with sandy wing coverts and body. Note white forearm-patches, greater upperwing coverts with line of dark blobs, underwing with fine line of white tips. Juvenile finely streaked all over.

DISTINCTIVE BEHAVIOUR

Roosts and nests in colonies on cliff faces, which become streaked with white droppings. Soars most often above hills and mountains, or glides over surrounding terrain in search of carrion. Descends in numbers to food, often with other vulture species, and feeds with long neck thrust into the carcass. Bathes regularly after eating.

E. Soder/NHPA

Subadult in gliding flight

ADULT Above pale fawn; greater wing coverts and scapulars have sooty-brown centres and fawn edges. Dense, white neck ruff. Below more rufous, with fine, white shaft streaks. Greater wing coverts often fawn or brown with white edges. Body and wing coverts often stained darker. Flight feathers and tail plain brownish-black, inner secondaries slightly paler. Head and neck covered with dense, white, hairy down, bare skin blue with pale pink shoulder patches. Crop down dark brown. Bill pale yellow with black cutting edges; cere black; eyes yellow; bare legs and long toes dark grey. Sexes similar in plumage, female about 4% larger (both sexes 6 200–8 500 g).

JUVENILE Dark reddish-brown overall, with white shaft streaks and white tips to greater upperwing coverts. All feathers have pointed tips. Head and neck covered with dense, woolly down; bare shoulder skin pale green. Long neck ruff usually streaked brown. Bill black, eye amber-brown.

V. G. Canseco/NHPA

Juvenile

DISTRIBUTION, HABITAT AND STATUS Now breeds only in the high mountains of northwestern Africa. A vagrant to the Red Sea coast from colonies in Arabia and Sinai. Non-breeding migrants, mainly juveniles, come from Europe during the northern winter and cross the Sahara to the northern Sahel and Ethiopia. Rare on the desert massifs, uncommon in arid scrub and savanna of the Sahel.

SIMILAR SPECIES Overlaps with only two other long-necked *Gyps* vultures, the similar-sized **Rüppell's Griffon** (p. 32) (black plumage scalloped with cream; dark orange bill; pale yellow eye; short, small ruff; juvenile with broad, pale streaks) and the smaller **African White-backed Griffon** (p. 30) (plain, dark greater upperwing coverts; white underwing coverts; black bill, cere, eye and neck skin; juvenile dark brown with fine streaks).

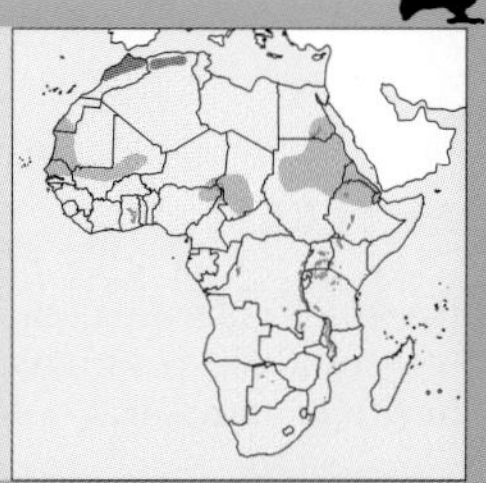

Large, long-necked, sandy-coloured vulture. Obvious white neck down and ruff. Adult eyes and bill yellow; juvenile eyes dark brown. Very large (about 95 cm tall, 250 cm wingspan). Restricted to northern Africa and Sahel region of sub-Saharan Africa.

Rüppell's Griffon, African White-backed Griffon

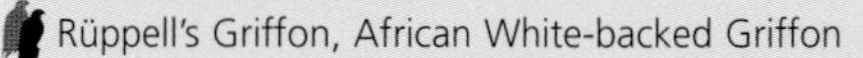

E. Soder/NHPA

White-headed Vulture

Trigonoceps occipitalis

AT PERCH

Angular head, high ruff and large wings notable. White head of adult obvious from far off, attractive bright facial colours at close range. Mottled band across greater coverts and, in female, white rear secondaries, notable in good light. Juvenile all dark with brown cap on head (except soon after fledging) and pale, washed-out head and leg colours.

IN FLIGHT

Broad, long, rectangular wings with long primary 'fingers' and short, square tail. Black above with pale band across centre of coverts. Below whiter than other vultures, with white crop, black breast-band, white belly and legs leading into white 'armpits' and white line along rear edge of coverts. Female has white inner secondary 'windows'. Juvenile all brown with dark wings and tail, a dark line at rear edge of upperwing coverts and a pale line at rear edge of underwing coverts. Colourful head and feet often obvious.

DISTINCTIVE BEHAVIOUR

Usually seen alone or in pairs, perched in trees or on the ground at carrion, small, helpless animals or other predatory birds with prey. Often first to discover carrion, descending from high, soaring flight. Very attentive to the moves of other scavengers and predators. Usually stands at the edge of gatherings of other vulture species, preferring to feed alone, but can dominate others for food when necessary. Nests most often on crown of tall, solitary thorn trees, often roosts nearby.

N. Dennis

Adult male (left) and female (right) sunbathing

ADULT Above black, including neck ruff and chest-band. White edges to feathers of median upperwing coverts. High, black ruff. Throat, crop, breast and thighs white, extending as white line along ends of greater underwing coverts. Flight feathers and tail black. Inner secondaries white in female. Broad, square head unfeathered, with tight, white down on crown and nape forming slight crest. Bare pale to bright pink skin on face and neck. Bill orange with black tip; small eye dull orange; cere and jawline sky blue; stout, bare legs pinky-orange. Sexes similar in size (both sexes 3 300–5 300 g).

JUVENILE Dark brown all over, apart from white mottling on mantle and pale edges to upperwing coverts giving scalloped effect. Flight feathers and tail dark sooty-brown. White down on head at fledging, soon turns dark brown. Bare skin and legs pale pink, cere pale blue and bill pale salmon. Moults into adult plumage over 6 years, paler secondaries of female evident from second year.

DISTRIBUTION, HABITAT AND STATUS Throughout open savannas of sub-Saharan Africa, avoiding arid treeless and moist forested areas. Adult pairs often sedentary and probably territorial. Sparsely distributed and nowhere common. Most abundant in western Africa.

SIMILAR SPECIES Most resembles larger **Lappet-faced Vulture** (p. 40) (adult has white leggings; broad, angular head; streaked underparts; naked pink head with lappets; black crop; short, black ruff; white forearm-patch; deep yellow or horn-coloured bill; pale grey legs; juvenile all-brown with plain black wings).

Handsome pied vulture. Black with white legs and belly, and blocky white head. Colourful orange bill and blue cere. Bright pinky-orange feet. In flight, white breast contrasts with dark wings with a thin, white line across their centre. Female has white inner secondaries. Juvenile all-brown with pale salmon bill. Very large (about 75 cm tall, 230 cm wingspan).

Lappet-faced Vulture

N. Dennis/SIL

Lappet-faced Vulture

Torgos tracheliotus

AT PERCH

Characteristic silhouette with rather upright stance; huge, dark wings; broad, square, naked head and deep, pale bill. In full sun often hangs head in body shade with lappets extended. Note white leggings, pale, bare legs, streaked breast, small, dark eye and black crop. Juvenile has brown leggings and dark bill.

IN FLIGHT

Long, broad, rectangular wings with very long primary 'fingers' and broad, short, wedge-shaped tail. White legs and forearms obvious on adult. Small, naked head and streaked breast evident at closer range. Juvenile all dark, apart from pale head and feet.

DISTINCTIVE BEHAVIOUR

Flies early on broad wings, often as slow, searching flights at low altitudes. Notable for bounding in among other vultures feeding at carrion, with wings half-open, head lowered and tail cocked, with white legs and forearms obvious. Usually stands around the edge at a carcass – to select large chunks to eat alone, or to sunbathe with wings outstretched. Colour of bare head and looseness of neck skin vary according to emotion and temperature. Often found in pairs, rarely congregates in larger numbers. Builds extensive stick-nest platform, often on top of a small, isolated tree, which may also serve in place of a dead tree as a roost.

M. Goetz

Adult in cooling posture

ADULT Above dark sooty-brown. Below white and downy, with long, dark brown feathers streaked across breast and vent. Long, very woolly thighs white in southern African birds or brown in many eastern and western African birds. Head and neck naked, flushing from pale pink to dark red, with loose lappets of nape and neck skin either dangling or contracted. Short, black neck ruff. Flight feathers and tail black, forearm and elbow covered with dense white down. Crop down black. Deep, heavy bill dull yellow (southern and central Africa) or blackish horn (eastern and western Africa, Arabia) with dark blue base, cere and jaw; small eye dark brown; short, bare legs and long toes pale blue. Sexes similar in plumage, female about 2% larger (both sexes 5 400–7 950 g).

JUVENILE Similar to adult, but browner; thighs and forearm down dark brown, head paler with covering of short, dark down on crown and nape, longer erectile ruff feathers, bill blackish-brown, cere pale yellow. Assumes adult plumage after 6 years.

DISTRIBUTION, HABITAT AND STATUS Widespread in sub-Saharan Africa and the Arabian Peninsula. Favours open, semi-arid desert, steppe and grasslands. Less common in moister and more wooded savanna and absent from forests. Patchily distributed, main concentrations in southern, eastern and northeastern Africa (and Arabia), extinct in southern parts of South Africa.

SIMILAR SPECIES Similar in basic colours and flight pattern to much smaller **Hooded Vulture** (p. 44) (white inner thighs; long, slender bill; unstreaked, brown breast; silvery patch at base of flight feathers; dark forearms; blue eyelids; juvenile with dark brown cowl). Similar to **White-headed Vulture** (p. 38) (white lower breast; white line across centre of underwing; square tail; female with white inner secondaries; juveniles of the 2 species very alike; red (adult) or pink (juvenile) bill distinctive; white down on head in adult). In size and structure, most like **Eurasian Black Vulture** (p. 42) (darker blackish-brown, downy head; black, downy throat; blue-grey facial skin) but only overlaps rarely in the upper Nile valley.

ALTERNATIVE NAMES: (African) Black Vulture, King Vulture, Nubian Vulture

Largest black vulture. Naked red head; white, fluffy leggings. Breast white, streaked black. Heavy black or yellow bill. In flight, white legs and forearms obvious against the broad, black wings and short, wedge-shaped tail. Juvenile browner, with brown leggings, pink head and black bill. Very large (about 100 cm tall, 258–280 cm wingspan).

White-headed Vulture (juv.), Hooded Vulture, Eurasian Black Vulture

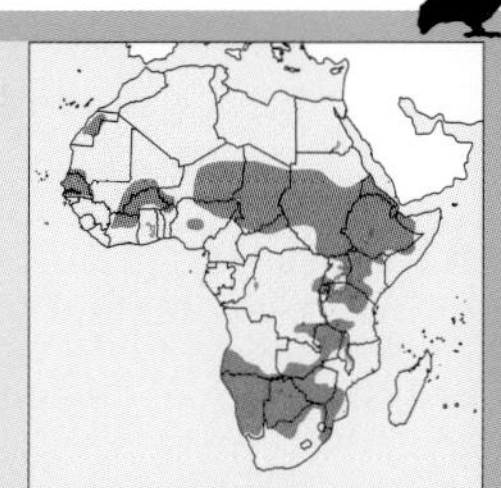

A. Shah/ABPL

Eurasian Black Vulture

Aegypius monachus

AT PERCH

Dark (almost black) overall colour, huge size, broad wing area and whitish legs are obvious. Blocky head and heavy bill appear relatively small, often withdrawn into the dense, shaggy ruff and back feathers. Note head and skin colours at close range. Adult has pale buff cap, dark horn-coloured bill and blue cere; juvenile has black bill separated by whitish cere from black head.

IN FLIGHT

Uniformly dark, almost black, overall, the pale feet and gape often evident in contrast. Flight feathers often appear slightly paler than body and wing coverts. Adult often has paler band of greater coverts across centre of wing. Long, broad, rectangular wings, with only slight secondary bulge on trailing edge and with long, upturned primary 'fingers'. Soars with wings held level and short, wedge-shaped tail often spread.

DISTINCTIVE BEHAVIOUR

Usually seen singly in Africa, probably mainly in immature plumage. Most often seen on the wing, soaring above hill and mountain slopes. Perches readily on trees, less often on rockfaces or the ground. Feeds at carrion, often with other vulture species.

F. Cahez/BIOS

Subadult in flight

Adult Dark, sooty-brown overall, flight feathers and tail slightly lighter with a silvery-grey wash below. Above, including dense neck ruff, mottled with dark fresh feathers and paler abraded ones. Greater coverts paler than rest of underwing. Vent dark grey. Head and neck have short, dense down. Buff cap on crown, ears and nape; cap surrounds black mask around eyes and on throat. Bill blackish-horn; eye dark brown; cere, gape and naked head skin blue to azure, showing through down; short, bare legs and long toes pale yellow. Sexes similar in plumage, female about 2% larger (male 7 000–11 500 g, female 7 500–12 500 g).

Juvenile Dark sooty-brown overall, almost matt black, except for off-white skin of cere, gape, head and feet. Moults into adult plumage over 6 years.

Distribution, habitat and status Was a very rare non-breeding migrant to the extreme northeast of Africa during the northern winter, recorded in Egypt in the Sinai desert and lower Nile valley. No recent records of breeding or vagrants in the mountains of Morocco or Algeria; linked previously to the Spanish population. Typically a bird of wooded mountains in its breeding range. Possibly extinct as an African species.

Similar species Easily confused with **Lappet-faced Vulture** (p. 40) (white leggings; streaky underparts; naked red head) which it closely resembles in overall size and proportions, and with which its range overlapped in drier areas of extreme northeastern Africa. Much darker than **Eurasian Griffon** (p. 36) and with different wing shape (secondary bulge; longer tail; wings in shallow V-shape when soaring), the only other large vulture within its range.

Alternative names: Cinereous Vulture, (European) Black Vulture

Huge, dark brown vulture. Broad, downy head with buff cap and dark mask and throat. Note blue cere, heavy, dark bill and pale, bare legs. Juvenile even darker, almost black, even on the head, with the bare skin areas dull white. Very large (about 90 cm tall, 250–295 cm wingspan). Was a very rare vagrant to northeastern Egypt.

Lappet-faced Vulture, Eurasian Griffon

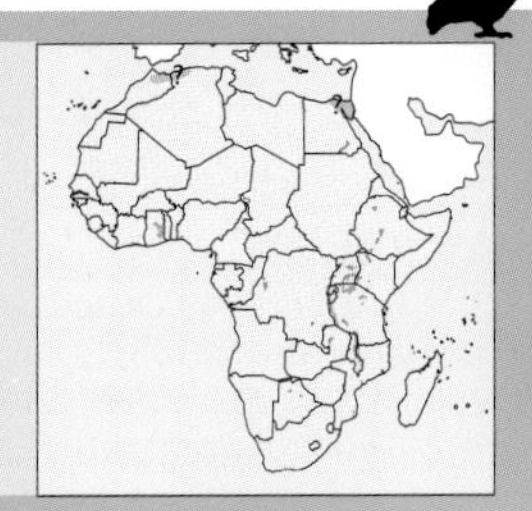

M. Gunther/BIOS

Hooded Vulture

Necrosyrtes monachus

AT PERCH

Sits upright with large, dark wings enclosing the body and the small head usually retracted in a hunched posture. On the ground, walks and runs energetically on long legs. Long, slender bill with small, oval nostril extends back into small, rounded head. Adult has bare, pink head, juvenile has dark cowl and blue-grey to pale pink face.

IN FLIGHT

Long, broad, rectangular wings with long primary fingers and short, square or slightly rounded tail. Pale, silvery 'windows' at base of flight feathers in juvenile and buff patch at base of forewing in adult. Note small, pink head and long bill at close quarters. White neck ruff and inner thighs often visible. Juvenile head, neck and thighs appear dark.

DISTINCTIVE BEHAVIOUR

Often seen perched along edges of forest and rivers, or soaring low and slowly over adjacent savanna. Frequently occurs in pairs. Soars earlier in the day than larger vultures and often first to discover carrion or small scraps of refuse and excreta. Uses the slender bill to pick at large carcasses after other scavengers have left, or to probe into dung and soft ground for insects. Nests on stick-platform built notably below the tree canopy.

L. Hes

Juvenile

D. Balfour

Adult showing head colours

ADULT Dark brown overall, except for white inner thighs and down on back of head, neck and front of throat. Feathers above finely edged with paler brown. Flight feathers black, with silvery wash below at base of inner primaries and outer secondaries. Crop down buff with dark spot at top. Long, slender bill dark horn colour; cere pale yellow; bare skin of head and neck flushing from pale to dark pink; eye dark brown; eyelids blue; slender, bare legs and long toes blue-grey. Sexes similar in plumage, female about 2% larger (both sexes 1 524–2 600 g).

JUVENILE Similar to adult, but darker and plainer, more clearly edged with paler brown above, and with dark brown crop and thighs. Flight feathers and tail all black. Cowl down more extensive with black crown and dark brown neck. Bare face pale blue-grey to pink with patches of dark down. Moults into adult plumage over 6–7 years.

DISTRIBUTION, HABITAT AND STATUS Widespread in sub-Saharan Africa with the exception of densely forested areas and treeless desert. Widely commensal with man and common in western and eastern to northeastern Africa, extending into a wider range of habitats including urban areas. In eastern and southern Africa uncommon and more restricted to wooded areas and game reserves.

SIMILAR SPECIES Most resembles juveniles of **Egyptian Vulture** (p. 46) (long, spiky, nape feathers; pointed wings; diamond-shaped tail; grey facial skin; slitted nostril) and **Palm-nut Vulture** (p. 48) (broad, rounded wings; broad, pale line at covert-flight feather junction; heavy, black bill; large, feathered head; narrow wedge of bare facial skin; vertical nostril slit; stout bill). In flight, similar wing and tail shape to much larger **Lappet-faced Vulture** (p. 40), but size not always easy to assess (adult has red head and heavy bill, white neck ruff and white leggings, but more extensive white below, streaked breast and patches along forewing).

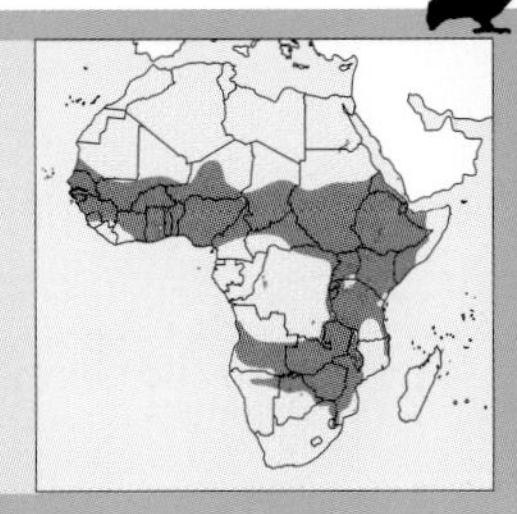

Small, dark brown vulture with woolly cowl. Bare face and neck. Long, thin bill. In adults, bare face flushes from pale to deep pink. Cowl, neck ruff and inner thighs white. In flight, note silvery wash below on flight feathers and short, square tail. In juveniles, cowl brown and facial skin blue-grey to pale pink with dark, downy patches. Large (about 60 cm tall, 170–176 cm wingspan).

Egyptian Vulture (juv.), Palm-nut Vulture (juv.), Lappet-faced Vulture

A. Wilson/Photo Access

Egyptian Vulture

Neophron percnopterus

AT PERCH

Often stands rather horizontally with short legs well forward and long wings and tail protruding behind. Often seen on the ground, looking rather like a large chicken as it walks about. Adult appears white, juvenile dark brown. At all ages the long, pointed face – backed by the spiky nape – and the pale legs are notable. Small, dark eyes and long nostril slit also obvious at close range.

IN FLIGHT

Note long, broad, rectangular wings with long primary fingers and short, diamond-shaped tail. Adult all-white except for black flight feathers, the upperwing with black edges to the coverts and trailing edge on either side of the whitish secondaries. Juvenile all dark brown but showing slight contrast between dark flight feathers and tail, and slightly paler coverts. Soars and glides with wings level.

DISTINCTIVE BEHAVIOUR

Flies long distance from cliff roosts to scavenge carrion or refuse. Also walks about to search for insects and birds' eggs, including in flamingo and pelican colonies. Renowned for ability to lift eggs in bill and smash them on the ground, or to use stones to break open large Ostrich eggs. Soars well, usually in pairs, rarely in small groups, including at roosts. Builds untidy stick-nest on rock ledge.

ADULT White overall, including tail, except for black flight feathers. Even the upper secondaries and inner primaries are washed with white. Sometimes becomes stained with dirt, especially on chest and wing coverts. Long, slender bill yellow with black tip; small eyes deep red; cere, bare face and throat deep yellow; stocky, bare legs and long toes pale grey, pink or yellow. Sexes similar in plumage, female about 2% larger (both sexes 1 584–2 200 g).

JUVENILE Dark brown overall, with black flight feathers and tail. Crop down white. Bill dark grey, eye dark brown, cere, face and throat pale dull yellow, legs grey. Moults into adult plumage by 4 years of age, becoming progressively paler.

DISTRIBUTION, HABITAT AND STATUS Widespread across northern and eastern Africa. European population extends to mountains along north coast of Africa, where it breeds, then migrates south. Breeding populations sedentary on Cape Verde, Canary and Socotra islands, across the southern Sahara, and into the eastern African highlands, rift valleys and grasslands. Extinct as breeding species in southern Africa, except possibly in northern Namibia; now only a very rare vagrant. Favours arid desert and steppe, but enters open savanna, grassland and farmlands on migration. Urban scavenger in some areas.

SIMILAR SPECIES Adult most like **Palm-nut Vulture** (p. 48) (black base to short, rounded tail; white primaries with black tips; broad, rounded wings; pale, heavy bill; bare, red face; yellow eye; vertical nostril slit). Juvenile most like juvenile **Palm-nut Vulture** (p. 48) (paler brown; more rufous underparts; heavy bill; small, bare facial area; broad, rounded wings; short, rounded tail) but even more like **Hooded Vulture** (p. 44) (back of head with downy cowl; short, square tail; adult with bare facial skin flushing bright pink; white thighs; oval nostril). Also resembles **White Stork** (*Ciconia ciconia*) in plumage, especially in flight (long, heavy, straight, red bill; long, red legs; long neck, including in flight; face feathered).

M&C Denis-Huot/BIOS

Adult

ALTERNATIVE NAMES: White Vulture, Scavenger Vulture

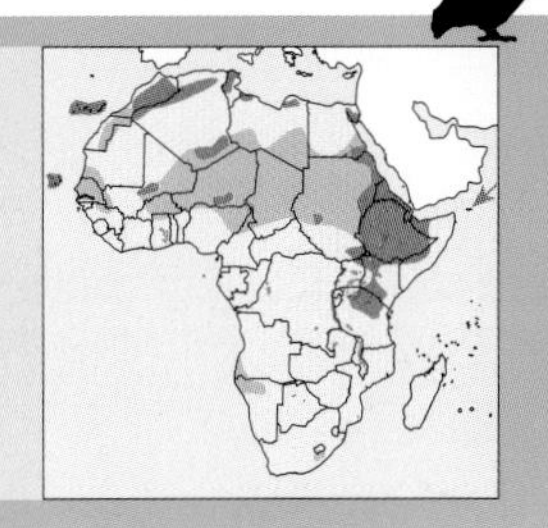

Small, white vulture with a long, thin bill. Extensive bare, yellow face. Long, spiky nape feathers; black flight feathers; white diamond-shaped tail. Juvenile plain dark brown. Large (about 50 cm tall, wingspan 155–170 cm).
Palm-nut Vulture, Hooded Vulture

L. Hes

Palm-nut Vulture

Gypohierax angolensis

AT PERCH

Stands upright on rather long legs. Appears white and thickset, with short, black wings and tail. Head large with long, loose nape feathers and long, heavy bill exaggerated by bare, red face. Juvenile dark brown, scalloped with buff above, but heavy, black bill and bare, pale yellow face still notable.

IN FLIGHT

Flaps broad, rounded wings with quick, stiff beats. Also soars with broad tail expanded to show distinctive flight pattern. Below white, with black wedges of secondaries and narrow, black tips to primaries. Above black, with white patches on primaries, white coverts and back, and white end of tail. Juvenile upperwing coverts slightly paler than flight feathers, rufous underparts separated by white line from dark flight feathers.

DISTINCTIVE BEHAVIOUR

Feeds mainly on fruits of raffia and oil palms, with some small carrion. Spends long periods perched near food trees, or walking about on beaches and river banks. Flies between feeding sites or, less often, soars over nesting area. Roosts in large trees, sometimes communally, and builds a large stick-nest in a tree, often near water. Congregates at good food sites and flies out to islands or boats to visit fishermen.

C. Paterson-Jones

adult feeding

C. Paterson-Jones

adult standing

ADULT White overall except for black secondaries, scapulars, greater upperwing coverts, primary tips and basal two-thirds of tail. Bill pale grey; cere yellow with vertical nostril slit; bare face and jaw red; eye yellow; stout, bare legs and long toes dull grey, pink or yellow. Sexes similar in plumage and size (male 1 361–1 710 g, female 1 712 g).

JUVENILE Dark brown above, with broad, pale edges to feathers. Paler brown and more rufous below, with white greater underwing coverts. Bill black, cere yellow, eye brown, bare face and legs pale yellow.

SUBADULT Pale brown body and wing coverts. Flight feathers and tail almost like adult's, with pale patch at base of primaries and white tail tip. Face becomes orange, eyes yellow and bill paler.

DISTRIBUTION, HABITAT AND STATUS Occupies edges of moist tropical and riverine forests of western and central Africa, including Bioko, extending eastwards and southwards along coastal forests and drier areas where palms planted. Very common and resident along major rivers through forest and tropical coasts. Uncommon and more local elsewhere; rare but regular vagrant far from usual habitats.

SIMILAR SPECIES Most resembles **Egyptian Vulture** (p. 46) (white adult; all-white diamond-shaped tail; all-black flight feathers; adult and brown juvenile have long, thin bill; bare yellow face; dark brown eye; long, horizontal nostril slit). Overlaps most in shape and habitat with **African Fish-eagle** (p. 50) (adult has small, white head; white chest; chestnut body and wing coverts; all-black flight feathers; all-white tail; black bill; bare, yellow face; juvenile brown with broad, black end to white tail; white mottling on head and chest). Brown juvenile also resembles **Hooded Vulture** (p. 44) (long, thin bill; downy cowl on back of head; long, rectangular wings; short, square tail) and juvenile **African Harrier-hawk** (p. 128) (much smaller and slighter; small bill; long, thin legs).

ALTERNATIVE NAME: Vulturine Fish-eagle

Dumpy white vulture with black wings and tail. Bare, red face; pale grey bill. In flight, has broad, rounded wings with black secondaries, black-tipped white primaries and short, black tail with broad, white tip. Juvenile brown with black bill and yellow face. Large (about 55 cm tall, 150 cm wingspan).

Egyptian Vulture, African Fish-eagle, Hooded Vulture

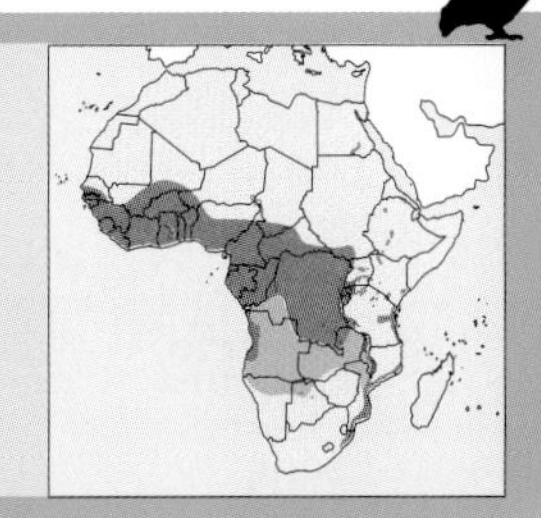

HPH Photography/Photo Access

African Fish-eagle

Haliaeetus vocifer

AT PERCH

White head and breast obvious from a distance, dark body colours only distinguishable at close range. Head relatively small with obvious bare yellow skin. Long, dark wings usually cover short, white tail. Immature plumages variable, with pale head and chest or untidy dark and light streaking most evident.

IN FLIGHT

Broad, rounded wings with dark flight feathers, and short, white tail form distinctive shape. Chestnut underparts and coverts add to contrast of white head, chest and tail. Soars with wings at slight dihedral; glides and dives with wrists bent; flaps with deep, slow beats.

DISTINCTIVE BEHAVIOUR

Spends long periods perched on prominent sites near water. Pair often soars above territory. Utters loud, yelping calls, tossing back head either at perch or in flight, often a pair together with sexual differences in pitch. Feeds mainly on fish taken from water surface in a dive, but also carrion, waterbirds and other small vertebrates, often stealing prey from other fish-eating birds. Builds large, obvious stick-nest in open fork of large tree or on rocks near water.

M. Goetz

Adult male

ADULT Head, neck, upper back and chest pure white. Body and wing coverts deep chestnut. Flight feathers and greater coverts black. Short tail pure white. Bill black; eyes pale brown; cere and lores bright yellow; stout, bare legs and feet pale yellow. **Female**, with deeper, more rounded, white bib, about 10% larger (male 1 986–2 497 g, female 3 170–3 630 g).

JUVENILE Brown with variable amounts of black and white streaking forming dark, streaked head, pale breast area and underwing coverts, and dark belly and thighs. Flight feathers a dark sooty-brown, primaries with white bases, secondaries all dark and tail mottled white with broad, dark terminal band. Cere, lores and legs grey.

SUBADULT Resembles adult with white head, neck and chest, but streaked with dark brown. Body and wings dark brown with white bases creating mottled effect; tail white with dark terminal band. Cere and legs paler yellow. Assumes adult plumage after about 5 years.

DISTRIBUTION, HABITAT AND STATUS Throughout sub-Saharan Africa, wherever there is open, shallow water at rivers, lakes, swamps, dams, estuaries or seashores. Widespread and common, locally even very common. Readily colonises new water bodies as they form.

SIMILAR SPECIES Adult unmistakeable. Juvenile and subadult resemble **Osprey** (p. 72) (wings barred; white underwing coverts; dark carpal patches; white body; wings narrower and usually bent; distinctive dark face pattern; yellow eyes), which also hunts mainly on the wing and often plunges underwater after fish. Juvenile also resembles juvenile **Palm-nut Vulture** (p. 48) (tail plain brown; head larger).

ALTERNATIVE NAMES: Fish Eagle, River Eagle

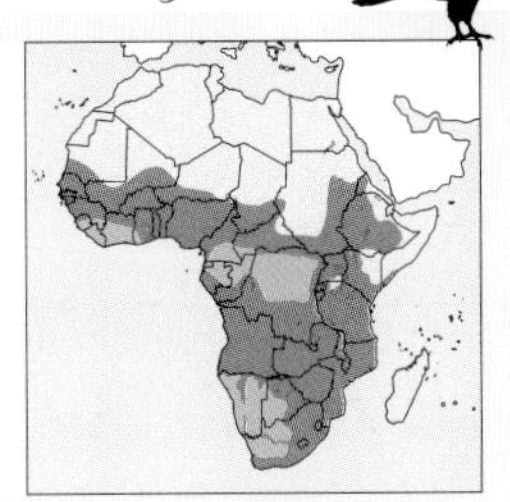

Pure white head and chest distinctive. Chestnut body and wing coverts. Black wings and short, white tail. Loud, yelping calls notable. Juvenile confusing and variable – mainly dark brown, streaked with black and white, and with broad, black terminal band to mottled tail. Large (about 60 cm tall, 191–237 cm wingspan).

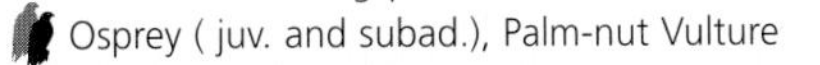

Osprey (juv. and subad.), Palm-nut Vulture

D. van Smeerdijk/Photo Access

Madagascar Fish-eagle

Haliaeetus vociferoides

AT PERCH

Large, brown eagle with relatively small, white head, short, white tail and pale, bare legs. Juvenile paler and browner and more streaky in appearance, with brown tail.

IN FLIGHT

All-dark wings and body with broad, upturned wings and short, white tail. Flies with deep, steady flaps but prefers to soar on extensive wing surfaces. Juvenile's pale rufous underwing coverts contrast with dark flight feathers and brown tail and with white primary 'windows'.

DISTINCTIVE BEHAVIOUR

Spends much time perched on prominent snags, or soars high over territory. Calls often and loudly with short, yelping calls, 'kyow-koy-koy-koy', uttered with head thrown back in flight or at perch; often a pair calls together. Hunts fish and a few crabs, taken from near the surface in a shallow dive. Also pirates from other waterbirds. Builds a large stick-nest in an open fork of a large tree or on a cliff near water.

D. Richards

Juvenile

ADULT Above dark dusky-brown. Rump has cream feather tips. Head, nape and sides of neck broadly streaked with cream, ear coverts white. Dark streak through eye and throat pale buff. Chest and underwing coverts brown with broad, deep rufous streaks, breast plain dark brown. Tail pure white. Flight feathers dark sooty-brown. Bill black; eyes dark brown; cere blue-grey; stocky, bare legs and feet pale yellow. Sexes similar in plumage, female larger.

JUVENILE Overall pale brown; feathers streaked with buff on head, neck and breast, and edged in buff elsewhere. Lower breast and vent dark brown and unmarked. Ears, throat and eyebrow-stripe unmarked grey and rufous; rufous wash to buff chest streaks. Tail brown, with some white mottling on inner webs of outer feathers. Tail feather shafts black in subadult plumage but exact plumage sequence undescribed.

DISTRIBUTION, HABITAT AND STATUS About 50 pairs remain in northwestern Madagascar (previously found all over island, and a vagrant to Mauritius, now probably the most endangered bird of prey in the world). Occupies seashore, islands, lakes and rivers, especially with wooded margins. Locally common at a few littoral sites.

SIMILAR SPECIES No similar large eagles on Madagascar. Similar in size, habits and body form, and especially in juvenile colour, only to mainland **African Fish-eagle** (p. 50) (call similar but longer, without abrupt ending on harsh, rasping yelp).

ALTERNATIVE NAME: Madagascar Sea-eagle

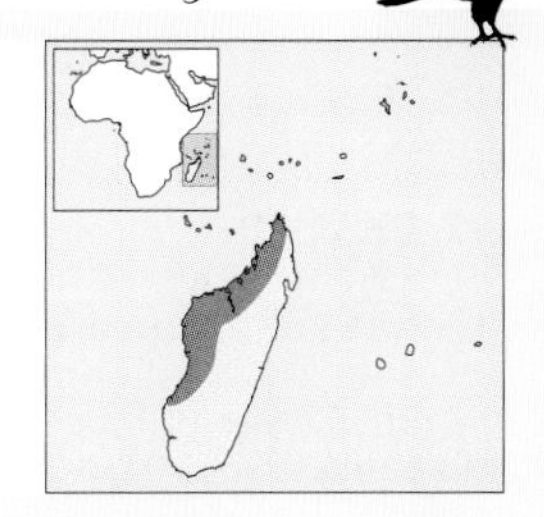

Short, white tail distinctive. Dark brown overall. Head and sides of neck streaked with cream. Loud, yelping calls, given with a toss of the head, also notable. Juvenile paler brown, more streaked with light brown, and brown tail mottled with white at the base. Yelping call similar to that of African Fish-eagle. Large (about 65 cm tall, 200 cm wingspan).

Only large eagle on Madagascar.

R. Thorstrom

Bateleur

Terathopius ecaudatus

AT PERCH

Appears stocky, large-headed and almost tailless, with long primary tips extending well below perch. Adult mainly black from the front or chestnut from behind, the grey or brown shoulder areas not always prominent. Juvenile brown all over. Red (adult) or dull turquoise or blue-grey (juvenile) areas of bare skin obvious. Male has silvery-grey 'shoulder', female browner with pale patch across secondaries.

IN FLIGHT

Characteristic bow-shaped wings with upturned tips, held at high dihedral. White undersides, rimmed black (widely in male, narrowly in female) and contrasting with black body. Grey and black upperside pattern also distinctive against chestnut back and notably short tail. Feet protrude past tail tip. Juvenile brown with longer tail, so that feet are concealed, and subadult brown and off-white in same basic pattern as adult.

DISTINCTIVE BEHAVIOUR

Spends long periods on the wing, where most often seen. Flies fast and low, usually within 100 m of the ground, without flapping but often rocking from side to side. When it does flap, it has notably short, fast, stiff wingbeats. When perched, large feathers contribute to fluffy, large-headed appearance. Pair often perch together before nesting, sometimes preening one another. Adopts special head-down posture, with bill tucked into breast, also when sunbathing with wings held fully extended and white undersides turned upwards. Feeds on small animals and carrion, including large carcasses, often descending from flight in a tight spiral. Builds large stick-nest in open fork of large tree, where it is sensitive to disturbance.

ADULT Jet-black overall, with chestnut back and tail (pale buff in uncommon cream-backed form). **Male** has silvery-grey lesser and median upper-wing coverts, forming pale 'shoulder', and all-black upper wings. Underwing white with broad, black trailing edge. **Female** has all upperwing coverts grey-brown and secondaries off-white with dark grey tips. Underwing white with narrow, black trailing edge. Bill yellow with black tip; eyes dark red-brown; cere, extensive bare facial skin, short, stout legs and long, thick toes with knobbly scales bright crimson. Female about 4% larger (both sexes 1 820–2 950 g).

JUVENILE Uniform chocolate-brown overall, with pale feather tips giving mottled effect. Underside of flight feathers paler grey-brown in contrast. Bill blue-grey with black tip; eyes dark brown; cere and facial skin dull blue-green and legs whitish.

SUBADULT Moults into distinct subadult plumage, predominantly brown but with pale grey-brown areas on underwing which will become pure white in adult. Later emergence of black feathers gives grizzled sooty effect that is matched by colouring of red skin areas. Only attains adult plumage at 7–8 years of age.

DISTRIBUTION, HABITAT AND STATUS Widespread over grassy and wooded savannas of sub-Saharan Africa, extending to Arabian Peninsula, and rarely to Israel. Absent from areas of forest and woodland with closed canopy, and also from arid steppe and desert habitats. Obvious and common in many areas but eliminated from areas of high human density and degraded habitats.

SIMILAR SPECIES Adults unmistakeable. Juvenile similar to brown snake-eagles, either juvenile **Black-chested Snake-eagle** (p. 58) or all ages of **Brown Snake-eagle** (p. 56) (stance more upright; eyes large and bright yellow; legs longer, toes shorter and both pale grey or buffy; dark brown underwing coverts; flight feathers white with bars (Black-chested) or all silvery-white (Brown)). Prefers more open, flat habitats but flight pattern very like **Jackal Buzzard** (p. 116) (black underwing coverts; white flight feathers with broad, black tips; body black, white and rufous below).

N. Dennis

Juvenile

ALTERNATIVE NAME: Bateleur Eagle

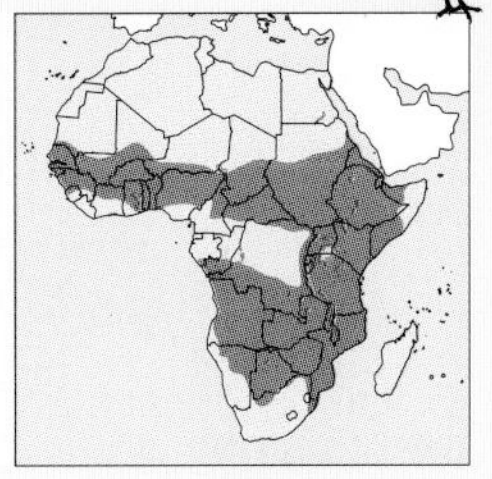

Stocky, black, large-headed eagle. Grey 'shoulders'. Prominent, bare, red face and legs. Chestnut or cream back and tail. Unique shape in flight, with long, pointed, bow-shaped wings and stubby tail. Sexes distinguishable by white in secondaries and width of black trailing edge. Juvenile brown with dull blue-green face and whitish legs. Large (about 50 cm tall, 180 cm wingspan).

Jackal Buzzard, Brown Snake-eagle, Black-chested Snake-eagle (juv.)

Brown Snake-eagle

Circaetus cinereus

AT PERCH
Dark brown overall appearance makes yellow eye and pale legs obvious. Large head and upright stance often diagnostic, even at a distance. Juvenile or subadult sometimes distinguishable only at close range.

IN FLIGHT
Contrast between dark brown underwing coverts and silvery-grey flight feathers distinctive from below. Tail bars also notable. Wings rather short and pointed against dark, heavy body with prominent head protruding. Wingbeats are fast and rather shallow. Soars at times but only rarely glides or attempts to hover unless quite windy.

DISTINCTIVE BEHAVIOUR
Spends much of time perched on top of a prominent emergent tree, often on a ridge or hilltop, even during the heat of the day. Feeds mainly on large snakes and lizards, as well as some other small vertebrates. Often strikes at prey from a long distance. Secretive when breeding in a small stick-nest on top of a densely foliaged tree, epiphyte or creeper.

M. Goetz

Subadult

I. Davidson

Subadult soaring

ADULT Dark brown overall. Flight feathers have white mottling at base of inner web and silvery-grey unbarred undersides. When fresh, tail has 3 narrow, white bars and fine, white tip. Bill black; large eyes deep yellow; cere, long bare legs and stubby feet pale grey. Sexes similar in plumage, female about 5% larger (both sexes 1 540–2 465 g).
JUVENILE Similar to adult, dark brown all over with silvery underwing and barred tail. Some individuals are slightly paler or have fine pale tips to the feathers of the upperside to give a faint scaled effect. White feather bases on head and breast often more obvious. Softparts like adult's.
SUBADULT Similar to adult and juvenile but at close range has fine, white bars and tips, or white mottling, on flank and vent feathers. White feather bases on head and breast frequently evident.
DISTRIBUTION, HABITAT AND STATUS Widespread in sub-Saharan Africa. Often common, but numbers fluctuate suggesting local movements that are poorly understood. Favours wooded thorn and broad-leafed savanna, even dense woodland, but absent from forest and more arid steppe and grassland habitats.
SIMILAR SPECIES Only really similar to juvenile **Black-chested Snake-eagle** (p. 58) (paler, more rufous-brown overall; darker above, mottled with white; below pale rufous-brown) and juvenile **Beaudouin's Snake-eagle** (p. 60) (pale brown above, white below). Sometimes confused with dark brown juvenile **Bateleur** (p. 54) (mottled with lighter brown; dark brown eye; short, whitish legs; blue-green cere; long, pointed wings; very short, unmarked tail; no white on underwing surface).

ALTERNATIVE NAME: Brown Harrier-eagle

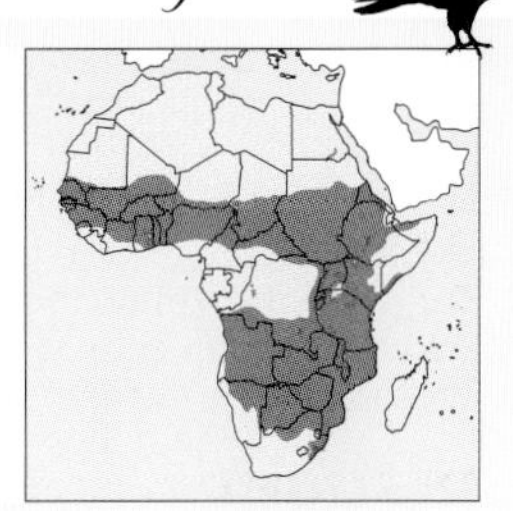

Dark, all-brown eagle. Glaring yellow eyes. Long, pale, bare legs. Large, block-like head. Upright stance when perched. In flight, silvery flight feathers contrast with dark brown underwing coverts. Juvenile and subadult similar. Large (about 65 cm tall, 164 cm wingspan).

Black-chested Snake-eagle (juv.), Beaudouin's Snake-eagle (juv.), Bateleur (juv.)

Black-chested Snake-eagle

Circaetus pectoralis

AT PERCH
Large head and bare legs obvious as often stands rather horizontally. White underparts contrast with predominantly black colour. Long primaries extend to tip of rather short tail. Juvenile appears rufous-brown, subadult notable for blobs on white chest and darker brown upperparts.

IN FLIGHT
Long, broad, square-tipped wings held stretched at slight dihedral. Underwings appear mainly white, with secondary and tail barring; black head and chest obvious. Juvenile appears rufous below with white flight feathers and barred tail.

DISTINCTIVE BEHAVIOUR
Hunts on the wing more than other snake-eagles, soaring, kiting when windy with wings bent, or hovering with awkward flapping. Strikes mainly at small reptiles, also other mammalian, avian or arthropod prey, often over long distances. Builds a small, inconspicuous stick-nest concealed on the crown of a low thorn tree, carrying reptiles to the nest with the tail dangling from the bill. When not breeding may gather at social roosts of up to 200 birds but still hunts solitarily.

R. du Toit

Adult hovering

ADULT Above black, including neck and chest. Breast and underwing plain white, apart from narrow black bars across secondaries and tail. Bill black; large eye bright yellow; cere and long, bare, legs pale grey. Sexes similar in plumage and size (both sexes 1 178–2 260 g).

JUVENILE Brown above with obvious pale feather edges and contrasting dark brown flight feathers. Paler rufous-brown below, including underwing coverts and head. Crown lightly streaked with darker brown and with grey wash over ears. Flight feathers white below, only faintly barred with brown. Eye, cere and leg colours as in adult.

SUBADULT Distinct subadult plumage similar to adult plumage but sooty-brown above. Below white or cream, marked with large, brown blobs on upper chest and broad, rufous-brown bars on flanks and flight feathers.

DISTRIBUTION, HABITAT AND STATUS Open steppe, savanna and woodland of eastern, central and southern Africa. Absent only from true desert and dense forest. Highly nomadic in many areas, possibly even migrant, but exact movements not known.

SIMILAR SPECIES Adult most like **Martial Eagle** (p. 74) (black underwings; spots on white breast and legs; distinct crest; feathered legs). Juvenile like **Brown Snake-eagle** (p. 56) (darker brown; shorter, more pointed wings; dark underwing coverts; unbarred silvery-white flight feathers).

A. Kemp

Subadult

ALTERNATIVE NAMES: Black-breasted Snake-eagle, Black-breasted Harrier Eagle

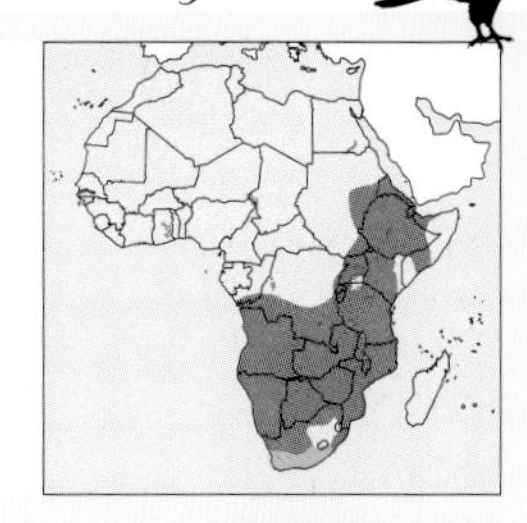

Black with plain white underparts. Black bars across secondaries and tail. Large head with prominent yellow eyes. Long, bare, pale grey legs. Juvenile pale rufous-brown with faintly barred flight feathers. Subadult brown above with pale scallops, below white with large, black spots on chest and faint, brown bars on breast, wing and tail. Large (about 55 cm tall, 178 cm wingspan).
Brown Snake-eagle (juv.), Martial Eagle (ad.)

Beaudouin's Snake-eagle

Circaetus beaudouini

AT PERCH

Grey-brown above, contrasting with dark brown flight feathers. Note barred tail. From the front, note the brown head with its large, yellow eyes, brown chest, barred breast and pale, bare legs. Juvenile mottled white and brown, with yellow eye and pale legs and pale rufous or white underparts slightly streaked. The large head is a feature at all ages.

IN FLIGHT

Long, broad, square-tipped wings; medium-length tail. From above, paler contour feathers contrast with darker flight feathers, of which the tail appears most obviously barred. From below, white underwings and tail are clearly barred. Wings have broad, dark grey trailing edge and primary tips and bands on tail. Juvenile similar, but browner below and with paler, less distinct barring.

DISTINCTIVE BEHAVIOUR

More often described perched than on the wing and only rarely reported to hang and kite in the wind, or to hover clumsily; this may explain why it favours less open habitats than its Short-toed and Black-chested relatives. Builds a small stick-nest on top of a low tree, but generally poorly known. Diet reported to consist of snakes and some other small vertebrates.

ADULT Dark grey-brown above, including head and chest. Below white, finely barred with brown. Vent white. Flight feathers dark grey-brown with broad, dark brown bars, visible on underside of tail as 2 narrow bars and a broad terminal bar. Bill black; eyes large, obvious and yellow; cere, long bare legs and stubby toes pale grey. Sexes similar in plumage and size.

JUVENILE Appears paler brown than adult, mottled with white, or dark brown above, with white bases to feathers. Head, neck and chest rufous to white, with brown centres to feathers creating streaked effect. Below pale brown, with extensive white feather bases. Softparts the same as adult's.

SUBADULT More grey-brown and usually darker above than juvenile, often including head. Below white, chest blotched and lower breast barred on flanks with brown and grey.

DISTRIBUTION, HABITAT AND STATUS Restricted to the open woodlands of western sub-Saharan Africa. Moves northwards in the rains and southwards in the dry season as its habitat, in effect, shifts with the seasons. Widespread, but nowhere common and little studied.

SIMILAR SPECIES Previously lumped together with **Short-toed** and **Black-chested snake-eagles** (pp. 62 and 58) as the same species, mainly through confusion of subadult plumages. Short-toed Snake-eagle most similar but overall paler, Black-chested Snake-eagle much darker and more solidly marked. Subadults similar, but juveniles quite different, neither like paler, less-marked adult Short-toed, nor like plain brown and rufous Black-chested.

ALTERNATIVE NAME: Beaudouin's Harrier-eagle

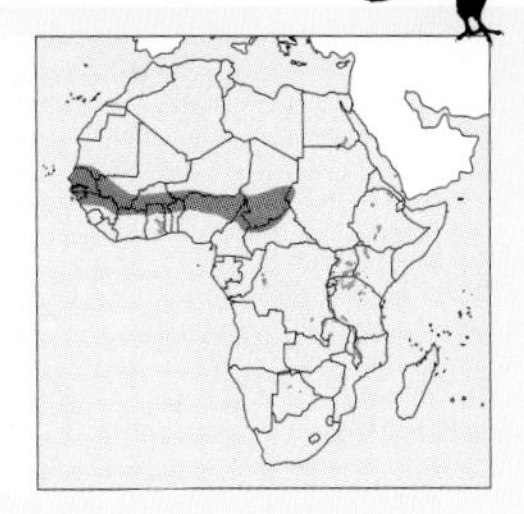

Grey-brown above, including head and chest. White below with narrow, brown bars. In flight, wings and tail grey-brown above, broadly barred with dark brown. Eyes yellow. Legs bare and pale grey. Juvenile browner above and pale brown and white below. Large (about 50 cm tall, 170 cm wingspan).

Short-toed Snake-eagle, Black-chested Snake-eagle

Short-toed Snake-eagle

Circaetus gallicus

AT PERCH

Perches rather upright. Yellow eyes obvious in large, dark head. Pale, bare legs conspicuous. Dark upperparts and chest contrast with white underparts. Breast barring evident only at close range. Pale cere often accentuated by white forehead and lores. Juvenile and subadult not easily distinguished from adult.

IN FLIGHT

Long, broad, square-tipped wings, always with dark tips to primaries, but with individually variable barring on the underside. Three tail bars, dark, protruding head and dark chest notable from below. From above, flight feathers obviously darker than contour feathers. When hanging or gliding, wing tips curved well back and wrist prominent. Hanging and hovering flight often distinctive.

DISTINCTIVE BEHAVIOUR

Spends much time on the wing, soaring on flat wings, kiting or gliding on bent wings, or hovering awkwardly. Flaps with strong, deep beats at take-off. Takes its main prey of reptiles and small mammals on the ground, often after a long stoop from a prominent perch or from in flight. Builds a relatively small stick-nest on top of a tree, usually well-concealed. Vocal before breeding but generally silent for much of the year.

ADULT Above dark grey-brown, including head and chest. Below white with narrow dark brown bars. Vent white. Flight feathers dark brown above, below white with narrow, brown bars, broad, brown tips and black primary tips. Broad terminal bar and 2 or 3 narrow bars of tail obvious from above and especially from below. Bill black; large eyes deep yellow; long, bare legs, stout, stubby feet and cere pale grey. Sexes similar in plumage, female about 2% larger (male 1 180–1 892 g, female 1 304–2 324 g).

JUVENILE AND SUBADULT Both similar to adult, but paler and markings not as obvious. Juvenile often has pale tips to upper contour feathers and is almost plain white below, including underwing.

DISTRIBUTION, HABITAT AND STATUS Widespread in southern Europe and Asia, small breeding population in northwestern Africa. Some migrate south in northern winter to sub-Saharan Africa, but not south of equator. Nests in small patches of woodland but prefers to hunt over open terrain, extending during migration to semi-desert. Uncommon in Africa, but formerly confused on migration with resident Beaudouin's Snake-eagle.

SIMILAR SPECIES Most resembles adult **Beaudouin's Snake-eagle** (p. 60) (usually darker above; less distinctly barred below and on the flight feathers; tends to frequent more wooded habitat). In flight, when wings bent and less barred below, resembles **Osprey** (p. 72) (narrower wings; distinctive dark facial pattern).

ALTERNATIVE NAMES: Short-toed Eagle

Dark brown above, including head and chest. Below white with dark brown bars. Tail has 3–4 dark bars. Large, yellow eyes. Bare, pale grey legs. In flight shows pale underwing with dark, narrow bars, dark trailing edge and black primary tips. Juvenile a slightly paler version of adult. Individual variation in colour and markings of adults and juveniles. Large (about 55 cm tall, 180 cm wingspan).

Beaudouin's Snake-eagle, Osprey

Southern Banded Snake-eagle

Circaetus fasciolatus

AT PERCH

Grey-brown above, white below and long, barred tail extending past wing tips distinctive. Note large head with pale yellow eye, cere and legs pale yellow. Sits rather upright, often with head retracted so that it appears squat and the tail length is exaggerated.

IN FLIGHT

Extensive barring of underparts, underwings and rather long tail obvious. Plain brown head and chest, black trailing edge and wing tips, and pattern of tail bars notable. Wings rather short and rounded with distinct notch. Flaps with rapid, shallow beats and soars when possible, but not an accomplished flier.

DISTINCTIVE BEHAVIOUR

Spends much time perched on a large, open branch below the canopy, watching the ground for the small reptiles and frogs that are its main prey. Calls frequently, especially in the early morning, with a fast, high crowing 'kok kok kok kwaaak' or 'kerk kowa' repeated at perch or in display flights. Otherwise seen infrequently in flight above the canopy, and sometimes perches in the open on forest edge. Makes short strikes at prey on the ground. Builds a small stick-nest, usually concealed among creepers.

ADULT Above dark grey-brown, feathers edged with pale rufous to give cinnamon effect. Rump darker with fine, white tips to feathers. Head and chest plain grey-brown. Below white, with relatively broad, brown bars except for vent. Upperwing dark grey-brown, underwing white with brown bars on coverts, and 4 or 5 black bars plus broad ends to flight feathers. Tail brown above, white below, showing 3 or 4 dark bars, a broad subterminal bar and, when fresh, a white tip. Bill black; eyes, cere, base of bill and long, bare legs with stubby toes pale yellow. Sexes similar in plumage, female about 2% larger (both sexes 908–1 100 g).

JUVENILE Dark brown above, streaked on head, mottled on nape and tipped with white on coverts. Below white or washed pale rufous, streaked with brown on throat and breast, barred on flanks. Wings and tail similar to adult, but with narrower dark bars in tail. Subadult plumage may be even paler below, with head and neck almost white or grey-brown, but not well-studied. Cere and legs deeper yellow than adult.

DISTRIBUTION, HABITAT AND STATUS Restricted to the east coast of Africa. Inhabits coastal forests and woodland, often in riverine forest or near water, including exotic plantations. Secretive and sedentary, but may be locally common once calls are recognised.

SIMILAR SPECIES Most resembles **Western Banded Snake-eagle** (p. 66) (greyer; almost unbarred below; short, black tail with a broad, white band across the middle; cere and legs deep yellow). Immature more difficult to distinguish but tail pattern quite different. Their ranges barely overlap in central Mozambique and northern Kenya. Both have pale yellow legs, unlike pale grey legs of other snake-eagles.

Adult

P. Pickford/SIL

ALTERNATIVE NAMES: Fasciolated Snake-eagle, Fasciated Snake-eagle, East African Snake-eagle

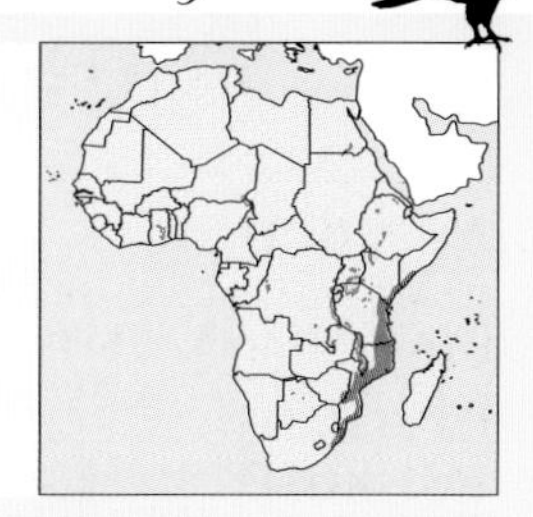

Plain cinnamon-brown overall. Grey-brown head and neck. Breast white with broad, brown bars. In flight, underwing and tail have narrow, dark brown bars, broad, black subterminal bars and white tips to all but outer primaries. Tail shows 3 or 4 dark bars. Large, pale yellow eyes. Cere and bare legs pale yellow. Juvenile pale brown and white, below more streaked than barred. Medium-sized (about 50 cm tall, 120 cm wingspan).

Western Banded Snake-eagle

Western Banded Snake-eagle

Circaetus cinerascens

AT PERCH

Appears squat, large-headed, short-tailed and plain grey-brown, apart from white tail band. Eye, cere and leg colours prominent at close range.

IN FLIGHT

Body and underwing coverts appear plain grey-brown against white underwings with narrow dark bars and dark trailing edges. Short tail with broad, black subterminal bar (and base in adult) distinctive at all ages.

DISTINCTIVE BEHAVIOUR

Spends much time perched below the canopy. Easily overlooked unless calling with its distinctive loud, descending, cawing notes. Hunts mainly snakes, and some other small vertebrates, dropping from its perch to trunks, foliage or the ground. Builds a small stick-nest, well-concealed within foliage or among creepers. Occasionally rises to soar and call above the canopy.

D. Richards

Adult

ADULT Plain grey-brown above and below, including underwing coverts. White tips to belly feathers and bars on flanks in all but a few unmarked individuals. Flight feathers dark grey (almost black) above, contrasting with paler upperparts. Below white with narrow, black bars and broad, black trailing edge. Short, black tail distinctive with broad, white, band across centre. Bill black; eyes pale yellow; cere, long bare legs and stubby toes deep yellow. Sexes similar in plumage and size (about 1 126 g).

JUVENILE Above paler and browner than adult, contour feathers with white edges and broad, white streaks on head. Below off-white streaked with pale brown, especially on belly and thighs. Flight feathers dark brown above, white below with brown bars and broader brown trailing edge. Tail has broad, black subterminal band; dark base almost concealed. Eye, cere and legs very pale yellow. Subadult plumage may be all dark grey-brown, without white markings on underparts, but remains to be properly distinguished.

DISTRIBUTION, HABITAT AND STATUS Moist woodlands of sub-Saharan Africa; patchy distribution from Senegal eastwards to Ethiopia and southwards to the Zambezi and Save river systems. Favours riverine forest patches with large, creeper-laden trees. Does not enter lowland rainforest. Locally common but secretive, sedentary and often only detected by calls.

SIMILAR SPECIES Most resembles **Southern Banded Snake-eagle** (p. 64) (more obviously barred below and on underwing; longer tail showing 3 or 4 narrow, more even dark bars; paler yellow cere and legs). Juveniles more difficult to distinguish but tail patterns remain quite different. Ranges adjoin only in central Mozambique and northern Kenya.

ALTERNATIVE NAMES: (Smaller) Banded Snake-eagle, Banded Harrier-eagle

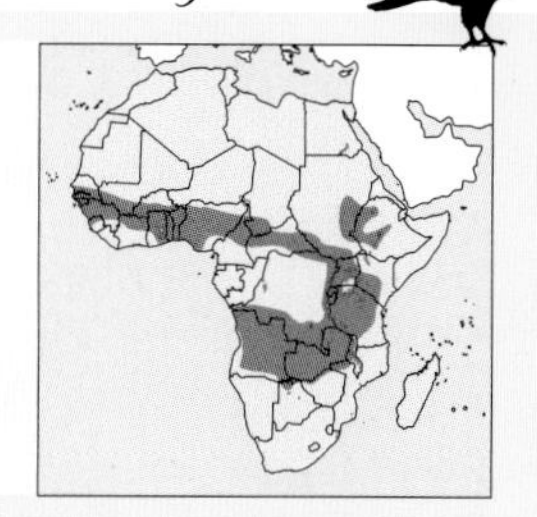

Stocky and pale grey-brown overall. Vestiges of white bars on lower breast and flanks. Eye pale yellow. Cere and bare legs deep yellow, almost orange. In flight, short, black tail with broad, white bar across centre distinctive. Underwings narrowly barred with broad, black trailing edge. Medium-sized (about 45 cm tall, 114 cm wingspan).

Southern Banded Snake-eagle

Congo Serpent-eagle

Dryotriorchis spectabilis

AT PERCH

Brown above and white below, with dark barring on flanks, stripes on throat and spots on breast. Upright stance with long, barred tail hanging below distinctive. Dark crown and large, loose nape feathers. Rufous wash on breast and nape notable in Congo basin subspecies.

IN FLIGHT

Barred wings, flanks and tail against white underparts obvious. Above plain dark brown, including rump. Long, rounded tail and short, rounded wings distinctive.

DISTINCTIVE BEHAVIOUR

Spends much time perched in dense mid-stratum of forest and 20–30 m up in secondary forest and on forest edge, searching for prey. Feeds mainly on reptiles and amphibians, catching them off trunks or foliage, or the ground, sometimes after hard, repeated strikes with the feet. Usually flies only a short distance between perches, but often seen flying above canopy or high across clearings. Utters low, crowing calls, not unlike those of Western Banded Snake-eagle.

ADULT Above grey-brown, darkest on crown and neck. Long nape feathers have white bases, creating mottled effect. Throat, chest and under-wing coverts pale rufous (West Africa) or white (Congo basin). Throat has black central and lateral 'moustache' stripes. Breast and underwing coverts have brown spots on feather tips. Lower breast white, broadly barred with brown and washed with rufous on flanks and thighs. Flight feathers grey-brown above with faint, darker bars and broad tips. Below white to pale grey with dark grey and brown bars and tips. Tail has 5 dark bars and a white tip. Bill black; large eyes yellow (male) or brown (female); cere, long, bare legs and stubby feet yellow. Sexes similar in plumage and size (about 700 g).

JUVENILE Rufous-brown above with white feather bases, especially on nape; darker ends and white tips create scalloped effect. Below white with large, dark spots, forming bars on flanks. Flight feathers similar to adult's but browner and with white tips. Eye grey, becoming yellow with age. Black streaks, rather than spots, below may indicate juvenile form or different subadult plumage.

DISTRIBUTION, HABITAT AND STATUS Lowland rainforest from Senegal to Cameroon (*D.s. batesi*) and throughout Congo basin (nominate subspecies). Appears restricted to large stands of primary evergreen forest. Widespread but generally considered uncommon unless calls are recognised.

SIMILAR SPECIES In shape and size most like **Long-tailed Hawk** (p. 146) (tail graduated with broad, white tips; grey above with prominent white rump; white bars across black tail and wings; below deep rufous (occasionally plain grey); eyes yellow; no loose nape feathers; juvenile most similar, with white underparts spotted brown, but white rump and tail bars prominent). Spotted underparts shared with juveniles of smaller, shorter-tailed forest accipiters (**Chestnut-flanked Sparrowhawk** (p. 174), **African Goshawk** (p. 172) and **African Cuckoo-hawk** (p. 190)).

ALTERNATIVE NAME: African Serpent-eagle

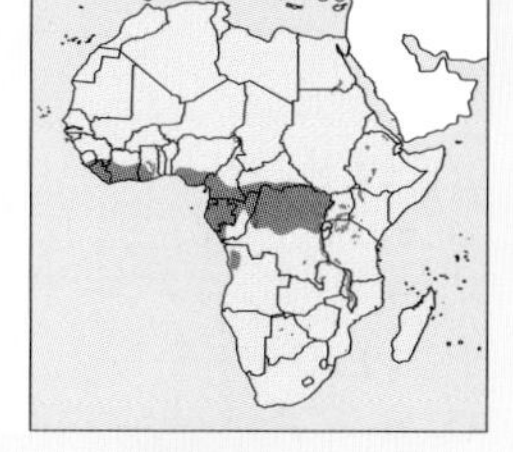

Long, barred tail most distinctive. Body dark brown above; below white or pale rufous, with broad, dark spots on breast and bars on flanks. Dark central throat and 'moustache' stripes. Bare, yellow legs. Eyes yellow (male) or brown (female). Short, barred, rounded wings. Juvenile has paler head and neck. Medium-sized (about 50 cm tall with long tail, 85 cm wingspan).

Long-tailed Hawk

Madagascar Serpent-eagle

Eutriorchis astur

AT PERCH

Dark brown above; white finely barred with brown below. Long, broad, narrowly barred tail obvious as bird perches upright on long, bare, yellow legs. Large head and crest notable.

IN FLIGHT

Long, broad tail and short, rounded wings with narrow, dark bars distinctive from below.

DISTINCTIVE BEHAVIOUR

Almost unknown until recently rediscovered. Located only in mid-stratum of dense forest, below canopy, where it spends long periods perched motionless on a branch. Large, loose nape feathers erected when excited. Flies from branch to branch and descends to forest floor, but generally reluctant to fly unless disturbed. Diet includes chameleons and maybe lemurs and birds. Call consists of a distinctive series of crowing notes 'pok pok caa caa', often the only clue to the presence of this secretive species.

R. Thorstrom

Juvenile

ADULT Above fuscous-brown with white feather bases and darker ends, creating barred effect. Below, ears and throat often paler and more streaked with brown on head and neck, but finely barred on rest of underside, including underwing. Flight feathers dark brown above and off-white below with narrow, darker brown bars. Tail has 6 dark bars and a narrow, white tip. Bill black; eyes yellow; cere grey but almost covered in bristles and feathered lores; long, bare legs and stout feet yellow. Sexes similar in plumage and size.

JUVENILE Similar to adult. Fine, white tips to large head feathers and voluminous crest feathers, mantle, coverts and rump give a scaly, scalloped effect. Below, dark bars more widely spaced. Eyes blue-grey.

DISTRIBUTION, HABITAT AND STATUS Confined to primary rainforest up to 1 000 m a.s.l. on north-eastern Madagascar, but also recorded at forest edge.

SIMILAR SPECIES Very similar in size and colour to **Henst's Goshawk** (p. 182) (shorter, narrower tail; white eyebrow; breast and throat more finely barred with dark grey).

ALTERNATIVE NAME: Long-tailed Serpent-eagle

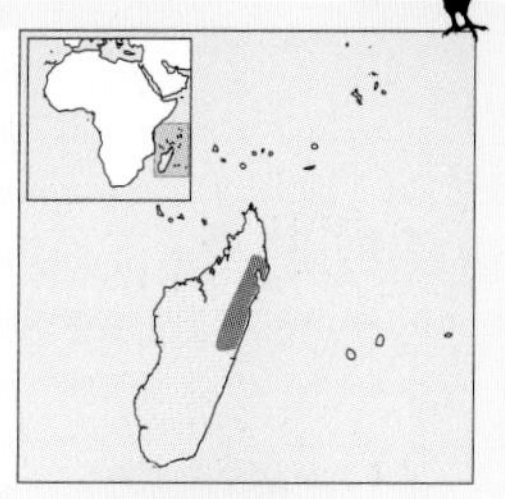

Long, very broad and floppy, multi-barred tail. Brown above, with faint, darker bars. Below and underwing white with fine, brown bars. Large head with large, loose nape. In flight, rounded wings and long tail distinctive. Flight feathers have numerous narrow, dark bars. Call is a distinctive series of crowing notes. Large (about 60 cm tall with long tail, 95 cm wingspan).

Henst's Goshawk

Osprey

Pandion haliaetus

AT PERCH

Usually perches rather horizontally on relatively short legs, appearing more kite-like than eagle-like. White head and obvious face-mask conspicuous. Large size of feet and (at close range) spiny soles and large, recurved claws are distinctive. Long primaries extend past tip of tail.

IN FLIGHT

Shape of narrow, bent, gull-like wings with long primaries is distinctive. Flies with wrists held higher than body, dropping to upturned primaries. Tail medium-length and usually held closed. Head relatively small and protruding on long, flexible neck. Mainly flaps, with deep, steady beats; only sometimes soars.

DISTINCTIVE BEHAVIOUR

Hunts with steady prospecting flight, working up and down a selected stretch of shore. Catches fish and some other prey by diving steeply into the water, sometimes becoming completely immersed. On emergence always hesitates in flight to shake off water and, if a fish is caught, to changes its grip so that the prey faces headfirst into the wind. Calls infrequently, mainly shrill whistles either drawn-out or in staccato bursts.

P. Ginn

Adult

ADULT Dark brown above. Head and neck white with broad, black stripe through eye and streaks on forehead. Below white with fine, black streaks on sides of chest. Flight feathers and tail dark brown above, white and grey below, with narrow, dark bars and black wing tips. Black carpal patch on white underwing coverts. Bill black; eyes yellow; cere, heavy bare legs and strong feet blue-grey. **Female** more darkly streaked on breast, and about 5% larger (male 1 220–1 600 g, female 1 250–1 900 g).

JUVENILE Similar to adult, but mask and hindneck streaked with white; feathers of upperparts have narrow, lighter tips, producing fine, scalloped effect on wing coverts. Moults directly into adult plumage at end of first year.

DISTRIBUTION, HABITAT AND STATUS May occur anywhere in Africa or on Socotra while on migration during northern winter. Breeds on the shore or offshore islands of the Red Sea and Mediterranean, and on the coasts of Canary and Cape Verde islands. Most are migrants from further north in Europe. Commonest along coasts, lagoons, swamps and large rivers of western Africa, elsewhere mainly found around sea coasts, large lakes and man-made dams. Immature birds often stay over in Africa.

SIMILAR SPECIES Unlike any other hawk. In similar aquatic habitats most often confused with juvenile **African Fish-eagle** (p. 50) (larger; heavily built; broad, rounded wings; short tail with mottled off-white base and broad, dark terminal band; eyes brown; head and neck streaked; brown below streaked with rufous and white). Resembles adult **Black-chested Snake-eagle** (p. 58) (solid black chest, no facial pattern).

 ALTERNATIVE NAME: Fish Hawk

Osprey

Dark face mask. Long, gull-like wings, usually held slightly bent. Powerful, blue-grey legs and feet. Yellow eyes. Lives around water and plunge-dives for fish. Large (about 45 cm tall, 140–170 cm wingspan).

African Fish-eagle (juv.), Black-chested Snake-eagle

Martial Eagle

Polemaetus bellicosus

AT PERCH

Appears tall and thickset when standing upright; height and build accentuated by broad, flat head and short tail. When bird stands horizontal, the large wings and long legs are notable. At a distance, adult looks black with a white belly. Spotted underparts, yellow eyes and pale cere and feet only notable at closer range. Juvenile looks very pale grey and white, the dark eye conspicuous, and blue-grey cere and feet. Subadult appears darker grey above.

IN FLIGHT

Distinctive flight silhouette of long, broad wings held at marked dihedral, short, fanned tail, and wide, upturned wing tips. Adult appears black with a white belly, juvenile white with dark grey barred flight feathers and tail.

DISTINCTIVE BEHAVIOUR

Often seen perched boldly on treetop. Spends long periods in flight, using the broad wings to soar, course, glide and (occasionally) to hover high overhead in search of its main prey of medium-sized birds, reptiles and mammals. Usually strikes after a long, shallow stoop, taking prey on the ground or in the air. Builds a massive stick-nest in the fork of a large tree or even a power pylon, which is often obvious at a distance.

L. Hes

Young adult perched

L. Hes

Adult in soaring flight

ADULT Above sooty-brown, including head, neck and chest. Breast, vent and feathered legs white, with small, dark brown spots. Flight feathers and tail black with narrow, dark grey bars. Underwing coverts sooty-brown. Bill black; eyes deep yellow; cere and long, bare feet dull yellow. Sexes similar in plumage, female about 9% larger (male 3 012–5 100 g, female 5 924–6 200 g).

JUVENILE Above grey, with white feather edges creating a scaled effect. Below (including underwing coverts) plain white. Flight feathers and tail black, finely barred with dark grey. Eye dark brown. Juvenile moults into subadult plumage after 2–3 years, attaining adult plumage after only 6–7 years.

SUBADULT Above dark grey with white feather tips, dark feathers on sides of neck extend forward to form broad collar. Eye dusky-yellow.

DISTRIBUTION, HABITAT AND STATUS Widespread across sub-Saharan Africa, largest resident breeding eagle in the flatter, more open habitats. Present in semi-desert to open woodland. Occurs at low densities; common in more extensive and undisturbed ranching and conservation areas.

SIMILAR SPECIES Adult most resembles adult **Black-chested Snake-eagle** (p. 58) (black head; yellow eye; white belly; bare legs; white underwings with narrow, black bars; narrower, more pointed wings; unspotted body) and, to a lesser extent, smaller adult **African Hawk-eagle** (p. 96) (streaked underparts; white cheeks; more slender; dark underwing coverts; white flight feathers with black trailing edge; pale 'windows' in primaries; long tail). Juvenile very similar to juvenile **Crowned Hawk-eagle** (p. 76) (washed rufous below, especially underwing coverts; grey eye and cere; deep yellow gape and feet; long, barred tail; short rounded wings with 3 or 4 obvious bars). Smaller **Long-crested Eagle** (p. 94) is black with white legs (but black belly; thin floppy crest; obvious white 'windows' in black wings).

ALTERNATIVE NAME: Martial Hawk-eagle

Large, powerful sooty-brown eagle. Breast white, finely spotted with dark brown. Long, white, feathered legs. Head broad and flat with slight crest. Fierce yellow eyes. In flight, the underwings, tail and head appear black in contrast to the white breast and legs. Juvenile grey above and plain white below, with dark grey flight feathers. Very large (about 70 cm tall, 212 cm wingspan).

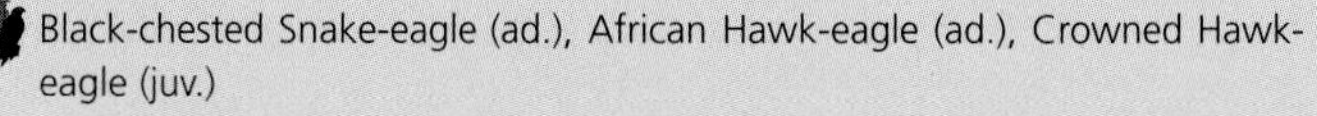

Black-chested Snake-eagle (ad.), African Hawk-eagle (ad.), Crowned Hawk-eagle (juv.)

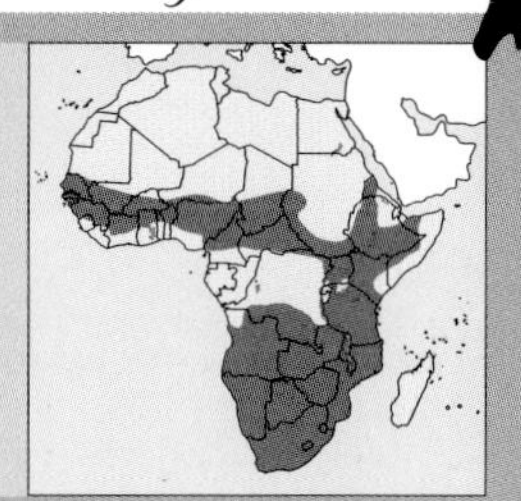

A. Weaving 'perched' and R. du Toit 'in flight'

Crowned Hawk-eagle

Stephanoaetus coronatus

AT PERCH

Broad chest, thick wings, broad crest, heavily spotted and barred legs, yellow feet and long, broadly barred tail are distinctive. Perches very upright with tail hanging vertically below. Adult appears black above and mottled black, white and rufous below. Heavy, black bill backed by narrowly spaced pale yellow eyes and wide, black crest is striking. Orange gape and yellow feet also notable. Juvenile appears white with a grey body and tail.

IN FLIGHT

Short, rounded wings with narrow bases, wide, bulging secondaries and long, rounded primaries, together with long, broad tail, are distinctive. Flight feathers of adult white with 2 (female) or 3 (male) black bars, broad, black ends and white tips, accentuating the chestnut underwing coverts. The tail appears black, and shows 3 or 4 broad, pale grey bars. Pale rufous underwing coverts of juvenile make pale grey flight feathers look even more heavily barred, with dark greater coverts, extra wing bar in each sex, and tail with dark subterminal band and white tip.

DISTINCTIVE BEHAVIOUR

Spends much of the time perched inconspicuously on a large branch within a forest tree. Most obvious when uttering its fast, repetitive yelping cries, especially when soaring over the forest and performing an undulating flight display. Takes mainly medium-sized mammals, either diving from a perch to the forest floor or snatching them on the wing from within the canopy. Builds a massive stick-nest in a major fork of a large forest tree, and mammalian bones often accumulate below.

J. Carlyon

Adult female perched, resting one leg

ADULT Above dark slate-grey, head and neck often paler rufous-brown. Crown has broad, black-tipped V-shaped crest at rear. Below black with broad white and rufous bars of variable intensity. Thick, feathered legs white finely spotted with black, sometimes forming bars. Underwing coverts bright chestnut, but greater coverts black with white bases to form a pied line demarcating the coverts. Flight feathers above dark grey, below white, with 2 (female) or 3 (male) dark bars and broad, dark tips. Tail dark grey above and white below with 2 black bars and a broad, black tip. Bill black; eyes pale yellow; cere grey; gape orange; heavy, bare feet deep orange-yellow. **Female** usually more heavily marked, always with one less dark wing bar, pale brown throat and about 9% larger (male 2 700–4 120 g, female 3 175–3 853 g).

JUVENILE Pale grey above with white feather tips producing scaled effect. Head, neck and underparts cream to white, the crown and crest with grey feather bases visible when feathers ruffled, the legs with fine, dark grey spots. Underwing coverts and upper breast washed to a varying extent with pale rufous. Flight feathers grey with 3 (female) or 4 (male) dark grey bars and broad, dark tips, greater underwing coverts also tipped dark grey. Eye grey, turning brown then yellow. Gape yellow. Moults to adult plumage over 4 years.

DISTRIBUTION, HABITAT AND STATUS Widespread in forests of sub-Saharan Africa, from lowland rainforest in western Africa and the Congo basin to montane forest in Ethiopia and eastern Africa, and riverine and coastal forest in central and southeastern Africa, including tall exotic plantations. Occurs at low densities but locally common in undisturbed habitat, even in suburbs and near dwellings.

SIMILAR SPECIES Adults are larger, darker and more heavily marked than any other African forest eagles, and they have distinctive broad crest, chestnut underwing coverts, pale yellow eye, grey cere and orange gape. Juvenile very similar to juvenile **Martial Eagle** (p. 74) (underparts plain white, including underwing coverts and legs; dark brown eye; pale grey cere and feet; short tail; long, broad, dark, faintly barred wings with upturned tips).

ALTERNATIVE NAME: (African) Crowned Eagle

Robust, long-tailed eagle with a broad crest. Powerful feathered legs. Dark grey above; below black or dark grey heavily barred with white and rufous. Legs white with black spots forming uneven bars. In flight, the barred underparts and chestnut wing coverts contrast with white flight feathers, with 2–4 dark bars and tail with broad, dark tip. Eye pale yellow. Cere grey. Juvenile grey above and pale rufous below with a grey eye. Very large (about 75 cm tall, 152–209 cm wingspan).

Martial Eagle (juv.)

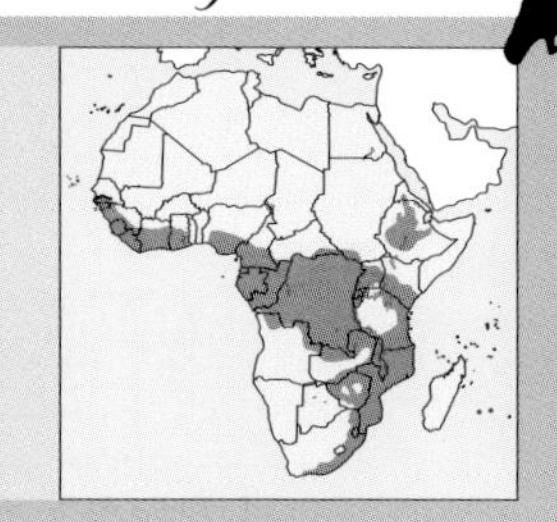

P. Ginn

Verreaux's Eagle

Aquila verreauxii

AT PERCH

Appears black overall, with only a narrow, white V-shape showing on the back. The pale bill, dark eye, and yellow face and feet are obvious in good light. Often stands rather horizontally, looking broad-shouldered, and showing thick, feathered legs and long wing tips reaching end of tail. Brown juvenile appears mottled at a distance, best recognised by buff, rufous and black head pattern. Small head and long neck give a snaky appearance.

IN FLIGHT

Distinctive wings with bow-shaped trailing edge due to short tertials, bulging secondaries, short inner primaries and long, broad outer primaries. Soars with wings at full stretch to show large, pale windows at base of expansive primaries. Black body and coverts distinct against grey-barred flight feathers. Juvenile appears mottled with black on chest and neck, but with characteristic wing shape, and black edge to wing coverts and white rump.

DISTINCTIVE BEHAVIOUR

Adults often found in pairs, perched on a prominent lookout or soaring and gliding around their rocky habitat. They make long stoops from perch or on the wing at their main prey of hyrax and hares, often a pair flying in tandem. Perform spectacular pendulum flight displays and occasionally utter a harsh yelping call. Build a huge stick-nest and roost on rockfaces, rarely in trees; sites usually well marked by 'whitewash'.

L. Hes

Adult perched

A. Froneman

Adult perched

ADULT All-black except for white 'V' on back and white rump. Flight feathers have narrow grey bars and broad, black tips; primaries have long, white bases; tail all-black. Bill blue-grey with black tip; eyes dark brown; cere, bare facial skin and large, bare feet deep yellow. Sexes similar in plumage; female about 7% larger (male 3 000–4 150 g, female 3 100–5 800 g).

JUVENILE Head patterned, with rufous-buff crown, black streaks on forehead, rufous nape and hindneck, and black cheeks, foreneck and upper breast. Rest of body sooty-brown, with broad, buff feather edges to give scaled effect on wing coverts and lower breast. Legs plain buff, rump and greater wing coverts black with only fine buff edges. Flight feathers black with faint, grey bars on underwing and pale patch at base of primaries. Moults into adult plumage over 4 years.

DISTRIBUTION, HABITAT AND STATUS Confined to rocky gorge, hill and mountain habitats, wherever hyrax occur, from the Sinai Desert in Egypt to Table Mountain in Cape Town. Extends to Israel and the Arabian Peninsula. Locally common, even on urban fringes. Main range along mountains and rift valleys from Ethiopia to South Africa, extending to highlands of Chad and Angola. Absent from western Africa.

SIMILAR SPECIES Not easily confused with any other species. Even the brown juvenile is larger and more mottled, and has a distinctive head and chest pattern that prevents confusion with other brown booted eagles.

ALTERNATIVE NAME: (African) Black Eagle

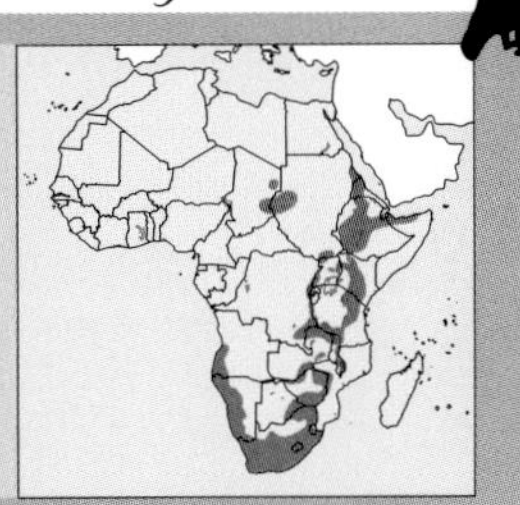

Robust, small-headed jet-black eagle with feathered legs; V-shaped white back; yellow facial skin and feet; dark eyes. In flight, note grey barring in flight feathers, white primary patches and narrow bases to the bow-shaped wings. Juvenile brown and black with rufous crown and nape and white rump. Very large (about 75 cm tall, 200 cm wingspan).

Distinct from all other eagles

A. Froneman

Tawny Eagle

Aquila rapax

AT PERCH

A stocky, reddish-brown eagle, the relatively short legs obvious in how close it sits to the perch. Note long, fluffy flank 'skirts' and short, thick tarsus. Adult appears more mottled than juvenile, especially on the breast, and female is more heavily marked than male. Yellow-brown eye (adult) and short gape visible at close range.

IN FLIGHT

Brown body and wing coverts contrast with grey-barred flight feathers, greater coverts and tail. Inner primaries paler, forming a pale wedge; outer primaries have broad black tips forming dark ends to the wings. Often soars, glides and flaps with wings slightly bent and downcurved. This exaggerates the sinuous trailing edge, close in to the body, bulging secondaries, shorter inner primaries and long outer primaries. Rounded tail quite long and often fanned.

DISTINCTIVE BEHAVIOUR

Often seen as a pair perched close together, usually on a treetop. Hunts small animals from perch or on the wing, usually after a fast stoop. Hunts less often on foot than Steppe Eagle. Often feeds at carrion and regularly pirates prey from other predators. Attracted to grass fires and other concentrations of food, such as termites, queleas or locusts. Utters throaty barking cries – 'kou' – especially when performing territorial pendulum flight display. Only eagle that regularly builds a flat stick-nest on top of a tree, like some vultures.

T. Carew/Photo Access

Juvenile pale form perched

N. Dennis

Adult male perched

ADULT Rufous-brown overall, especially on head, neck and thighs. Wing coverts have dark brown centres and breast has dark streaks, creating grizzled, tawny effect. Flight, greater wing covert, shoulder and tail feathers dark brown above, below grey-brown with narrow, paler grey bars. Bill black; eyes tawny yellow-brown; cere (with oval nostril) and bare feet are yellow. **Female** usually darker, more marked, more streaky, and about 6% larger (male 1 696–1 954 g, female 1 572–3 100 g).
JUVENILE Uniformly pale chestnut, with dark brown flight and tail feathers above, grey-brown barred with paler grey below. Body plumage fades to pale cream or 'blond', often leaving face rufous or blotched with new rufous feathers. Greater wing coverts and rump tipped with white to form thin, pale lines. Eye brown. Moults into adult plumage by 4 years of age.
DISTRIBUTION, HABITAT AND STATUS Widespread breeding resident throughout sub-Saharan Africa and in northeastern Africa. Occupies most habitats except true desert and dense forest. Common in openly wooded savanna and steppe, more patchy amongst denser woodland.
SIMILAR SPECIES Most like migrant eagles that are present only during southern summer. Most similar to slightly larger **Steppe Eagle** (p. 82) (more uniform colouring; pale nape-patch; longer gape; brown eyes; juvenile has obvious white wing and rump lines), but also confused with smaller **Lesser Spotted Eagle** (p. 84) (paler; long, thin legs; yellow eye; rounded nostril; unbarred flight feathers; loose nape feathers), **Greater Spotted Eagle** (p. 86) (generally darker; long, slender legs; long, spiky nape feathers; rounded nostril) and much smaller **Wahlberg's Eagle** (p. 106) (dark and pale forms; slimmer; long, narrow tail; long, slender legs).

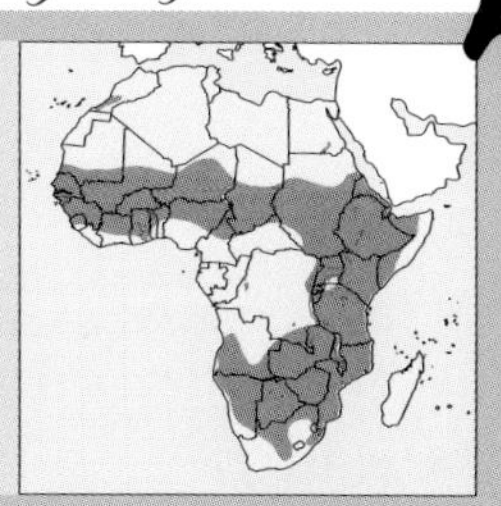

Resident large, reddish-brown eagle with stocky, feathered legs. Variable in colour and extent of tawny markings. Eye tawny yellow-brown. Yellow gape extends back only to level with front of eye. Juvenile plain pale chestnut, fading to light 'blond' plumage, with faint white lines along greater wing coverts and across rump. Large (about 60 cm tall, 182 cm wingspan).

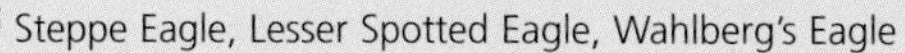

Steppe Eagle, Lesser Spotted Eagle, Wahlberg's Eagle

Steppe Eagle

Aquila nipalensis

AT PERCH

Large, dark, stocky eagle; long, fluffy leggings and short, stout legs often obvious. Wing tips reach tip of tail. Note dark eye, elongated yellow gape, oval nostril and pale nape-patch at close range. Paler juvenile has obvious white lines along edge of wing coverts and secondaries.

IN FLIGHT

Long, broad wings with extensive primaries, sinuous trailing edge formed by bulge of secondaries, and quite long, rounded tail. Contour feathers appear dark against barred flight feathers and tail. Wings held forward and downcurved when gliding, and flat when soaring. Juvenile distinctive with white lines along coverts, white tips to secondaries and tail, white rump and pale patch at base of primaries.

DISTINCTIVE BEHAVIOUR

Hunts alone, mainly for small mammals, but often gathers in small groups at food sources such as termite emergences, quelea colonies or locust swarms. Also eats carrion. Often associates with Lesser Spotted and Tawny Eagles. Takes most small-animal prey on the ground, pulling it from within bushes or often taking it while walking around. Rarely reported hunting in flight except when breeding. May roost in groups in trees, or on the ground in semi-desert areas.

ADULT Uniformly dull, dark brown overall. Most have a pale rufous nape-patch. Flight and tail feathers dark brown with black tips, narrow, dark grey bars and pale patch above at base of primaries. Bill black; eyes dark brown; cere (with oval nostril), large gape (extending backwards to level with back of eye) and bare feet yellow. Two subspecies enter Africa, of which *orientalis* is by far the commoner; the slightly larger and darker *nipalensis* is very rare. Sexes similar in plumage, female about 6% larger (male 2 250–3 110 g, female 2 600–3 800 g). **JUVENILE** Cinnamon-brown overall with pale buff feather edges creating scaled effect. Broad, pale cream tips to dark brown greater wing coverts, rump, secondaries and tail, together with all-cream greater underwing coverts, form a series of pale and dark lines above and below on wings and rump. Covert and body feathers much paler than flight feathers, except for pale bases to primaries. Retains pale greater covert stripes and pale rump for several years; only reaches adult-hood at 4 years.

DISTRIBUTION, HABITAT AND STATUS A non-breeding migrant to Africa during the northern winter, entering Africa via the Arabian Peninsula where part of the population remains to over-winter. Spreads into eastern and southern Africa, mainly in open savanna and woodland. Widespread and locally common but makes extensive, unpredictable movements.

SIMILAR SPECIES Most similar (especially *orientalis*) to slightly smaller **Tawny Eagle** (p. 80) (more mottled and with tawny streaks; no pale nape-patch; shorter gape; tawny yellow-brown eye), but also confused with smaller **Lesser Spotted Eagle** (p. 84) (generally paler; long, slender legs; yellow eyes; rounded nostril; loose nape feathers; unbarred flight feathers; juvenile with small, pale wing-covert spots and pale nape-patch), **Greater Spotted Eagle** (p. 86) (generally darker; long, slender legs; long, spiky nape feathers; rounded nostril; juvenile has large, white wing-covert spots) and much smaller **Wahlberg's Eagle** (p. 106) (darker and lighter forms; slimmer; long, narrow tail; long, slender legs).

D. Robers/Photo Access

Adults

Squat, dark, plain brown eagle with stocky feathered legs. Ginger nape-patch. Note large gape extending back to level with rear edge of eye. Fluffy leggings and stubby feet rather small for the large body. Juvenile paler, with broad, white lines around edge of wing coverts and white rump. Large (about 65 cm tall, 174–260 cm wingspan).

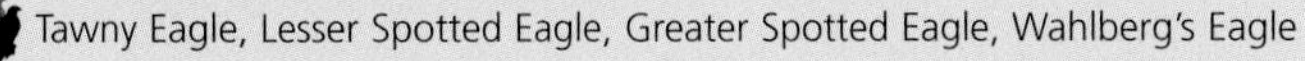

Tawny Eagle, Lesser Spotted Eagle, Greater Spotted Eagle, Wahlberg's Eagle

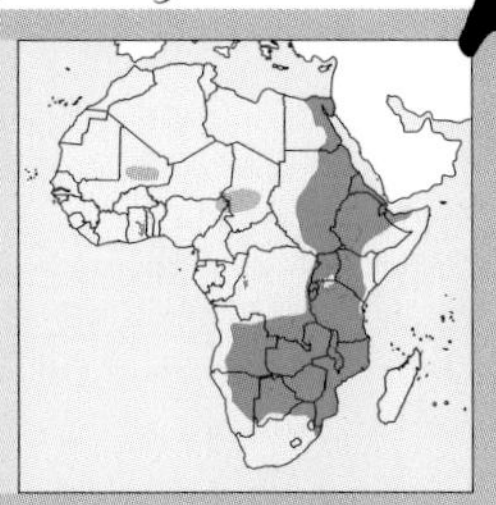

Lesser Spotted Eagle

Aquila pomarina

AT PERCH

Long, slender, upright brown eagle with a rather short tail. Long legs and tightly feathered tarsus give a 'stovepipe' effect. Head distinctive, with long, loose, pointed nape feathers, yellow eye and sharp face, formed by rather small bill and wide gape. Unmarked tail; pale rump-stripe rarely visible. Juvenile obvious with spotted wing coverts.

IN FLIGHT

Brown eagle with long, rectangular, faintly barred wings and straight trailing edge, the relatively short, rounded tail usually fanned. Coverts and body usually paler than flight feathers. Pale rump distinctive, forming narrow horseshoe-shaped base to tail and contrasting with dark back. Note pale patches at base of inner primaries and, in juvenile, pale tips to greater coverts and secondaries.

DISTINCTIVE BEHAVIOUR

Usually solitary, but hundreds may gather at swarms of winged termites or near flocks of breeding queleas, often with Steppe and other local eagles. In Africa, appears to hunt mainly from a perch, descending to take small-animal prey from the ground, sometimes walking about or reaching into bushes. Not often seen hunting on the wing as it does when breeding. Roosts in trees near current food source, sometimes in small groups.

ADULT Brown overall, with individual variation in tone. Pale buff on the rump. White bases to long, pointed nape feathers sometimes evident. Flight feathers, including tail, only faintly barred. Pale patches at base of inner primaries. A rare (5%) pallid form, and intermediates mainly pale buff with dark flight feathers. Bill black; eyes tawny-yellow; cere with round nostril yellow; legs tightly feathered, bare feet yellow. Sexes similar in plumage, female about 3% larger (male 1 053–1 509 g, female 1 195–2 160 g).

A. Froneman

Juvenile

JUVENILE Usually slightly darker than adult, with buff patch on nape. Upperwing coverts tipped with small, pale spots, creating speckled effect. Rump, vent and tips to greater upperwing coverts and secondaries white, forming distinct pale bands on tail and wing. Below streaked with paler brown. Dark, narrow bars more obvious on tail. Eyes brown. Soon moults pale nape patch and vent; only adult in 4th year.

DISTRIBUTION, HABITAT AND STATUS A non-breeding migrant to Africa during the northern winter, crossing via the eastern Mediterranean or the Arabian Peninsula. Most move south to the tall woodlands of central and southeastern Africa, but a few remain in eastern Africa. Widespread, and locally common at food sources, but probably overlooked due to confusion with other brown eagles. Possibly a vagrant to Seychelles.

SIMILAR SPECIES Most similar to slightly larger **Greater Spotted Eagle** (p. 86) (bushier nape; brown eye; underwing coverts darker than flight feathers; lacks pale nape patch; large spots in juvenile form white lines across coverts and rump), but also confused with larger and heavier **Tawny** and **Steppe eagles** (pp. 80 and 82) (usually paler; barred primary and tail feathers; short fluffy legs; pale rump wider and more diffuse; tawny-brown eye; oval nostrils), and smaller **Wahlberg's Eagle** (p. 106) (darker and lighter forms; slimmer; long, narrow tail; body often darker than flight feathers; brown eye; legs also long and slender).

Dark brown eagle with long, thin, feathered legs. Long, pointed nape feathers. Tawny-yellow eyes. Wide gape. In flight, note pale rump, pale edges to wing coverts, and pale patches at base of inner primaries. Juvenile has obvious buff spots on feather tips above and buff stripes below. Rare pallid forms and some intermediates occur. Large (about 55 cm tall, 134–170 cm wingspan).

Greater Spotted Eagle, Steppe Eagle, Tawny Eagle, Wahlberg's Eagle

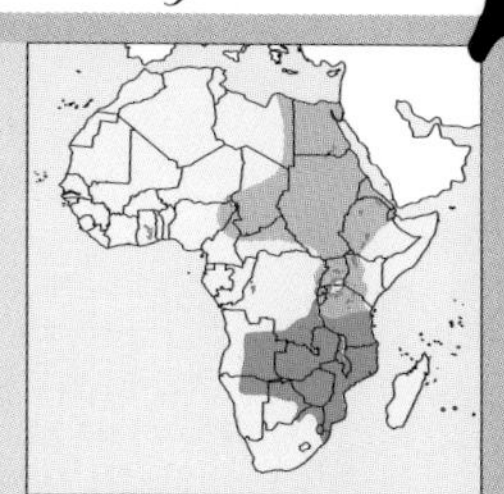

C. Haagner/ABPL

Greater Spotted Eagle

Aquila clanga

AT PERCH

A large, dark, stocky brown eagle with long, slender legs. At close range note dark brown eye, spiky nape feathers, round nostril and purple gloss above. Juvenile distinctive with large, white spots across wing coverts. Wings reach tip of short tail. Dark brown eye colour at all ages, and large cream to white spots on wing coverts and rump feathers in juveniles, are only distinguishing features, especially compared with Lesser Spotted Eagle (adult has amber eyes, juvenile has smaller spots).

IN FLIGHT

Uniformly dark brown eagle with broad wings, distinct primary 'fingers', curved trailing edge from long secondaries and indented tertials, short, slightly wedge-shaped tail. Juvenile distinctive, with pale rump and pale lines across upperwing coverts. Soars with wings bowed forward, flaps with deep, heavy beats.

DISTINCTIVE BEHAVIOUR

Little studied in Africa. Mainly found perched and solitary, although enters Africa in small flocks and sometimes gathers in small numbers at termite or locust emergences, often with Steppe and other local brown eagles. Seems to favour semi-arid open thorn savanna but also overwinters in tall woodland, and must cross desert and mountain habitats en route.

ADULT Dark brown overall, almost blackish, including unbarred flight feathers. Nape feathers especially long, loose and pointed. Purple gloss on feathers above. Below duller with paler vent. Legs long and tightly feathered. Very rare pale buff form with dark flight feathers. Bill black; eyes dark brown; cere (with round nostril) and bare feet yellow. Sexes similar in plumage, female about 5% larger (male 1 537–2 000 g, female 2 150–3 200 g).

JUVENILE Even darker brown than adult, with large, pale buff or white tips to upperwing and rump forming pale U-shaped rump and broken lines of 'pearls' along upperwing. Secondaries and tail also have narrow, pale tips when fresh.

DISTRIBUTION, HABITAT AND STATUS A non-breeding migrant to Africa during northern winter, entering mainly along the Bosporus and Arabian Peninsula. Uncommon at crossing points, only rarely reported in Africa, mainly from northeast and along Rift Valley. Recently satellite-tracked south to Zambia, so may be more widespread than thought and probably often overlooked.

SIMILAR SPECIES Most similar to, but still difficult to distinguish from, slightly smaller **Lesser Spotted Eagle** (p. 84) (slimmer; contour feathers paler than flight feathers; tidier nape; yellow eye; juvenile with small spots on coverts and pale nape-patch), but also confused with larger and heavier **Tawny** and **Steppe eagles** (pp. 80 and 82) (barred primary and tail feathers; short, fluffy legs; pale rump wider and more diffuse; oval nostril), and smaller **Wahlberg's Eagle** (p. 106) (has darker and paler forms; slimmer; long, narrow tail; but legs also long and slender).

Juvenile

ALTERNATIVE NAME: Spotted Eagle

Bulky, chocolate-brown eagle with long, thin, feathered legs. Long, spiky nape feathers. Purple sheen above. Juvenile even darker, with large, white spots forming broken lines across wing coverts and shoulders, and white rump. Very rare pale form. In flight, covert and flight feathers equally dark, wings broad, tail short and both unbarred. Large (about 55 cm tall, 155–182 cm wingspan).
Lesser Spotted Eagle, Steppe Eagle, Tawny Eagle, Wahlberg's Eagle

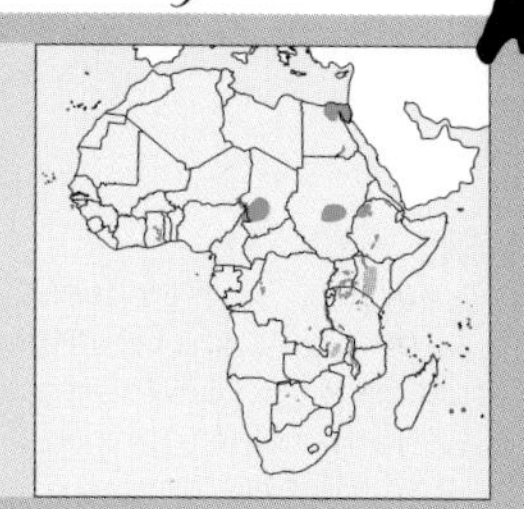

Hellio & Van Ingen/NHPA

Spanish Imperial Eagle

Aquila adalberti

AT PERCH

A large, very dark brown eagle with voluminous leggings paler than body. The pale crown and nape, white shoulder and front of wing and, at close range, the pale eyes, heavy bill and wide, yellow gape are diagnostic. Juvenile appears streaked with dark brown and light rufous; note buff edges to upperwing coverts and rump, and to black flight and tail feathers.

IN FLIGHT

Adult dark overall with prominent buff top to head and white leading edge to wing, also paler patch and wedge at base of primaries. Juvenile's pale body and coverts contrast with dark, white-tipped flight feathers; pale patch and wedge at the base of juvenile's primaries. All have long, broad wings, wide secondary row, long, rounded tail and protruding head distinctive.

DISTINCTIVE BEHAVIOUR

Hunts from a perch or on the wing. Main prey is rabbits, but also eats other small to medium-sized vertebrates.

V. Canseco/NHPA

Juvenile with prey

ADULT Plain very dark brown overall, except for cream or buff crown and nape and large, white shoulder patches extending along leading edge of wing. Flight feathers almost black, paler grey barring only evident as pale patch at base of primaries and wedge on inner primaries. Bill black; eyes pale brown; cere and heavy bare feet yellow. Sexes similar in plumage, female about 8% larger (both sexes 2 500–3 500 g).

JUVENILE Individually variable. Dark brown body and wing coverts, but feathers heavily streaked and tipped with light rufous so that it appears pale rufous with dark streaks. Legs and vent almost plain pale rufous. Pale body and coverts contrast with the black, white-tipped flight feathers and tail; lighter rufous bars only obvious where they form a pale base to the outer primaries and a wedge on the inner primaries. Buff feather tips form pale line at edge of upperwing and rump. Dark outer primary coverts form a black carpal arc. Eyes dark brown. Moults to adult plumage over at least 5 years, turning paler initially and then appearing darker and blotchy in the intermediate stages.

DISTRIBUTION, HABITAT AND STATUS Main breeding range on the Iberian Peninsula used to extend to northern Morocco. Now a rare vagrant to northwestern Africa. Confined to uninhabited slopes with good bush cover and patches of forest.

SIMILAR SPECIES Overlaps only with similar-sized **Golden Eagle** (p. 92) (also dark brown; crown dark; pale golden nape more streaky; dark, narrow-based wings; white legs; juvenile dark brown, unstreaked, with white base to black-tipped tail) and, less commonly, with much smaller **Tawny Eagle** (p. 80) (paler brown; lacks white shoulder patches and grey tail). Widely separated geographically from very similar **Eastern Imperial Eagle** (p. 90).

ALTERNATIVE NAME: Adalbert's Eagle

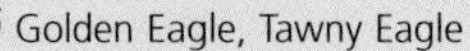

Heavily built, very dark brown, booted eagle. Buff crown and nape, white leading edge to wing, grey tail with dark tip. Juvenile heavily streaked with dark brown and rufous, and has black, white-tipped flight feathers and tail. Large (about 65 cm tall, 177–220 cm wingspan). Rare vagrant to northern Morocco from Spain.

Golden Eagle, Tawny Eagle

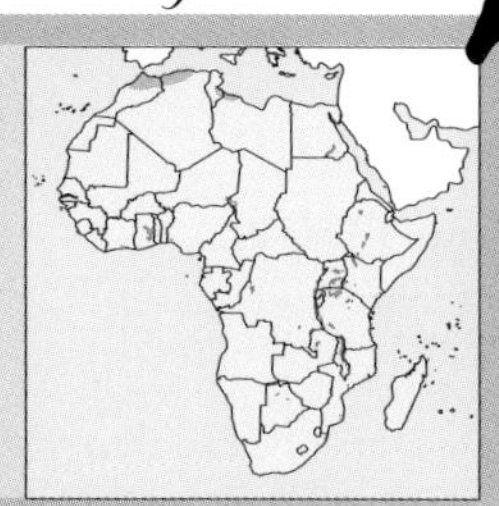

Eastern Imperial Eagle

Aquila heliaca

AT PERCH

A large, dark brown eagle with voluminous leggings paler brown in colour. The pale crown and nape, white shoulder patches and, at close range, the pale eyes, heavy bill and wide, yellow gape are diagnostic. Dark grey, banded tail not often visible beneath the long wings. Juvenile appears dark brown with light streaks or light brown with dark streaks, depending on the individual. Note white edges to upperwing coverts, rump, and black flight and tail feathers.

IN FLIGHT

Very long, broad wings with extensive secondaries, long, rounded tail (usually fanned) and protruding head create distinctive silhouette. Adult dark overall with pale crown and front of shoulder, and paler patch at base of primaries and tail. Dark band at end of tail not always obvious. Juvenile has pale body and coverts contrasting with black flight feathers. Flight feathers have white tips and pale patch and wedge at the base of the primaries.

DISTINCTIVE BEHAVIOUR

Unstudied on its wintering range in Africa. Feeds mainly on small mammals and also attracted to carrion. On breeding grounds hunts from a perch or while soaring.

ADULT Plain dark brown overall except for heavy, cream streaks on the long, pointed crown and nape feathers, small, white patch at base of shoulders, and pale brown flanks and vent. Flight feathers almost black, paler grey barring only evident as a pale patch at base of primaries and a wedge on the inner primaries. Tail dark grey, mottled with dark brown, with a broad, dark subterminal band. Bill black; eyes pale brown; cere and heavy, bare feet yellow. Sexes similar in plumage, female about 8% larger (male 2 450–3 950 g, female 2 800–4 250 g).

JUVENILE Individually variable. Dark brown body and wing coverts, but feathers heavily streaked and tipped with light buff to appear pale brown with dark streaks. Legs and vent pale rufous and rump plain buff to white. Pale body and coverts contrast with black, white-tipped flight and tail feathers; lighter brown bars only obvious where they form a pale base to the outer primaries and wedge on the inner primaries. Buff feather tips form a pale line at edge of upperwing coverts and pale rump. Dark outer primary coverts form a black carpal arc. Eyes dark brown. Moults to adult plumage over at least 5 years, appearing even paler initially and then darker and blotchy in intermediate stages.

DISTRIBUTION, HABITAT AND STATUS Rare, non-breeding migrant during the northern winter to the Nile Valley, Sudan, Ethiopia and Kenya. May be overlooked among other large, brown, migrant booted eagles. Most records are from lowland grasslands which are similar to their traditional breeding areas of open, lightly wooded or cultivated plains. Often reported in the company of other brown eagles.

C. Ruoso/BIOS

Adult

SIMILAR SPECIES Much larger than other large, dark brown **Greater Spotted** and **Steppe eagles** (pp. 86 and 82) (lack the pale crown, nape and shoulder-patch and grey tail base). Streaky juveniles resemble smaller **Tawny Eagle** (p. 80) (more blotched than streaked; tawny eyes) and juvenile **Steppe Eagle** (p. 82) (white lines along wing coverts and rump more prominent, especially on underwing). Range widely separated from that of very similar **Spanish Imperial Eagle** (p. 88).

Heavily built, dark brown booted eagle. Prominent cream streaks on crown and nape. Small, white 'epaulettes' on shoulders. Grey tail with broad, dark tip. Juvenile appears heavily streaked with dark and pale brown, and has black, white-tipped flight feathers and tail. Large (about 65 cm tall, 180–215 cm wingspan). Rare migrant to Nile Valley and eastern Africa.

Greater Spotted Eagle, Steppe Eagle, Tawny Eagle

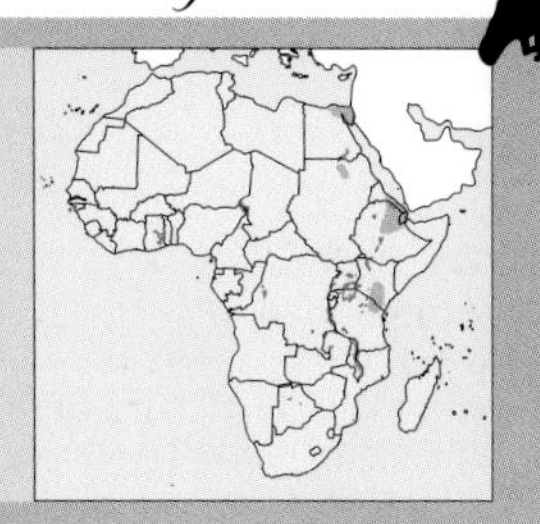

C. Ruoso/BIOS

Golden Eagle

Aquila chrysaetos

AT PERCH

Appears dark brown, long and slender with, at close range, golden patches on nape and across wing coverts. Relatively small head on a long, snaky neck and long flight feathers reaching end of tail enhance the effect of elongation. Juvenile lacks golden areas of plumage, best distinguished by white tail with broad, black tip.

IN FLIGHT

Slender body with a long tail, short inner secondaries and long, extensive primaries form a distinctive shape in flight. Pale bases to flight feathers (especially primaries) and to the dark-tipped tail are notable. Juvenile has pale areas even more exaggerated.

DISTINCTIVE BEHAVIOUR

Usually solitary; perches inconspicuously, but more likely to be seen gliding along hill ridges or soaring overhead. Moves onto lowland flats in winter. Hunts mainly small mammals and medium-sized birds, which it takes on the ground or in flight. Utters loud, yelping calls but not very vocal. Builds a large stick-nest on a rock ledge or in a large tree protruding from a steep slope.

ADULT Above uniformly dark brown except for golden tips to long, pointed nape feathers and median upperwing coverts. Below slightly paler with long, white, feathered legs. Flight feathers and tail black mottled and banded with grey, especially at base of flight feathers, and forming broad, dark terminal band to tail and flight feather tips. Bill black; eyes pale brown; cere and bare feet yellow. Sexes similar in plumage, female about 9% larger (male 2 840–4 450 g, female 3 630–6 665 g).

JUVENILE Darker brown than adult, without golden nape and wing patches. Base of flight feathers and tail broadly mottled and barred with white, to form prominent pale wing areas and an off-white tail with a broad, black terminal bar. Eye dark brown. Attains adult plumage by 5 years of age.

DISTRIBUTION, HABITAT AND STATUS Resident breeding species in mountainous areas of north-eastern Africa, including massifs well south into the Sahara Desert. Rare vagrant, usually immatures, to western parts of sub-Saharan Africa.

SIMILAR SPECIES Range overlaps only with that of similar-sized **Spanish Imperial Eagle** (p. 88) (dark brown; long, broad, rectangular, all-dark wings with pale patch in primaries; crown and nape uniformly pale; buff legs and vent; juvenile streaked dark and has flight feathers pale rufous with dark brown) and, less commonly, with much smaller **Tawny Eagle** (p. 80) (paler brown body; dark brown, barred flight feathers).

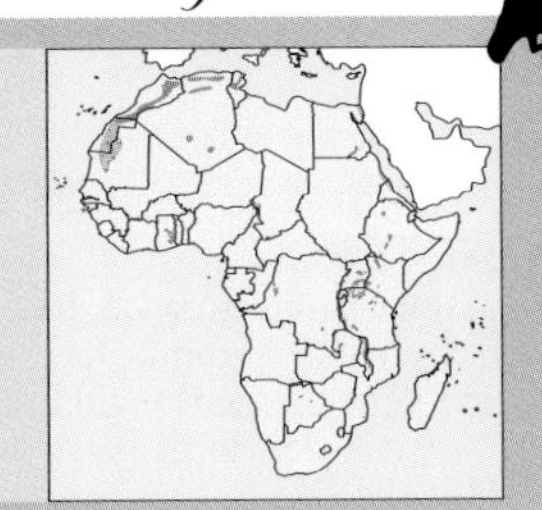

Dark brown eagle with golden nape and forewing. White feathered legs. Slender, broad-shouldered build. In flight, pale grey bases to primaries and dark tip to grey tail distinctive. Juvenile darker brown without pale nape, but bases of primaries and tail white. Very large (about 65 cm tall, 190–227 cm wingspan). Breeding resident only in northeastern Africa.

Spanish Imperial Eagle, Tawny Eagle

Long-crested Eagle

Lophaetus occipitalis

AT PERCH

Black silhouette with erect crest unmistakeable. Note yellow eye, white (usually male) or brownish (usually female) legs, and grey-barred tail at close range. Small bill combines with wide gape and broad head to produce frog-like face.

IN FLIGHT

At a distance, wings appear black with large, white primary 'window' patches above and below. From closer by, the black underwing coverts with white 'commas' at the end, the grey-barred secondaries and tail, and their broad, black trailing edges become visible. Broad, rounded wing shape, with long, outer secondaries and primaries, and long tail are also distinctive. Soars with wings widely spread, and flaps with slow, shallow wingbeats.

DISTINCTIVE BEHAVIOUR

Most often seen perched alone on a prominent lookout at the edge of woodland. Changes perches at intervals or takes off to soar overhead, usually not very high. Often calls in flight with high, protracted scream. Hunts mainly for large rodents, usually taken among long grass and swallowed whole. Builds a well-concealed stick-nest near the woodland edge.

I. Davidson

Adult soaring at full stretch

I. Carlvon

Adult male

Adult Sooty-brown overall, except for long, slender, white (male) or brownish (female) feathered legs. Long, thin feathers at back of crown usually held as an erect crest. Flight feathers and tail black, barred with pale grey and with broad, black tips. Base of primaries, and median underwing primary coverts, white. Bill black; eyes, cere, wide gape and bare feet deep yellow. Female about 9% larger (male 912–1 363 g, female 1 367–1 523 g).

Juvenile Very similar to adult, with white legs and shorter crest at fledging. Eyes grey.

Distribution, habitat and status Widespread throughout sub-Saharan Africa except for the arid treeless areas in the northeast and southwest. Most common where damp, marshy areas occur among patches of woodland and forest, including agricultural plots and exotic plantations. Sedentary in areas of high rainfall but an unpredictable vagrant where local rains or drought alter the suitability of habitat.

Similar species Long crest differentiates this species from any other. Recalls **African Hawk-eagle** (p. 96) (black above; white legs; yellow eye; but white below and on cheeks, streaked with black; pale yellow cere; narrow gape; dark wings with pale primary 'windows'; but coverts without white 'commas'; secondaries almost unbarred; wings long and rectangular). Also resembles much larger, heavier adult **Martial Eagle** (p. 74) (white breast spotted with brown; crest broad and short; all-dark wings; short tail).

Long, erect, floppy crest obvious. All-black with long, white (male) or brownish (female) feathered legs. Yellow eye. Wide, yellow gape. In flight, large, white 'windows' in wings and white bars across tail. Juvenile like adult. Medium-sized (about 50 cm tall with crest, 115 cm wingspan).

African Hawk-eagle, Martial Eagle

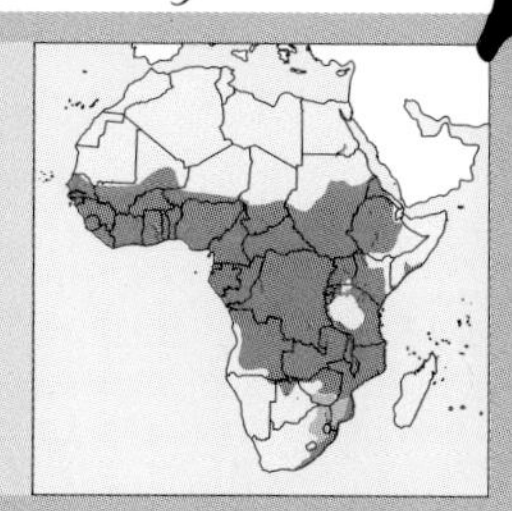

African Hawk-eagle

Hieraaetus spilogaster

AT PERCH

Tall, slender eagle with long tail reaching well below wing tips. Black above and white below, with obvious black streaking on the breast and long, white, feathered legs. Yellow eye looks fierce within the black cap. Juvenile dark brown above and chestnut with dark streaks below, the rufous eyebrow and cheeks, brown eye and pale cere most notable at close range.

IN FLIGHT

Pale primary 'windows' most obvious, together with heavily spotted wing coverts and broad, black trailing edge to wing and tail. Juvenile with chestnut underparts, the coverts edged in black. A long-winged, long-tailed eagle with narrow wing base, broad, bulging secondaries and rounded tips. Fast, agile flier with fast, deep wingbeats, like a very large goshawk.

DISTINCTIVE BEHAVIOUR

Secretive. Adults usually seen in pairs, most often when perched on a protruding dead branch, soaring high overhead or making low, aerial searches. The main prey is gamebirds and small mammals, taken from the ground or in dashing aerial pursuit. Builds a notably large stick-nest in a tall tree, often along well-wooded river banks.

I. Davidson

Adult female

Adult male

ADULT Black above, including crown and ear coverts. Below white, including throat and foreneck, with bold black stripes on upper breast, underwing coverts and flanks. Legs and vent white. Flight feathers and tail black above, below white with broad, black tips. Primaries have broad, white bases above, secondaries and tail have narrow, pale grey bars above and below. Bill black; eyes deep yellow; cere pale yellow; long, slender, bare feet pale greenish-yellow. **Female** more heavily streaked below and about 4% larger (male 1 150–1 300 g, female 1 444–1 640 g).

JUVENILE Dark brown above, with pale rufous and white bars on feather bases often showing on wing coverts and scapulars. Eyebrows, ears, throat, underparts and underwing coverts rufous, with dark brown streaks on breast and coverts. Wing and tail patterns dull version of adult plumage with barring much more obvious and without the broad trailing edges. Greater underwing coverts distinct black. Attains adult plumage by 3 years of age.

DISTRIBUTION, HABITAT AND STATUS Widespread throughout sub-Saharan Africa except for areas of extensive lowland forest and treeless desert. Extends into arid scrub where there are trees along watercourses, but most common in open woodland and savanna bushveld.

SIMILAR SPECIES Most resembles the smaller, stockier **Ayres's Hawk-eagle** (p. 100) (crested; adult more heavily streaked below with black head, or paler grey with extensive white markings; wings and tail heavily barred; lacks pale 'windows' at base of primaries; legs spotted; juvenile has pale brown scaling above, rufous head and neck with black crest, paler rufous and less streaked below, yellow eye). Similar in colour to much smaller **Black Sparrowhawk** (p. 180) (bare yellow legs; adult has black bib and thighs, unmarked white (or all-black) breast, deep red eye; juvenile has grey-brown eyes; darker rufous thighs).

ALTERNATIVE NAME: African Eagle

Black above, below white with obvious black streaks. Long, white, feathered legs. In flight, large white primary 'windows' and dark coverts. Broad black trailing edge to wings and long tail. Yellow eye. Juvenile dark brown above, below chestnut with dark brown streaks, eye brown. Large (about 55 cm tall, 142 cm wingspan).

Ayres's Hawk-eagle, Black Sparrowhawk

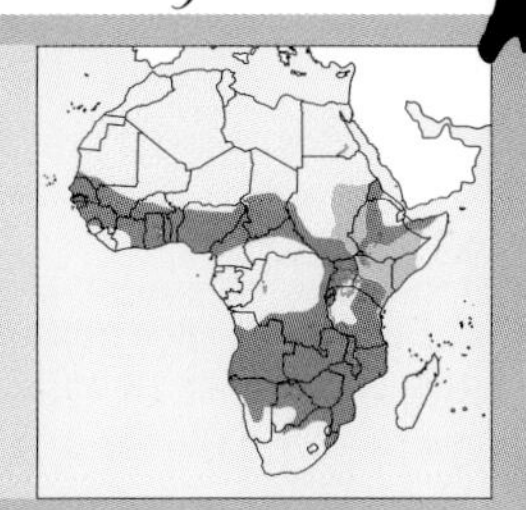

Bonelli's Hawk-eagle

Hieraaetus fasciatus

AT PERCH

Stands tall with long, grey tail protruding well below wing tips. Appears blackish-brown above and white streaked black below, with a dark cap enclosing the yellow eye. Juvenile brown above and rufous below with brown eye.

IN FLIGHT

Combination of black underwing coverts and white leading edge, body and flight feathers distinctive. Pale grey primary 'windows' obvious above and below; broad, dark trailing edge to wing and tail also notable. Pale mottling on mantle often prominent as white back patch. Juvenile with pale rufous underparts. Long wings with narrow bases, broad secondaries and rounded tips. Long, tail square when closed, rounded when fanned. Flies with fast, deep wing-beats, quick and agile, like a very large goshawk.

DISTINCTIVE BEHAVIOUR

Often hunts on the wing, coursing along hillslopes or soaring overhead. Makes fast dashing pursuits after its main prey of small to medium-sized mammals and birds. Builds a large stick-nest, often on a cliff face or on a tree growing from a hillside.

V. Canseco/NHPA

Adult perched on a rockface

A. Manzanares/Bruce Coleman Ltd

Subadult female 'mantling' over prey

ADULT Sooty-brown above, with pale, barred feather bases to nape, mantle and upperwing coverts. Below and cheeks white streaked with black, especially on upper breast. Heavily blotched on median and greater underwing coverts; barred and often pale rufous on vent. Legs plain pale brown, almost white. Flight feathers dark blackish-brown above and pale grey below, bases of primaries pale grey. Secondaries and tail barred with pale grey with broad, dark tips. Bill blue-grey with black tip; eyes deep yellow; cere and long, bare feet yellow. **Female** darker, more heavily marked and about 5% larger (male 1 500–2 160 g, female 2 000–2 500 g).

JUVENILE Dark brown above; head and underparts pale rufous streaked with dark brown. Individual variation in intensity of colours. Flight feathers dark brown above, pale grey below, with pale patch at base of primaries and pale bars on secondaries. Tail grey-brown above, pale grey below with narrow, darker bars. Eye brown. Attains adult plumage by 4 years of age.

DISTRIBUTION, HABITAT AND STATUS Breeding resident in the mountains of northwestern Africa. Favours rocky wooded foothills but extends up to 2 000 m a.s.l. Generally uncommon. Vagrant to, but possibly breeds in, extreme northeastern Africa along Red Sea Mountains.

SIMILAR SPECIES Only resident booted eagle of similar size in range is **Tawny Eagle** (p. 80) (paler brown; mottled (adult) or plain (juvenile); stockier; shorter-legged; tawny, yellow-brown eyes). Others, much larger, all-dark, and with long, dark, rectangular wings and shorter tails, are **Spanish Imperial Eagle** (p. 88) (rare vagrant; white 'epaulettes'; white to buff crown and nape; juvenile streaky) and **Golden Eagle** (p. 92) (breeds in similar habitat; golden nape and wing-patch; juvenile plain brown with white tail base and pale primary wing patches).

ALTERNATIVE NAME: Bonelli's Eagle

Bonelli's Hawk-eagle

Tall, slender, booted eagle. Dark sooty-brown above with pale, streaky mantle. Below white with dark streaks. In flight, note black underwing coverts with white leading edge, and pale primary 'windows'. Tail long and grey with broad, black tip. Eyes yellow. Juvenile dark brown above, rufous and streaked below. Large (about 55 cm tall, 150–180 cm wingspan). Only in northern Africa.

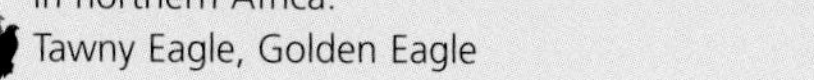
Tawny Eagle, Golden Eagle

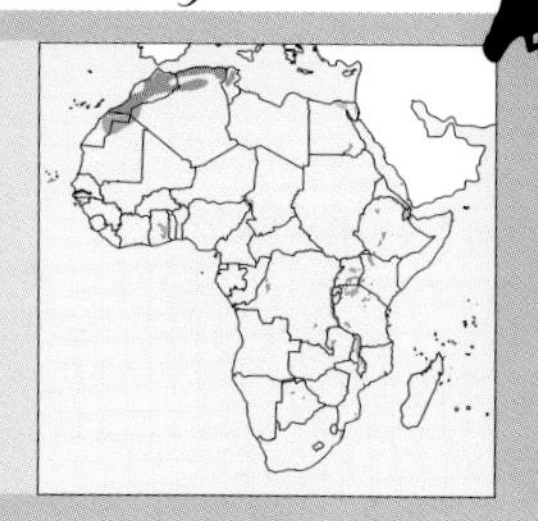

Ayres's Hawk-eagle

Hieraaetus ayresii

AT PERCH

Appears dark, squat and thickset, with long, strong feathered legs and obvious crest. Note heavy streaking on breast and black head in dark adults. Pale birds appear scaly above and white below. Pale rufous head and underparts of juvenile, with dark crest patch, most obvious; yellow eye, greenish-yellow cere and feet, and white shoulder-patches also distinctive. Moults directly into adult plumage.

IN FLIGHT

Heavily barred flight feathers and tail with broad black trailing edges and tips obvious at all ages. Adults appear black above with white, variably streaked, underparts. Juvenile brown above and rufous below. Soars on straight, broad, rectangular wings with long, square tail. Flaps with deep, fast wingbeats, or dives with wrists apart and wing tips touching, and with white 'headlights' especially obvious.

DISTINCTIVE BEHAVIOUR

Usually seen on the wing, soaring and gliding in search of its main prey of small birds, most of which are taken in a fast stoop in heart-shaped conformation. Often perches well-concealed inside the tree canopy, emerging to make fast dashes after prey, including pursuit among branches. Builds a small stick-nest in a high fork of a tall, woodland tree.

ADULT Most are black above; black head and face give a capped, falcon-like effect. Short, dense crest. Below, including throat and underwing coverts, white with heavy black streaks becoming spots on the legs. Flight feathers and tail also black above and white below, underside with heavy, black bars and broad, black tips. Others variously paler, often with white forehead and eyebrows. Palest are dark grey overall, above feathers broadly edged with white, especially on the nape, to give a scaled effect. Below white, lightly spotted or finely streaked with grey. Rare all-black form occurs. All have prominent white patches at the base of the wings. Bill blue-grey with black tip; eyes deep yellow; cere and bare feet pale greenish-yellow. Female usually darker, about 14% larger (male 714 g, female 879–940 g).

JUVENILE Dark brown above, feathers broadly tipped with pale rufous, creating scaled effect. Head, underparts and underwing coverts pale rufous, crown heavily streaked with dark brown, crest black and sides of chest finely streaked with dark brown. Flight feathers, greater wing coverts and tail dark brown above, below white with same black barring as adult. White shoulder patches and softpart colours also same as adult.

DISTRIBUTION, HABITAT AND STATUS Widespread in forests and woodlands of sub-Saharan Africa but generally rare. Most frequent in tall, deciduous miombo woodlands of south-central Africa, where it breeds during the southern winter. Many then disperse to drier areas to the south, and maybe to the the north and west, during the southern summer, when they enter cities and exotic woodlands. Species may be more sedentary in parts of northeastern and western Africa.

P. Pickford/SIL

Adult, dark form

SIMILAR SPECIES Most like the larger **African Hawk-eagle** (p. 96), especially paler birds (more slender; crestless; less streaked; adult often has white forehead, legs white, pale 'window' at base of primaries, broad, black trailing edge to secondaries and tip of tail, darker underwing coverts; juvenile plainer brown above and on head, deeper chestnut, more heavily streaked below, brown eye). Similar in size, colour and habitat to **Black Sparrowhawk** (p. 180) (crestless; bare yellow legs; adult has black bib and thighs, unmarked white breast, deep red eye; juvenile deeper rufous, more streaked below, brown thighs, grey-brown eye). Juvenile similar in build and habits to pale juvenile **Booted Eagle** (p. 104) (same white 'headlights'; unbarred dark flight feathers; plain grey tail; paler brown and rufous overall; no crest; brown eye; yellow cere and feet).

ALTERNATIVE NAME: Ayres's Eagle

Dark, stocky little eagle with feathered legs. Note short, tufted crest. Most adults black above, and white heavily streaked with black below. Others paler; the palest is dark grey scaled with white above, below white lightly streaked with grey. Also a rare all-black form. Juvenile brown above, pale rufous on the head and below with dark brown streaks. At all ages, flight feathers heavily barred, white 'headlights' on shoulders, yellow eyes and pale greenish cere and feet. Medium-sized (about 40 cm tall, 124 cm wingspan).
African Hawk-eagle, Black Sparrowhawk, Booted Eagle

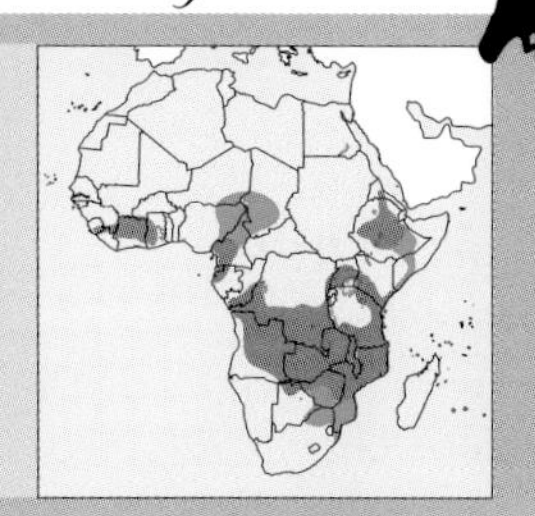

Cassin's Hawk-eagle

Spizaetus africanus

AT PERCH

Appears very dark grey from behind, with long, barred tail exaggerated by short wing tips. From in front, white except for incomplete black collar and flank patches. At close range, note yellow or brown eye, barring on flight feathers and long, feathered legs. Juvenile dark brown above and pale rufous below with dark spots and brown eyes.

IN FLIGHT

Short, rounded wings with black underwing coverts and pale primary 'windows', together with long, barred tail with broad, dark subterminal bar, are obvious. Black coverts connect with collar wedges and flank patches on sides of white body. Juvenile appears dark below with faint brown and rufous covert streaks and faint flight feather bars.

DISTINCTIVE BEHAVIOUR

Little studied. Usually seen circling low over forest canopy or soaring high along forested ridges. Sometimes utters a long, high-pitched scream. Probably spends much time perched within the canopy, where it is thought to take prey of birds and squirrels. Builds a stick-nest high in a forest tree.

ADULT Above sooty-brown to black; greater upperwing coverts, scapulars, back and rump have dark brown tips and dark rufous-brown bars. Below white, including throat. Dark brown feather tips form incomplete collar on side of breast. Flanks and vent boldly spotted with black. Leg feathers often tipped brown, to appear faintly mottled. Flight feathers sooty-brown above, with dark brown tips and dark rufous-brown bars. Below white, with narrow, dark grey bars and black tips to flight feathers. Underwing coverts black with white bases to primary coverts. Long tail dark charcoal grey above with dark bars and broad, dark tip; below pale grey with darker grey bars and subterminal bar. Bill black; eyes yellow (male) or brown (female); cere and bare feet pale yellow. **Female** usually more heavily spotted on flanks and legs, and larger (male 938–1 049 g, female 1 153 g).

JUVENILE Dark brown above, with white feather bases on nape, neck and back creating mottled effect. Head and neck pale rufous with dark brown streaks on crown and neck. Below pale rufous with dark brown spots, brightest rufous and heaviest spots on upper breast, fading to pale rufous with few or no spots on the legs and vent. Flight feathers and their greater coverts brown above with pale bars at base and wide, dark brown tips; below white with faint, grey bars and rufous-streaked underwing coverts. Tail has narrow, dark bars and tip. Eyes brown.

DISTRIBUTION, HABITAT AND STATUS Restricted to primary lowland rainforest, from Togo eastwards to western Uganda, the Congo basin and northern Angola. Rarely seen but may be overlooked due to secretive habits and dense habitat. Call recognition may be useful.

SIMILAR SPECIES Most resembles and overlaps in range with **Ayres's Hawk-eagle** (p. 100) (slightly crested; adult heavily streaked below and has black head, or is paler grey with extensive white markings; legs spotted; wings and tail heavily barred; lacks pale 'windows' at base of primaries; juvenile has pale brown scaling above, rufous head and neck with black crest, yellow eye) and smaller **Black Sparrowhawk** (p. 180) (crestless; longer tail and wings; bare yellow legs; adult with black bib and thighs, unmarked white breast, deep red eye; juvenile more streaked below and has brown thighs, grey-brown eyes).

Black above and white below. Small, stocky eagle with spotted flanks and feathered legs. Long, barred tail. Eye yellow (male) or brown (female and juvenile). Slight crest. Faint brown barring on wings. In flight, rounded wings with black underwing coverts distinctive. Juvenile brown above, rufous below with dark spots. Medium-sized (about 50 cm tall, 120 cm wingspan).

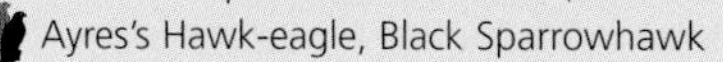

Ayres's Hawk-eagle, Black Sparrowhawk

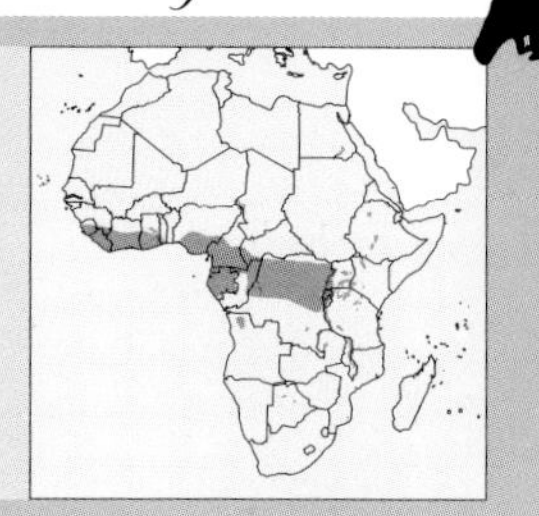

Booted Eagle

Hieraaetus pennatus

AT PERCH

Broad, stocky little eagle with short, heavily feathered legs, a large, rounded head and dark cheeks. Note pale patch across wing coverts and, at close range, pale brown eyes. White shoulder-patches and rump rarely visible, but plain grey tail sometimes evident.

IN FLIGHT

Above distinctive in both forms, with pale wing coverts, scapulars and rump forming a wide, pale V-shape across wings. Tips to greater coverts form a thin, pale line. White rump and plain grey tail. Note pale form with black tips to white underwing coverts and browner head. Dark form with sooty-brown greater coverts. Pale wedge of inner primaries distinctive to both forms. Pale juvenile has streaked breast and underwing coverts. Soars on flat wings, flaps deep and fast, with rounded tail spread and often flexed in flight although appears square when closed.

DISTINCTIVE BEHAVIOUR

Inconspicuous and not often seen perched. Usually spotted flying high overhead, showing distinctive flight patterns, or stooping in heart-shaped posture, with wrists apart, wing tips touching and the white 'headlights' glaring. Feeds mainly on small birds and a few mammals or reptiles. Calls with high, piping trill, especially during courtship display flight. Builds a small stick-nest on a rockface or in a tree growing thereon.

ADULT Occurs in 2 main colour forms. Both brown above, feathers edged with pale brown and forming a pale band across the median upperwing coverts. Both have white rump and white patches at base of leading edge of wing. Pale form (about 75%) is white or pale grey below, sometimes washed rufous with fine, dark streaks on sides of chest. Underwing coverts tipped with black. Dark form is brown below. Cheeks and throat always darker, streaked with grey or black. Flight feathers sooty-brown, the inner primaries grey, forming a pale wedge, the secondaries with narrow grey bars below. Bill blue-grey with black tip; eyes light brown; cere and small, bare feet yellow. Sexes similar in plumage, female about 10% larger (male 510–770 g, female 840–1 250 g).
JUVENILE Dark form similar to adult, but slightly more rufous below with fine, black streaks. Pale form washed rufous below with dark brown streaks on breast and pinstripes on abdomen; legs plain rufous. Eyes dark brown.
DISTRIBUTION, HABITAT AND STATUS Resident breeding species in northwest Africa, and in southwestern Africa and Namibia, where moves north during southern winter. Most enter sub-Saharan Africa from all across Europe as non-breeding migrants during the northern winter. The 2 populations are indistinguishable in the field. May be encountered in any habitat on the continent, but nowhere common. Most frequent in broken terrain (southern Africa) or flat plains with open woodland (migrants). Vagrant to Seychelles.
SIMILAR SPECIES Most similar in size and behaviour to **Wahlberg's Eagle** (p. 106) (also has dark and pale forms; plain dark flight feathers and tail, dark rump, long, slender, feathered legs, slight crest, dark eyes). **Black Kite** (p. 186) has similar pale upperwing coverts (plain dark flight feathers and tail; bare legs; forked tail; dark eyes). Much smaller than other brown booted migrants (**Lesser Spotted**, **Greater Spotted** and **Steppe eagles** (pp. 84, 86 and 82)) or resident **Tawny Eagle** (p. 80). Pale form juvenile similar to juvenile **Ayres's Hawk-eagle** (p. 100) (darker; more streaked; crested; yellow eye, pale greenish cere). Pale form adult in flight recalls much larger **Egyptian Vulture** (p. 46) (wedge-shaped, white tail; long, pointed wings; long bill; naked yellow face).

N. Dennis/ABPL

Adult pale form perched

Small eagle with heavily booted legs. Occurs in 2 colour forms, both dark brown above. Dark brown flight feathers; pale grey tail; below white (commoner pale form) or dark brown (dark form), including underwing coverts. Both have white 'headlights' on the shoulders. Dark cheeks; pale brown eyes. Juvenile has darker streaks below. Medium-sized (about 40 cm tall, 110–132 cm wingspan).

Wahlberg's Eagle, Ayres's Hawk-eagle (juv.)

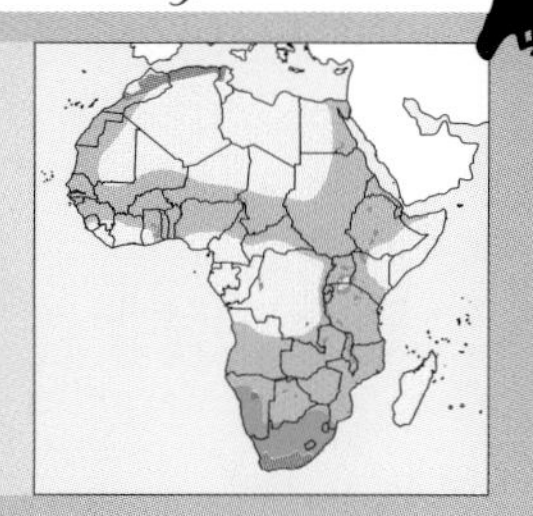

Wahlberg's Eagle

Aquila wahlbergi

AT PERCH

A small, slender eagle with a small bill, small face, dark eye and (most distinctive) a tiny, pointed crest on the back of the head. When it stands horizontally, long, slender, booted legs obvious; when vertical, the rather long tail is evident.

IN FLIGHT

When soaring, the long, rectangular wings with square tips and straight trailing edge and the long, square tail (usually held closed) are distinctive, like 2 planks nailed together in a cross. Folds wings when gliding, and fans tail when seeking lift in a thermal. Underwing coverts dark compared with flight feathers in dark brown forms.

DISTINCTIVE BEHAVIOUR

Usually seen perched on edge of or within a tree or, more often, soaring and gliding overhead. Hunts from either position, making a fast stoop or slow parachute descent to its prey of small animals, especially birds, lizards and rodents. Often calls with distinctive long whistle. Builds a small stick-nest in an upper fork of a large tree, often along a watercourse or foothill.

L. Hes

Pair perched, pale and dark forms

L. Hes

Dark form in flight with straight wings and long, narrow tail

ADULT Dark brown, cream and white forms, plus various intermediates. Commonest form dark brown all over, varying from deep chocolate to dull ochre, sometimes with light and dark browns mixed, especially on head and wing coverts. Less common form has cream underparts and broad, cream edges to darker brown feathers above. Rarest form has white underparts and broad, white edges to grey feathers above. Flight feathers and tail of all forms dark brown above, pale grey below with narrow, paler bars becoming more prominent the paler the body plumage. Intermediates differ most around the head, which may have areas of white in dark forms or areas of brown or grey in paler forms. Bill black; eyes dark brown; cere and bare feet yellow. Sexes similar in plumage, female about 3% larger (male 437–±900 g, female ±800–1 400 g).

JUVENILE Apparently indistinguishable from adult. Fledges directly as its particular form.

DISTRIBUTION, HABITAT AND STATUS Widespread in wooded savanna and woodlands of sub-Saharan Africa. A small, resident, population in equatorial east Africa, but mainly an intra-African migrant. Breeds in southern woodlands in the southern summer, then returns to northeastern and western woodlands in northern summer, possibly to breed again. In many areas the commonest eagle when it is present.

SIMILAR SPECIES Most similar in size and habits to **Booted Eagle** (p. 104) (also has dark and pale forms; pale form with dark head and dark spots on white underwing coverts; both forms have white rump, pale upperwing coverts, plain pale grey-brown tail, white shoulder spots, light brown eyes). Smaller and more slender, and has a small crest and dark eye, compared with the other brown booted eagles that are in Africa during the northern summer (**Lesser Spotted**, **Steppe** and **Tawny eagles** (pp. 84, 82 and 80)) and with the even larger eagles that reach only as far as northeastern Africa (**Greater Spotted** and **Eastern Imperial eagles** (pp. 86 and 90)). Flight silhouette always distinctive.

Small, brown eagle with long, booted legs. Slight, pointed crest; dark brown eyes. Characteristic flight silhouette of long, rectangular wings and a long, narrow tail. Plumage variable, mostly all-dark chocolate-brown, otherwise pale grey above, white below and other combinations. Juvenile indistinguishable from adult. Medium-sized (about 45 cm tall, 141 cm wingspan).

Booted Eagle, Tawny Eagle

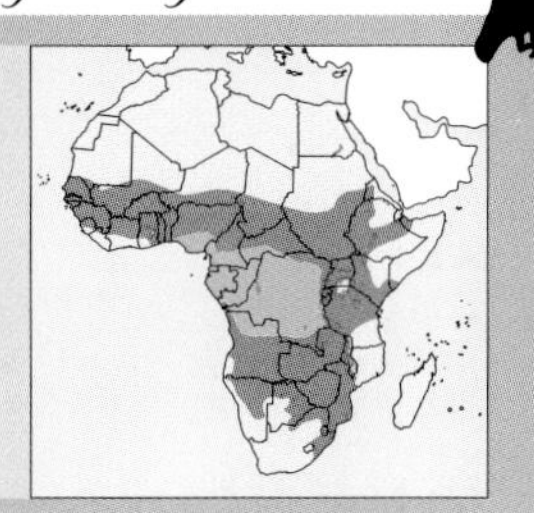

Eurasian Buzzard

Buteo buteo

AT PERCH
Best identified by typical dumpy buzzard shape and relatively large, rounded head. Plumage varies from pale to dark brown; underparts most distinctive with mixture of streaks, blotches and bars; dark thighs; often a pale band across centre of breast.

IN FLIGHT
White underwing with black trailing edge and tips, dark coverts and black carpal patch typical of several buzzard species. Darkness of coverts and bars in secondaries varies with plumage type. Upperwing coverts and body usually paler than flight feathers. Pale patch at base of primaries. Wings somewhat broader, tail shorter, darker and more evenly barred, and head broader, than in some other buzzards. Often soars on slightly raised, forward-held wings with tail fanned, or flaps with shallow, stiff beats. Only rarely hovers.

DISTINCTIVE BEHAVIOUR
Frequently on the wing, including during migration when it may be seen in large flocks. Otherwise solitary, often spending long periods on open perches, searching the ground below for small-animal prey.

ADULT Two subspecies; considerable individual variation. Migrant Steppe Buzzard (*B.b. vulpinus*) from northeastern Europe brown to dark brown above, often with paler streaks on head and rump. Below white or buff, streaked brown on throat, blotched on breast, often with solid dark chest and pale band across centre. Barred on belly, underwing and vent, with dark rufous thighs. Varies below from white to uniform chocolate brown. Flight feathers dark brown above, white below, with broad, black tips, bars on secondaries obvious in darker forms. Sooty-brown carpal patch. Tail rufous-brown above and grey below, with dark brown bars ending in broad sub-terminal bar. European form (*B.b. buteo*) generally less rufous, more brown, more white below, with less obvious tail bars. Bill black; eyes pale to dark brown; cere, long, bare legs and stubby toes yellow. Sexes similar in plumage, female about 4% larger (male 525–1 183 g, female 625–1 364 g).
JUVENILE Similar to adult but generally paler, feathers of upperparts tipped with rufous, under-parts and underwing coverts whiter, streaked rather than blotched or barred. Thighs and vent barred brown. Tail browner, with narrower, more evenly spaced dark bars. Often with more white streaks on crown. Eyes pale grey-brown.
DISTRIBUTION, HABITAT AND STATUS Non-breeding migrant to Africa during northern winter. Some southern European *B.b. buteo* birds migrate to the North African coast (rarely as far south as Liberia in the west), entering at Gibraltar and other Mediterranean crossings. Most are northern Steppe Buzzards (*B.b. vulpinus*) that enter via the Middle East and Arabia and travel via eastern and central Africa to the main wintering grounds in southern Africa. Some remain along the Nile and in eastern Africa. May be encountered in any habitat, but most prefer open steppe, grassland and savanna with scattered hunting perches, including agricultural land. Very common in southern Africa, uncommon further north and scarce in northern Africa except during migration. Other forms resident and breeding on Madeira, Canary, Cape Verde and Socotra islands sometimes considered separate, isolated species.

A. Weaving

Adult

SIMILAR SPECIES Most resembles resident **Mountain Buzzard** (p. 110) (less rufous above, especially on the tail; tail more obviously barred; whiter below; large, dark spots below, rather than streaks, blotches and bars; juvenile has yellow eyes, both species have streaked breasts and narrow tail-bars). Underwing pattern also similar to **Long-legged Buzzard** (p. 114) (plain pale rufous tail), or **Western Honey-buzzard** (p. 124) (3 irregularly spaced tail bars).

ALTERNATIVE NAMES: Common Buzzard (*B.b. buteo*), Steppe Buzzard (*B.b. vulpinus*)

Small, brown buzzard. Wide range of pale, medium and dark plumages. Most are brown above, often with dark chest and pale band across breast. Below brown mottled with white. Tail evenly barred and usually more rufous-brown. In flight, dark coverts, dark carpal patch and dark rim to wings distinctive. Medium-sized (about 45 cm tall, wingspan 102–128 cm).

Mountain Buzzard, Long-legged Buzzard, Western Honey-buzzard

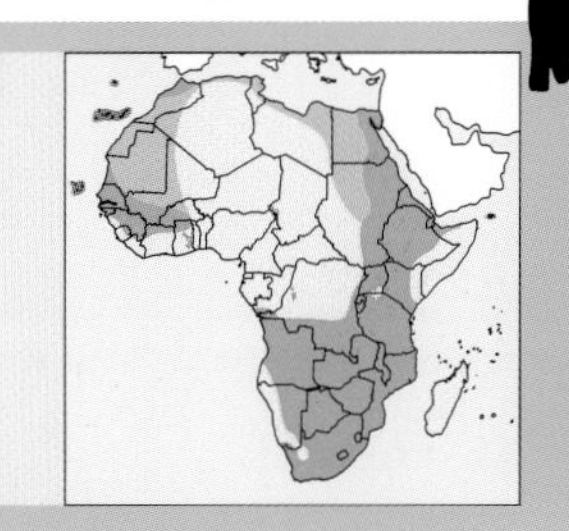

HPH Photography/Photo Access

Mountain Buzzard

Buteo oreophilus

AT PERCH

Sits rather vertically, with typical dumpy buzzard shape and relatively large, rounded head. Spotted underparts and thigh markings the only really distinctive features.

IN FLIGHT

Pale underwing with dark borders, dark tips, dark (*B.o. oreophilus*) or spotted (*B.o. trizonatus*) coverts and dark carpal patch. When soaring, often fans tail to expose bars, which indicate age (less barred in adult).

DISTINCTIVE BEHAVIOUR

Spends long periods perched, often at the edge of forest patches or exotic plantations. Waits for prey to emerge below before gliding down to catch it; main prey consists of small mammals, reptiles and insects. Rarely hunts over open grasslands near forests. Sometimes soars overhead, especially prior to nesting, and may utter shrill mewing calls. Builds stick-nest in upper fork of large forest tree.

I. Sinclair

B.o. oreophilus adult (Ethiopia)

ADULT Plumage individually variable. Two subspecies exist. In northern and eastern Africa (*B.o. oreophilus*) brown above with white feather bases on nape; below white, finely streaked on throat, rest with large, dark brown spots, but thighs and vent more rufous and barred. In southern Africa (*B.o. trizonatus*), brown feathers above with rufous edges; below, small, dark, broad streaks rather than spots, often with unmarked white band across centre of breast. Thighs usually heavily barred with dark brown. Flight and tail feathers have dark bars, including broad terminal bar, grey-brown above and pale grey below. Flight feathers contrast with dark brown (*B.o. oreophilus*) or white-spotted (*B.o. trizonatus*) underwing coverts and sooty carpal patch. Tail bars barely visible from below in *trizonatus*. Bill black; eyes pale brown; cere, bare legs and stubby toes yellow. Sexes similar in plumage, female about 4% larger (both sexes about 700 g).

JUVENILE Similar to adult, but with fine, brown streaks rather than spots or broad streaks on underparts. All-dark tail bands narrow. Eyes dull yellow.

P. Steyn

B.o. trizonatus adult at nest (southern Africa)

DISTRIBUTION, HABITAT AND STATUS Patchy distribution in hill and montane forests from Ethiopian highlands down Rift Valley and adjacent mountains to Malawi (*B.o. oreophilus*); separate population in southern and eastern South Africa (*B.o. trizonatus*). Apparently sedentary, except in southern South Africa where they move north along Drakensberg mountains during southern winter from southern winter-rainfall areas.

SIMILAR SPECIES Difficult to distinguish during southern summer, especially in South Africa, from migrant **Eurasian Buzzard** (p. 108) (adults generally browner below; breast more mottled; less barred on flanks; less evenly spotted on breast; tail more rufous with bars more visible; juvenile has brown eyes; both species have streaked breasts and narrow tail bars). Problem compounded by plumage variation in both species, but only Eurasian Buzzard has very dark or chestnut forms. Resident status, forest habitat and display calling may assist identification, but not infallible.

ALTERNATIVE NAMES: Forest Buzzard, Woodland Buzzard, African Buzzard

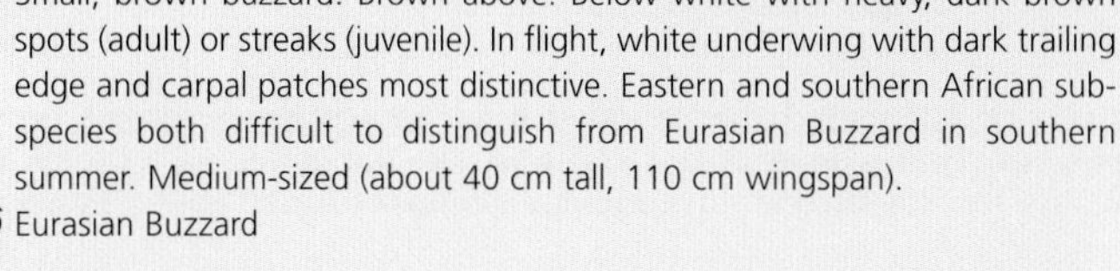

Small, brown buzzard. Brown above. Below white with heavy, dark brown spots (adult) or streaks (juvenile). In flight, white underwing with dark trailing edge and carpal patches most distinctive. Eastern and southern African subspecies both difficult to distinguish from Eurasian Buzzard in southern summer. Medium-sized (about 40 cm tall, 110 cm wingspan).

Eurasian Buzzard

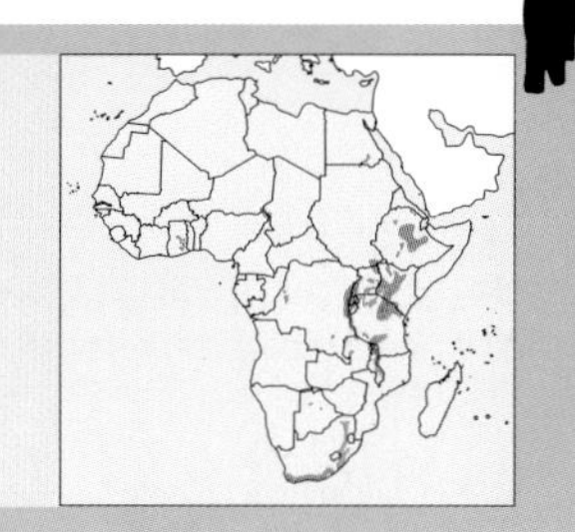

C. Paterson-Jones

Madagascar Buzzard

Buteo brachypterus

AT PERCH

Typical dumpy buzzard shape with relatively large, rounded head. Brown above, broad, brown chest and heavy streaking on sides of white belly below. From behind, tail barred with broad sub-terminal band. Pale eye, cere and legs distinctive. Juvenile has all-streaky breast and dark eye.

IN FLIGHT

Rather broad, rounded wings and rather long tail for a buzzard. Dark underwing coverts below, edged in grey, and white centre to body distinctive from below. Pale tail base obvious above. Flaps with stiff, shallow beats, or soars with wings slightly upturned and pulled forward.

DISTINCTIVE BEHAVIOUR

Spends much time perched within or on edge of forest patches. Sometimes soars overhead, often a pair together, but takes most of its small-animal prey from a perch. Builds a stick-nest in the upper fork of a large forest tree.

P. Oxford/Planet Earth Pictures

Subadult

O. Langrand

Juvenile soaring

ADULT Individually variable. Above dark brown with darker back contrasting with pale bars on rump and, in many, grey wash to head and white base to tail. Below cream, streaked with brown on throat, almost solidly streaked with rufous-brown on chest and underwing coverts. Heavily streaked with brown on flanks, but centre of belly and vent plain white. Thighs dark brown faintly barred white. Flight and tail feathers brown above, white below, with dark bars. Greater primary underwing coverts tipped black, secondaries barred grey or brown. Bill black; eyes pale yellow, cere blue-grey, bare legs and stubby feet very pale yellow. Sexes similar in plumage, female slightly larger.

JUVENILE Similar to adult, but with fine, paler tips to feathers above, and all streaked brown below without solid colour on chest. Obvious narrow, dark tail bars and little contrast between rump and tail. Eye brown.

DISTRIBUTION, HABITAT AND STATUS Endemic to Madagascar. Widespread and common in range of forested, open wooded or secondary habitats throughout the island. Only uncommon on deforested central plateau.

O. Langrand/BIOS

Adult

SIMILAR SPECIES Only buzzard on Madagascar. Resembles **Bat Hawk** (p. 198) (also has dark brown plumage, white underparts, yellow eye, pale legs; but white eyelids, white nuchal spots, black tail narrowly barred with white), but Bat Hawk built and flies quite differently, like a large falcon with long, pointed all-dark wings. Most like **Mountain Buzzard** (p. 110), but ranges differ.

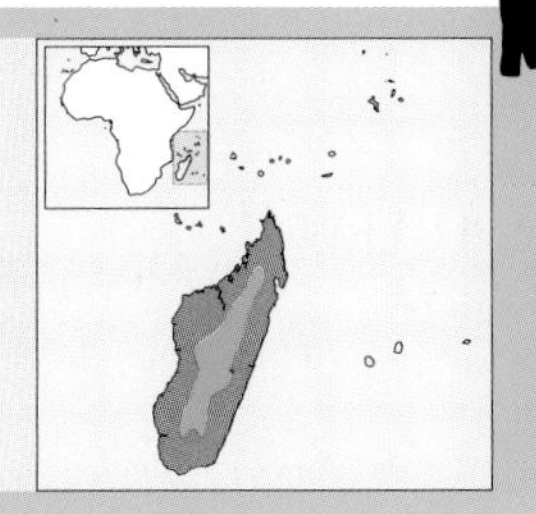

Small, dark brown buzzard. Note pale yellow eyes and legs, and blue-grey cere. Variable plumage. Chest usually brown, flanks heavily streaked with brown, vent white. In flight, wings and tail appear only lightly barred from below; from above, pale base to tail is often distinctive. Medium-sized (about 40 cm tall, 100 cm wingspan).

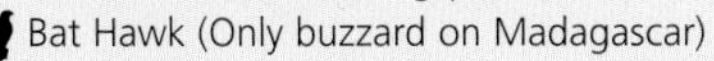
Bat Hawk (Only buzzard on Madagascar)

K. Wothe/Bruce Coleman Ltd

Long-legged Buzzard

Buteo rufinus

AT PERCH

Long, yellow legs obvious against dark thighs. Large, rounded head (typical buzzard shape) and chest usually very pale but much individual variation. Large, almost white tail obvious from behind.

IN FLIGHT

Above usually paler than dark brown flight feathers, except in sooty form. Always some pale grey barring at primary bases. Underwing white, with black trailing edge and tips, dark coverts, black carpal patch and only faintly barred secondaries. Large, pale tail always obvious, with barring in juvenile and especially in dark form. Flaps with steady, deep, elastic wingbeats. Often soars on level, rectangular wings with large secondary area. Glides with long primaries pointed back, sometimes hovers clumsily when kiting against the wind.

DISTINCTIVE BEHAVIOUR

Spends long periods at perch, on prominent rocks or trees or on the ground. Usually seen in soaring flight, sometimes stopping to kite or rarely to hover. Hunts mainly small mammals and reptiles but takes many large insects when not breeding.

Seitre/BIOS

Juvenile soaring

Hellio & Van Ingen/NHPA

Juvenile feeding

ADULT Pale, medium (most common) or dark forms. Above pale, dark or sooty-brown, the feathers edged with paler rufous. Head, neck and chest white or cream with faint rufous streaks. Underwing plain deeper rufous, except for all-sooty dark form. Rump and thighs always dark, contrasting with plain white tail that grades into rufous tip (narrowly barred black in dark form). Flight feathers sooty-brown above, with pale grey barring at primary bases; below white with broad, black tips, carpal patches and faint bars across secondaries (barring heavy in dark form). Bill black; eyes pale brown; cere, long bare legs and stout toes yellow. Sexes similar in plumage, female about 6% larger (male 590–1 281 g, female 945–1 760 g).

JUVENILE Similar to adult of same plumage type, best distinguished by black bars across end of tail.

DISTRIBUTION, HABITAT AND STATUS Subspecies *B.r. cirtensis* (small, generally pale, without dark form) breeds from Mauritania eastwards to Egypt and the Arabian Peninsula, rare vagrant south to Senegal and Cape Verde Islands. Nominate subspecies is a non-breeding migrant from Europe to sub-Saharan Africa during the northern winter, mainly via the Nile Valley and Arabian Peninsula to northeastern Africa, with scattered sightings further south, questionably as far as South Africa. Most reports from arid desert steppe, savanna and grassland. Nowhere common, rare further south.

SIMILAR SPECIES Rufous form resembles juveniles of **Black-chested Snake-eagle** and **Jackal Buzzard** (pp. 58 and 116) (lack pale head and chest and black carpal patches; tails clearly barred; in Snake-eagle cere and legs pale grey). **Eurasian Buzzard** (p. 108) smaller, with dark head and pale brown abdomen, but dark forms only easily distinguishable on broad subterminal tail bar.

Large, brown buzzard. Range of pale, medium and dark plumages. Eagle-like, with long, broad, rectangular wings and large tail. Above brown to rufous. Below white to cream, rufous or sooty. Head and chest always palest. Underwing coverts and flanks dark. Tail plain pale rufous, juvenile's with dark end-bars, all barred in dark form. In flight, underwings white with dark coverts, barred secondaries, and black carpal patch, trailing edge and tips. Medium-sized (about 45 cm tall, 126–155 cm wingspan).

Black-chested Snake-eagle (juv.), Jackal Buzzard (juv.), Eurasian Buzzard

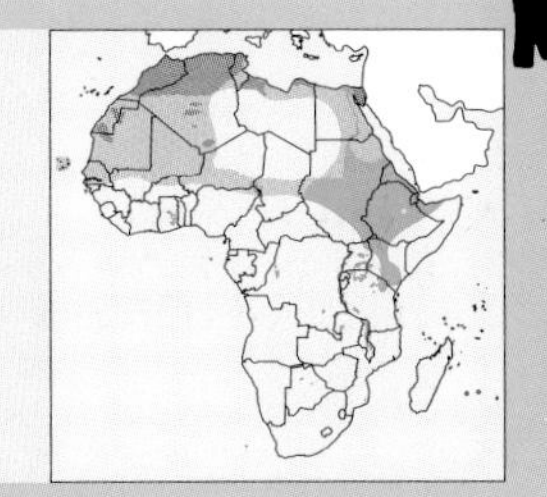

E. Janes/NHPA

Jackal Buzzard

Buteo rufofuscus

AT PERCH

Black upperparts, grey-barred secondaries and rufous tail obvious from behind. From front, chestnut chest and black and white barred underparts distinctive. Juvenile brown, distinguished by contrast between dark back and paler, plain rufous-brown breast. Appears stocky with large, rounded headed and rather short, strong legs.

IN FLIGHT

Long, broad wings held at marked dihedral. Black wing coverts and dark body (brown in juvenile) contrast with white underwing or grey-barred upperwing, and with rufous tail. Broad, black trailing edge and wing tips notable at all ages. Flaps with deep, strong strokes, soars with rocking flight, or hangs with wrists bent.

DISTINCTIVE BEHAVIOUR

Spends much time perched on prominent lookout for its main prey of insects, small reptiles, mammals and birds. Often on utility poles in open habitat, or rocks and trees if available. Hunts regularly on the wing, soaring or kiting in search of food and then parachuting to the ground; only rarely makes dashing pursuit after prey. Utters loud, deep, yelping calls, reminiscent of jackal's. Builds a large stick-nest, usually on a rockface, less often in a tree.

A. Wilson/Photo Access

Adult

ADULT Above black, including head, neck and throat. Below variable. Chest usually rufous, often with white upper edge and black marks below, rarely all-white or all-black. Lower breast and underwing coverts black, barred on coverts, flanks, thighs and vent with white or rufous. Flight feathers black above, with narrow, pale grey bars and broad, black tips, below white with barring only faintly evident. Tail plain rufous. Bill black; eyes dark red-brown; cere, stout bare legs and stocky feet yellow. Sexes similar in plumage, female about 7% larger (male 865–1 080 g, female 1 150–1 370 g).

JUVENILE Dark brown above, with pale edges to nape, back and upperwing coverts creating scaled effect. Head and sides of neck often washed with slate-grey. Below paler, plain and more rufous-brown. Flight and tail feathers brown with narrow, darker bars, paler base to primaries and rufous wash on tail. Eyes, cere and legs dull yellow.

DISTRIBUTION, HABITAT AND STATUS Range restricted to South Africa and southern Namibia; vagrant to southern Botswana and Zimbabwe. Favours rocky outcrops as roosts and nest sites, with open hunting areas nearby. Occurs from high Drakensberg mountains to small hills in the Namib Desert. Common in small-stock farming areas of drier Karoo steppe and short, highveld grasslands.

SIMILAR SPECIES In build, most resembles **Augur Buzzard** (p. 118) (all underparts white, not just rarely chest area; black collar in female; juvenile almost white below with dark streaks), but ranges only overlap in central Namibia and rarely in southern Zimbabwe. Resembles **Bateleur** (p. 54) (upturned wings; black trailing edge; dark body, but underwing coverts white; red (adult) or blue-green (juvenile) face and legs.) Juvenile very like uncommon rufous form of **Long-legged Buzzard** (p. 114) (longer, narrower wings; dark carpal patch) and similar to smaller **Eurasian Buzzard** (p. 108) (less rufous, more clearly marked with brown and white).

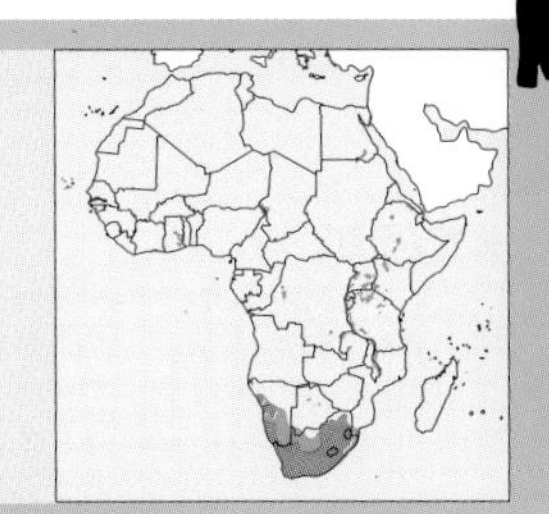

Chestnut chest, rarely black or white. Rufous tail. Black upperparts and head. Lower breast barred black, rufous and white. Secondaries barred grey. In flight, underwing black, with broad, white band across flight feathers. Juvenile dark brown above and plain paler rufous-brown below. Medium-sized (about 40 cm tall, 125 cm wingspan).

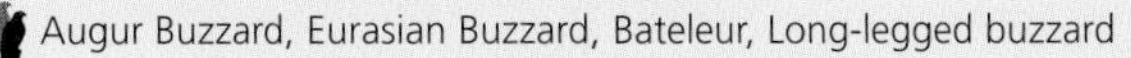

Augur Buzzard, Eurasian Buzzard, Bateleur, Long-legged buzzard

Augur Buzzard

Buteo augur

AT PERCH

Appears black from behind and white from the front except for rufous and black forms. Grey-barred secondaries and rufous of tail and (in some individuals) throat evident at closer range. Juvenile appears brown above and buff below. Usually sits rather vertically and appears dumpy with large, rounded head and long primaries extending well past tail.

IN FLIGHT

Characteristic wing shape with very broad secondaries, long, pointed primaries and short tail. Below white with black trailing edge and wing tips, also black 'comma' at wrist. Body and underwing coverts rufous or black in respective forms. Above, grey bars obvious on secondaries and primary bases. Often on the wing, usually soaring on raised wings with upturned tips and rocking flight, or hanging in the wind with bent wrists, but can flap fast with shallow, stiff wingbeats.

DISTINCTIVE BEHAVIOUR

Spends much time perched in the open, on rocks, mounds, trees or utility poles. Also hunts on the wing for long periods, soaring or kiting, from where it descends slowly on the main small-animal prey of reptiles and rodents. Also capable of fast, rapacious hunting of hares, hyrax and gamebirds. Calls with loud, yelping notes at perch or on the wing. Builds a large stick-nest on a rockface or in a tree.

ADULT Above black, including head. Below white, including underwing coverts (except for black carpal patch). Flight feathers black above, barred with pale grey, especially on secondaries. Below white with broad, black tips; barring only faintly visible. Tail deep rufous, sometimes with faint, dark grey bars. **Female** has black throat and/or collar on sides of neck. Form common in Somalia (Archer's Buzzard) rufous below with black streaks on white throat and chest and rufous wash over shoulders. Form common (10–55% of population) in wet forested areas of northern and eastern Africa is black below. Bill black; eyes brown; cere and stout, bare legs and feet yellow. Female about 4% larger (male 880–1 160 g, female 1 097–1 303 g).
JUVENILE Above brown; feathers have slightly paler tips, producing scalloped effect. Below pale buff with brown streaks on throat and sides of breast. Density of streaking individually variable from fine pale brown to heavy dark brown. Flight feathers grey-brown, wings barred with darker grey, as in adult, tail rufous-brown with faint, narrow, dark bars and broader subterminal band. Eye, cere and legs pale yellow.
DISTRIBUTION, HABITAT AND STATUS Patchy distribution from Ethiopia southwards to Namibia and Zimbabwe. Occupies hilly and mountainous country, from sea level to the highest peaks. Favours areas with rockfaces, wooded slopes, or exotic plantations for roosting and nesting, interspersed with more open grasslands for hunting. Apparently sedentary; commonest buzzard and widespread in eastern Africa, only locally common elsewhere.

Adult landing at nest

A. Weaving

SIMILAR SPECIES In flight, most resembles adult **Bateleur** (p. 54) (white underwing, black trailing edge and short, rufous tail similar; but body always black; back rufous or cream; face and legs red (adult) or blue-green (juvenile); longer, narrower wings) or longer-winged adult **Black-chested Snake-eagle** (p. 58) (barred underwing and tail; black head and chest; often hovers). Juvenile paler below and more streaked than other brown buzzards, more like juvenile hawk-eagles which have feathered legs and smaller, more crested heads. Overlaps in Namibia with identically shaped **Jackal Buzzard** (p. 116) (mixed black, white and rufous below; black underwing coverts).

ALTERNATIVE NAME: Archer's Buzzard (*B.a. archeri*)

Widespread normal form black above and white below. Distinctive in flight, the broad wings white below with broad, black trailing edge and tips. Short, bright rufous tail. Rufous-breasted form in northern Somalia and black-breasted form in northern and eastern Africa. Juvenile brown above, below pale buff to white with variable amount of dark streaks. Medium-sized (about 45 cm tall, 132 cm wingspan).

Bateleur, Black-chested Snake-eagle, Jackal Buzzard

D. Balfour

Red-necked Buzzard

Buteo auguralis

AT PERCH

From behind, rufous head, neck and tail contrast with grey-brown body and wings. From front, rufous head and white underparts, together with rufous to dark brown chest-band and spotted breast, distinctive. Juvenile's breast paler than back and has a few rufous and dark brown spots.

IN FLIGHT

Brown wing coverts and grey flight feathers, with pale patch at base of primaries and dark trailing edges and tips, contrast with white underparts and dark breast-band. Rufous tail with dark subterminal bar and white tip distinctive. Juvenile appears all brown with slight barring in secondaries and on tail. Wings broad and rounded. Flaps with deliberate short strokes, spends much time soaring with tail fanned.

DISTINCTIVE BEHAVIOUR

Most often seen perched along forest edge or soaring over clearings, into which it dives in pursuit of small-animal prey, especially insects and small reptiles and mammals. Calls with shrill, mewing screams. Builds stick-nest in upper fork of large forest tree.

I. Sinclair

Juvenile

ADULT Above dark grey-brown, with rufous crown and broad rufous edges to nape, neck and mantle feathers. Below white, heavily spotted with dark rufous or brown, especially on chest where spots form a solid dark band. Flight feathers dark grey with broad, black tips, primaries white basally below and secondaries barred with darker brown. Tail deep rufous with narrow, black subterminal band. Bill black; eyes brown; cere and bare legs yellow. Sexes similar in plumage, female about 7% larger (male 560–620 g, female 660–890 g).

JUVENILE Above grey-brown with obvious rufous tips to feathers. Below pale rufous-brown or white with a few dark brown spots on chest and flanks. Flight feathers grey-brown with darker tips, tail rufous-grey with narrow, darker bars and broader subterminal band. Eye pale creamy-yellow, cere and legs pale yellow.

DISTRIBUTION, HABITAT AND STATUS Western and central Africa. Favours forest edge and broadleafed woodland, including secondary and gallery forests. Absent from large tracts of lowland rainforest. Locally common, especially during migration between wooded savanna in the north in the wet season (May–June), and forest edge to the south in the dry season (Oct.–Nov.).

SIMILAR SPECIES Other buzzards uncommon in range; rufous head and neck and white underparts with heavy spotting and dark chest-band make it distinct from **Eurasian Buzzard** (p. 108). Similar to much larger **Beaudouin's Snake-eagle** (p. 60) (grey head, very white underwings; tail obviously barred with brown; yellow eye; pale grey legs).

Small, brown buzzard of lowland woodland. Rufous head and neck. Rufous tail with black subterminal band. White below, spotted with brown to form breastband. Medium-sized (about 35 cm tall, 95 cm wingspan).

Beaudouin's Snake-eagle, Eurasian Buzzard

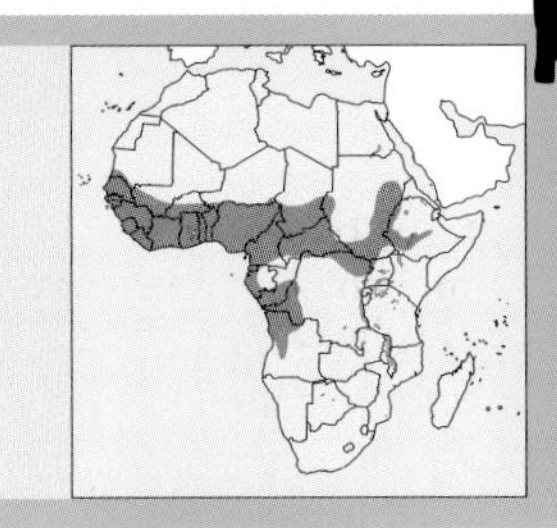

I. Sinclair

Grasshopper Buzzard

Butastur rufipennis

AT PERCH

Dark grey-brown above, white throat and rufous underparts most notable. Head grey. Glimpse of black-tipped rufous primaries, white line along tips of secondaries, yellow bill or pale yellow eye distinctive. Often sits rather horizontally, the long, pointed wings extending at least to the tip of the tail.

IN FLIGHT

The pale rufous primaries edged with black are unique among African raptors. Otherwise appears uniform dark grey-brown above; below, the rufous underparts contrast with the grey flight feathers. Grey tail appears tipped with black. The long, broad, pointed wings, long, square tail and buoyant, swerving flight suggest a kestrel, kite or harrier.

DISTINCTIVE BEHAVIOUR

Often gregarious in small flocks, especially when not breeding. Also spreads out to hunt alone, but quick to congregate at grass fires or insect emergences. Hunts mainly from a perch for the large insects that are its main prey, taking most prey on the ground, or less often making short, aerial pursuits. Breeding almost unstudied, but known to become vocal and to build a substantial stick-nest in a low tree, though apparently not colonial.

D. Richards

Juvenile

ADULT Above grey-brown, feathers thinly edged with rufous on mantle and upperwing coverts. Head dark grey with fine, black streaks. Below rufous with dark brown streaks, especially on chest. Throat white, edged and bisected by black stripes. Vent white and underwing coverts white or heavily spotted with dark grey. Upper primaries and their upperwing coverts rufous with black tips; below primaries white to very pale rufous with grey tips. Secondaries dark grey-brown with broad rufous bases and white tips. Tail grey with dark subterminal bar but other dark, narrow bars incomplete. Bill yellow with black tip; cere, bare legs and feet yellow. Sexes similar in plumage and size (male 310–342 g, female 300–383 g).

JUVENILE Head and neck pale rufous streaked with dark brown. Feathers above have broader, rufous edges than those of adult. Below darker rufous, retaining throat-stripes, but dark stripe through eye also evident. White tips to secondaries and primaries more obvious, tail unbarred and bill black. Eye deeper yellow.

DISTRIBUTION, HABITAT AND STATUS Across Africa just south of Sahara. Favours open, dry thorn savanna for nesting, but uses various open grasslands, savannas, floodplains, burnt areas, and even woodlands during the dry season. Migrates south in the dry northern winter season (Sept.–March) and north again in summer with the rains (April–Sept.) to breed. Ranges between about 9°N and 15°N when breeding, south to 5°N in the dry season, or to well south of the equator in eastern Africa. Exact movements, timing and nesting areas vary according to local and seasonal rainfall patterns. Widespread, locally sometimes very common, but regional status changes from year to year.

SIMILAR SPECIES Very distinctive, especially with rufous wing-patches. Somewhat like **Black Kite** (p. 186) in sociable behaviour and in often perching horizontally, but the latter more uniformly coloured and has forked tail. Flight recalls harrier or kestrel, but size and coloration quite different.

Slim, small, long-winged buzzard. Note white throat with black centre and 'moustache' stripes. Pale yellow eyes obvious. Dark grey-brown above with grey head, below pale chestnut streaked with dark brown. White tips to secondaries and yellow base to bill obvious at close range. In flight, pale rufous wing-patch and black tips to wings and tail distinctive. Juvenile has pale streaky head and dark eye-stripe. Medium-sized (about 30 cm tall, 90 cm wingspan). Very distinctive and often social.

Black Kite

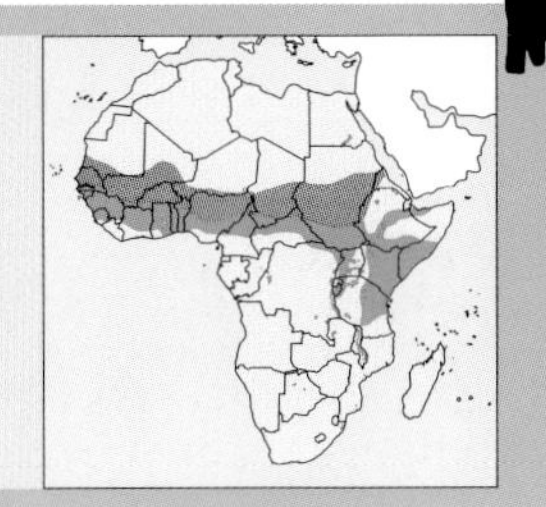

Western Honey-buzzard

Pernis apivorous

AT PERCH

Stands rather horizontally on relatively short legs, rather like Black Kite. Long tail extends well past wing tips, showing broad terminal band. At close range note pale yellow eyes, lack of eyebrow, feathered lores, small, sharp face and only slightly curved claws.

IN FLIGHT

Distinctive silhouette with small, protruding head, long, rectangular wing pinched in at body and long, barred tail. Wings held flat when soaring. Flaps with slow, deep wingbeats. Black trailing edge and broad, black tips to white wings distinctive in most plumages, with black carpal patch and dark or barred underwing coverts. Pale primary bases usually obvious from above.

DISTINCTIVE BEHAVIOUR

Spends much time within cover, perched or flying from one tree to the next in search of the bee and wasp nests that are its main food source. Sometimes descends to the ground to dig for nests. Often returns to same area over several days. Soars high and well, sometimes with flapping, especially when migrating.

ADULT Above usually grey-brown, with dark brown subterminal bar and white feather-tips creating scaled effect. Crown and face often grey, including feathered lores. In pale form, head, neck, chest and rump almost white. Underwing coverts similar to breast, which varies from white to sooty, plain or variously spotted to barred with darker brown. Flight feathers dark brown above with broad, black subterminal bar and pale tip. Below white with black tips, other dark bars variable but usually most evident on secondaries. Black carpal patch. Bill black; eyes pale yellow; cere grey; short, stout legs yellow. **Female** usually browner, about 2% larger than male (both sexes 440–1 050 g).

JUVENILE Similar to adult, but head usually paler and head and breast darkly streaked in both white and sooty-brown forms. Eye brown; cere yellow. Upperparts and flight feathers with broader pale brown or white tips.

DISTRIBUTION, HABITAT AND STATUS Non-breeding migrant to Africa during northern winter. Enters in tens of thousands on a broad front from Gibraltar to the Arabian Peninsula. May turn up anywhere, but considered uncommon to rare within Africa. Most sightings from forests and dense woodland of western, central and eastern Africa; uncommon further south to South Africa and Namibia. Vagrant to Bioko and Seychelles.

SIMILAR SPECIES Very similar to **Crested Honey-buzzard** (p. 126) (larger; loose crest; eye red-brown; brown chest gorget enclosing white throat with black rim; shorter tail with 3 equal wide bars – 2 dark and 1 white; black trailing edges to wings). Also similar to smaller **Eurasian Buzzard** (p. 108), same-sized **Long-legged Buzzard** (p. 114) and even larger **Short-toed** or **Beaudouin's snake-eagles** (pp. 62 and 60). Pattern of barring on tail and to an extent on wings are main consistent differences, unless seen close up with smaller head, shorter legs, pale yellow eyes and feathered lores.

Adult dark form sunbathing

ALTERNATIVE NAMES: European Honey-buzzard, Eurasian Honey-buzzard

Small-headed, grey-brown buzzard. Range of pale, medium and dark plumages. Tail always has 2 dark basal bars and 1 broader, dark terminal bar. At close range, feathered lores, pale yellow eyes, lack of brow ridge, weak bill and short legs distinctive. In flight, black carpal patches and black tips and trailing edges to wings. Generally brown above, often with grey face and pale rump, below white, rufous or sooty, and plain, spotted or barred. Medium-sized (about 45 cm tall, 130–150 cm wingspan).

Crested Honey-buzzard (rare), Eurasian Buzzard, Beaudouin's Snake-eagle

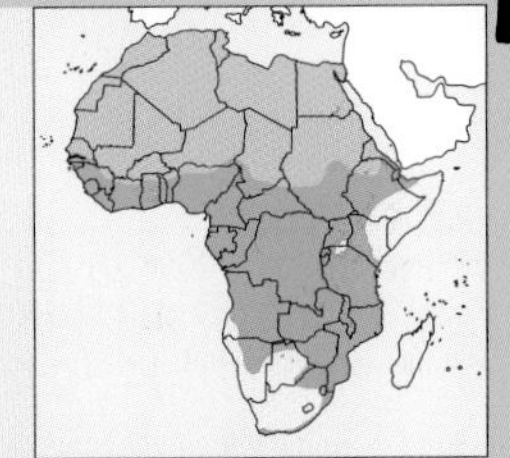

G. Robbrecht/Wildlife Pictures

Crested Honey-buzzard

Pernis ptilorhyncus

AT PERCH

Stands rather horizontally on relatively short legs, rather like Black Kite. Tail extends past wing tips, showing broad terminal band. At close range note red-brown eyes, lack of eyebrow, feathered lores, loose crest, black-rimmed white throat, and small, sharp face.

IN FLIGHT

Distinctive silhouette with small, protruding head, long, rectangular wing pinched in at body and barred tail. Wings held flat when soaring; flaps with slow, deep wingbeats. Black trailing edge to pale flight feathers distinctive in most plumages.

DISTINCTIVE BEHAVIOUR

Behaviour in Africa virtually unknown – deduced from better-known behaviour of Asian populations. Probably spends much time within cover, perched or flying from one tree to the next in search of bee and wasp nests, its main food. Sometimes descends to the ground to dig for nests. Soars high and well, sometimes with flapping, especially when migrating.

ADULT Individually variable. Above buff to dark grey-brown; fine, rufous edges to feathers and dark ends creating scalloped and spotted effect. Face and crown often grey, in contrast to loose, black crest at nape and white throat with black rim and sometimes black centre streak. Below most variable, usually with plain rufous chest, and breast plain rufous or white with rufous spots or dark brown bars. Flight feathers and tail sooty-brown with 2 broad, grey to buff bands. Bill black with grey base; grey cere; eyes red-brown; short, stout, bare legs and long, yellow toes with straight, black claws. Sexes similar in plumage, female about 2% larger (both sexes 740–1 490 g).
JUVENILE Similar to adult, but often browner; usually more streaked below but individually very variable, lacks dark chest of adult, throat usually not white, head brown and streaked, tail with narrow, evenly spaced grey and dark brown bands.

DISTRIBUTION, HABITAT AND STATUS A non-breeding migrant to Africa during the southern winter from far eastern Asia. So far recorded only once, in 1996, in Egypt, when observed migrating north out of Africa with flocks of Eurasian Buzzard and Western Honey-buzzard. May be more frequent visitor than realised but identification difficult.
SIMILAR SPECIES Very similar to smaller **Western Honey-buzzard** (p. 124) (no loose crest; eye pale yellow; throat plain and same colour as head; longer tail with 2 dark median bars and 1 broad, dark terminal bar; same black trailing edges to wings) and only recently recorded in Africa. Also similar to smaller **Eurasian Buzzard** and same-sized **Long-legged Buzzard** (pp. 108 and 114), but different pattern of barring on tail, and to an extent on wings, are main consistent differences, unless seen close up with smaller head, shorter legs, pale yellow eyes and feathered lores.

ALTERNATIVE NAMES: Oriental Honey-buzzard, Eastern Honey-buzzard

Dark grey-brown, small-headed buzzard. Range of pale, medium and dark plumages. Loose crest. Brown chest around black-rimmed white throat. Tail has dark bars across centre and tip. Black trailing edges to wings. Most are dark brown above with grey face, below plain rufous, spotted with white or barred with brown. At close range note feathered lores, red-brown eye, lack of brow ridge, weak bill and short legs. Medium-sized (about 45 cm tall, 135–150 cm wingspan).

Western Honey-buzzard, Eurasian Buzzard, Long-legged Buzzard

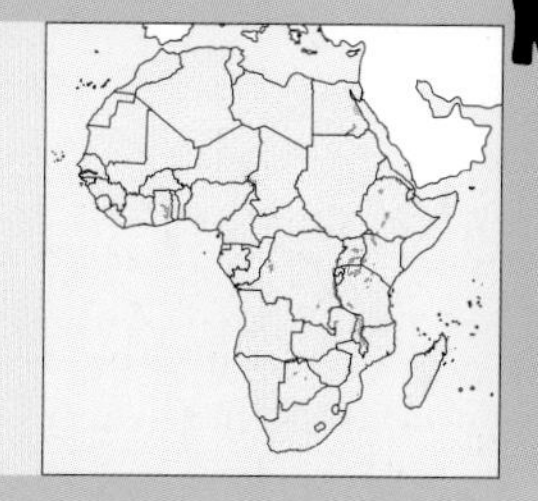

African Harrier-hawk

Polyboroides typus

AT PERCH

Appears as a large, grey hawk when standing upright, with large wing area extending to tip of long tail. Barred underparts not always obvious. Yellow face with small, black bill and eyes, and long, thin, yellow legs conspicuous. Black tail with white band and tip conclusive when visible. Juvenile plumage confusingly variable, but note odd build and bare face.

IN FLIGHT

Long, very broad, rounded wings, exaggerated by slender body and tail, make for very buoyant flight. Below, black ends to flight feathers separated from grey bases, and from grey barred coverts and body by thin white line. Black tail with obvious white band across centre. Juvenile with brown wing coverts and dark brown flight feathers and tail broadly banded with grey-brown.

DISTINCTIVE BEHAVIOUR

Obvious when hunting in trees, on rockfaces or on the ground, where it clambers about at odd angles, even hanging upside down, with wings flapping for balance. Looks into holes and crevices with its small, bare face, or inserts its long, slender 'double-jointed' legs. Also hunts on the wing, floating and flapping slowly between trees or soaring low overhead before floating down on small-animal prey. Often attracts hostile mobbing behaviour from other birds, whose nests it frequently robs. Eats some fruit, especially oil palm fruit in lowland forests. Usually perches within cover. Obvious display flight with drawn-out whistling calls. Builds stick-nest on rocks or trees, often on hillside or in riverine trees.

ADULT Grey overall appearance, with black blobs on scapulars and greater upperwing coverts. Rump barred black and white. Breast and underwing coverts finely barred black and white, rarely plain grey. Flight feathers grey (finely vermiculated with black), and with long, black ends and white tips to secondaries. Tail black with broad, white band across centre and white tip. Bill black; eyes dark brown; cere yellow; bare facial skin yellow, but flushes pink or red; very long, thin, bare legs and small feet yellow. **Female** less spotted above than male, about 3% larger (male 500–720 g, female 580–950 g).

JUVENILE Individually variable. Brown overall, from pale buff to dark chocolate, unmarked or spotted or barred. Flight feathers and tail dark brown with broad, paler grey-brown bars. Distinctive proportions, bare face, voluminous nape, very long, thin legs and hunting habits are the most certain identification features.

DISTRIBUTION, HABITAT AND STATUS Widespread in sub-Saharan Africa, in all types of habitat from semi-desert scrub to lowland rainforest, usually in hilly terrain in more open habitats. Common in woodland and forest habitats, more sparse in drier areas.

SIMILAR SPECIES Proportions, form and colouring very distinctive. Most similar to smaller **Dark, Pale** and **Eastern chanting-goshawks** (pp. 148, 150 and 152) (cere and legs red or orange; face feathered; broad white rump; tail with narrow black and white bars; juvenile eyes yellow).

Adult hunting in tree crevices

Juveniles resemble larger, stockier large-headed **Brown Snake-eagle** (p. 56) and juvenile **Black-chested Snake-eagle** (p. 58) (pale grey cere; thick, pale grey legs; feathered face; large, yellow eye; flight feathers mainly white below).

ALTERNATIVE NAMES: Gymnogene, Banded Harrier-hawk

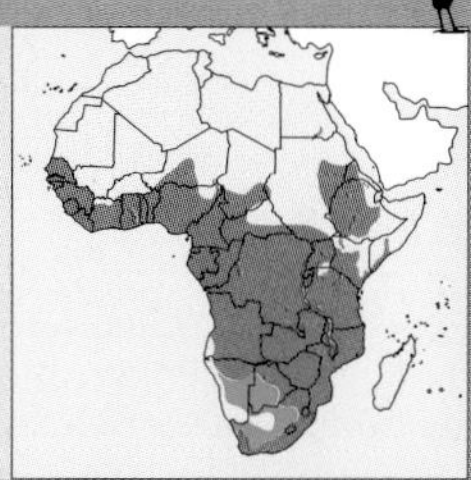

Slender, grey hawk. Broad, black outer wing. Long, black tail with broad, white centre band. Large, loose nape feathers. Small, bare, yellow face that flushes red. Long, thin, bare, yellow legs with small feet. Breast finely barred with black and white. Clambers about in trees. Juvenile variable shades of brown; variation can lead to confusion. Large (about 60 cm tall, 160 cm wingspan).

Chanting-goshawks, Black-chested Snake-eagle (juv.), Brown Snake-eagle

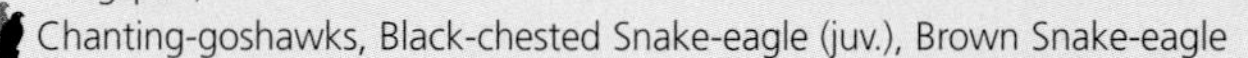

N. Dennis

Madagascar Harrier-hawk

Polyboroides radiatus

AT PERCH

Large, upright hawk with large wings almost reaching tip of tail. Note black tail with pale grey centre band. Adult appears pale grey, with dark bars below. Obvious yellow face, small, black bill and eyes, and long, thin, yellow legs. Most juveniles look pale brown or white with dark brown markings; distinctive build and bare face remain most notable features.

IN FLIGHT

Slender body and tail, with slow, buoyant flight on long, broad, rounded wings. Broad grey area of underwing coverts rimmed by black flight feathers. Tail also black, with obvious pale band across centre; square when closed, rounded when fanned. Juvenile wing coverts mottled brown and white against dark brown flight feathers and tail; tail has regular grey-brown bands.

DISTINCTIVE BEHAVIOUR

Clambers about with wings flapping when hunting on trees, cliffs or the ground. Peers into holes and crevices with the small, bare face, or inserts the long, slender 'double-jointed' legs. When hunting on the wing, floats slowly between trees or soars, looking down for small-animal prey; eagle-like appearance in flight distinctive. Robs bird nests, including nests of other raptors. Builds a stick-nest in a tall tree. Calls often with distinctive, drawn-out buzzard-like 'keeeoow'.

ADULT Dove-grey overall, often with brown wash above. Rump white with black bars. Feathers of scapulars and greater coverts have black blobs, and end in fine, black line and white tip. Breast and underwing coverts white with fine, black bars. Flight feathers grey, finely vermiculated with black and with long, black ends and white tips. Tail black with broad, white band across centre and white tip. Bill black; small eyes dark brown; cere yellow; bare facial skin yellow, pink or red; long, thin, bare legs and small feet yellow. **Female** with fewer or no black blobs on coverts, and slightly larger.

JUVENILE Above brown, with spots on head and neck and mottling on back and wings formed by white bases and tips to feathers. Rump often has broad, white bars. Below appears white with broad, brown spots, streaks and bars, including on the underwing coverts. Least marked on legs and vent. Flight and tail feathers dark brown, with even, broad, paler grey-brown bars. Individually variable in coloration and markings, but less so than African Harrier-hawk; most much paler.

DISTRIBUTION, HABITAT AND STATUS Throughout Madagascar, but mainly in forests and woodlands in the north and east of the island, including stands of exotic trees. Uncommon on central plateau.

Juvenile

SIMILAR SPECIES Quite different shape from any other Madagascan raptor. In plumage pattern, most resembles **Henst's Goshawk** (p. 182) (stocky build; short wings; feathered face; yellow eyes; white eyebrow-stripe; tail with evenly spaced dark and lighter brown bars). Juveniles recall brown and white **Madagascar Buzzard** (p. 112) (feathered face; grey cere; yellow eyes; pale yellow legs; stocky build; large, rounded head; heavily streaked on flanks).

ALTERNATIVE NAME: Banded Gymnogene

Slender, pale grey hawk. Breast white with fine, black bars. Bare yellow or red face. Long, loose nape feathers. Very long, thin, bare, yellow legs with small feet. In flight, broad, black edge to grey wings. Long, black tail with broad, white band across centre. Juvenile dark brown above and pale brown below, flight and tail feathers dark brown barred with pale grey-brown. Large (about 60 cm tall, 155 cm wingspan).

Henst's Goshawk, Madagascar Buzzard

African Marsh Harrier

Circus ranivorus

AT PERCH

Stands rather upright for a harrier, on long, strong, bare yellow legs. Large and heavily built for a harrier. Appears dark brown from behind; paler from in front, with brown, buff and rufous streaks. Rufous thighs often notable. Barred tail, greyer in male, sometimes evident. Bright yellow eye. Juvenile with pale chest-band, crown and shoulder-bar, and with rufous-barred tail.

IN FLIGHT

Typical floating harrier flight, with wings held up in shallow V-shape, especially when gliding. Also soars on level, extended wings with long tail usually closed. Upperwing coverts appear paler than dark flight feathers, the evenly spaced dark and light bars especially evident below. Pale underwing contrasts with much darker body. Rufous belly and thighs notable at close range. Juvenile has pale chest-band, leading edge to wings and, sometimes, crown.

DISTINCTIVE BEHAVIOUR

Most often seen coursing low over reeds and sedges along the water's edge, less often over adjacent scrub-, grass- and farmlands. Hunts its main prey of rodents and small birds after sharp twists to the ground, less often after a short aerial chase. Often soars high overhead, sometimes performing floppy aerial dances with high, squealing calls. Roosts and nests on the ground or on a platform of vegetation above the marsh.

ADULT Dark grey-brown overall. Above, feathers edged and streaked with rufous, white bars on wing coverts. Head, neck and underparts usually heavily edged and streaked with buff, cream and white. Lower breast, thighs and vent plain rufous. Underwing coverts buff with heavy, dark brown streaks. Flight and tail feathers sooty-brown above and grey-brown below, with paler, evenly spaced bars (bars rufous above and pale buff below). Bill black; eyes, cere and long, bare legs yellow. Flight feathers, especially base of primaries and upper tail, greyer in **male**; **female** browner and about 6% larger (both sexes, 382–590 g).

JUVENILE Dark grey-brown overall, with prominent, pale buff tips to shoulder and upperwing covert feathers, especially on lesser coverts. Crown and rim of ear coverts also streaked with buff. Broad cream streaks form band across chest, also some black and rufous streaks on breast. Underwing coverts dark brown heavily streaked with cream, and with dark carpal patch. Flight feathers dark brown above. Below, primaries have dark tips and cream bases, secondaries dark grey with only faint barring. Tail evenly barred with dark brown and rufous. Eye red-brown, cere and legs pale yellow.

DISTRIBUTION, HABITAT AND STATUS Patchily distributed across eastern, central and southern Africa where suitable wetlands exist. Locally common and often obvious, but declining in many areas as wetlands become degraded. Extensive movements, especially of juveniles, may bring birds to isolated waterbodies far from normal range.

N. Dennis

Adult with fish carrion

SIMILAR SPECIES In behaviour and habitat preference very like larger, heavier **Western Marsh Harrier** (p. 134) (overall less streaky; flight and tail feathers unbarred on adults; adult male has pale buff head, plain grey flight feathers and tail; adult female and juvenile have cream crown, throat and leading edges to wings, and dark 'mask', very like some juvenile African Marsh Harriers but never with pale chest-band). Resembles brown adult females and juveniles of smaller **'dry-country' harriers** (pp. 138–144) (all paler and more rufous; with more or less white rumps; lacking paler leading edge to wing).

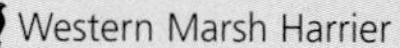

Large, dark, rufous-brown harrier with few distinguishing features. Streaked with rufous, buff and cream. Rufous thighs and belly. In flight, wings and tail clearly barred with dark grey and light brown. Juvenile has buff crown, cream band across chest and along leading edge of wing. Medium-sized (about 40 cm tall, 110 cm wingspan). Usually associated with wetlands.

Western Marsh Harrier

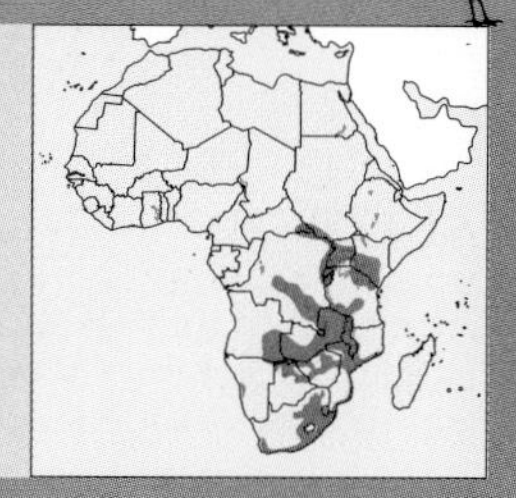

N. Dennis/SIL

Western Marsh Harrier

Circus aeruginosus

AT PERCH

A relatively heavily built harrier with fairly short, stout legs. Male's grey secondaries, greater coverts and tail, pale head and streaked chest, against the dark rufous underparts and black primaries, are notable. Female and juvenile appear plain dark brown apart from the pale crown and throat, dark eye-mask and pale shoulder-bar.

IN FLIGHT

Buoyant flight with deep, steady wingbeats, interspersed with some gliding when the wings are held in a shallow V-shape, or more rarely soaring on spread and slightly raised wings. Flight feathers and tail unbarred in all plumages. Male has black wing tips, grey upperwings and tail, white underwings with dark trailing edge, brown body and white underwing coverts with a little brown speckling. Pale head and streaked chest often evident. Female and juvenile appear plain dark brown apart from pale patch at base of primaries, cream leading edge to wings, and cream head with dark mask. Paler, more rufous outer feathers notable on female when tail spread.

DISTINCTIVE BEHAVIOUR

Spends much time coursing slowly and low, back and forth, over vegetation. Glides less than other harriers, but like them hunts by fast twists and dives into cover. Roosts amongst tall, dense groundcover, alone or in small, communal groups. Breeds amongst tall marsh vegetation. Calls and performs aerial dance over breeding areas.

ADULT Dark brown overall. Individual variation in colour and extent of markings, becoming paler with age. **Male** has head, neck, chest and underwing coverts broadly edged with buff or white to appear pale and streaked. Upperwing coverts tipped with rufous, but greater coverts pale grey, to match pale grey flight feathers. Black ends to primaries. Underwing white, with black primary tips and broad, grey secondary tips. Tail plain pale grey, often with faint darker bars on outer feathers. Belly and thighs plain dark brown with rufous wash. **Female** plain dark brown above, with fine, rufous edges to wing coverts and flight feathers and paler rufous outer tail feathers. Below paler brown with some rufous streaking and paler patch at base of primaries. Crown, nape and throat pale cream or white, divided by dark brown ear coverts around and behind the eyes. Lesser upperwing coverts pale cream. Some all-dark individuals. **Both sexes:** eyes, cere and long, bare legs pale yellow; bill black. Female about 5% larger (male 405–667 g, female 540–800 g).

JUVENILE Both sexes similar to adult female, overall darker brown except for paler rufous tips to wing coverts, and with brown streaking in cream crown and lesser upperwing coverts. Eye brown.

DISTRIBUTION, HABITAT AND STATUS Small resident breeding population, distinct from European population as subspecies *harterti*, in northwestern Africa. Most birds are non-breeding migrants to the Mediterranean basin and sub-Saharan Africa from Europe during the northern winter. Commonest in western and northeastern Africa, where the African Marsh Harrier is absent, extending into eastern Africa, and straggling as far south as northern South Africa. Migrates on broad front and may occur in any habitat, but favours marshy areas and hunts over adjacent dry habitats. Reaches Socotra, and rare vagrant to Madeira, Canary, Cape Verde and Seychelles islands.

Adult female

SIMILAR SPECIES In behaviour and habitat preference, very like slightly smaller and lighter **African Marsh Harrier** (p. 132) (more streaky; flight feathers and tail barred dark and light brown; no obvious difference between sexes; juvenile has pale chest-band). Resembles brown adult females and juvenile of the smaller **'dry-country' harriers** (pp. 138–144) (all paler and more rufous; with more or less white rumps; lacking paler leading edge to wing), a few adult males even having paler rumps.

ALTERNATIVE NAMES: Eurasian Marsh Harrier, European Marsh Harrier, Marsh Harrier

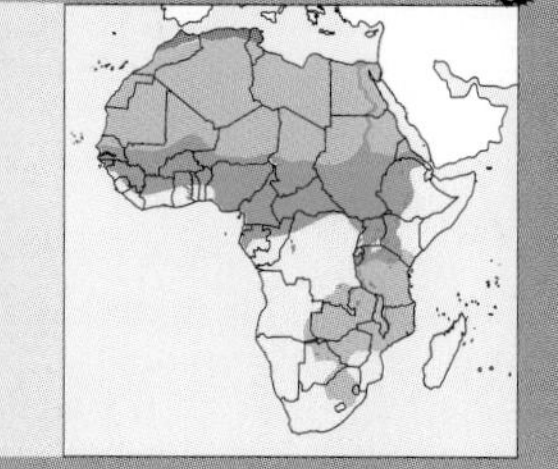

Large, dark brown harrier. Adult male has pale grey head; adult female and juvenile have cream crown and throat. Streaked chest. Plain grey flight feathers, black primaries and grey tail (male) or dark brown flight and tail feathers (female and juvenile). Dark face 'mask'. Cream leading edge to wings. Individually variable. Medium-sized (about 40 cm tall, 110–130 cm wingspan).

African Marsh Harrier

P. van Gaalen/Bruce Coleman Ltd

Madagascar Marsh Harrier

Circus maillardi

AT PERCH

Long flight feathers and tail, small head with typical harrier facial disc, and long legs all notable features. Adult male appears black from behind but heavily streaked with white on nape, and with pale grey patches on wings and grey tail. White underparts and and streaking on throat and chest contrast with black crown and face. Adult female and juvenile appear brown, paler and more rufous below, with pale, streaky head and chest.

IN FLIGHT

Flies buoyantly with steady beats and intervals of gliding or soaring with wings raised in a shallow 'V'. Pale rump obvious at all ages. Adult male appears white below with black wing tips, tail tips and trailing edges; above black with grey flight feathers and tail plain or narrowly barred and tipped black. Black face with yellow eye, and streaked neck and chest, notable at close range. Adult female and juvenile dark brown with broad light and dark bars across wings and tail.

DISTINCTIVE BEHAVIOUR

Hunts main prey of small vertebrates by coursing low over vegetation, twisting down to snatch at food with the long legs. Also hunts over forest canopy, dropping into foliage or snatching prey from off the canopy. Soars and calls with high whistling notes above the breeding areas. Nests and roosts on the ground amongst tall vegetation, often in marshes. Hunts more over mountain forest on Réunion (race *maillardi*) than on Madagascar and Comoros (race *macrosceles*).

ADULT Male black above, including crown and ear coverts. Fine grey tips to lesser wing coverts; rump white. Nape, neck, throat and chest white with heavy black streaks. Breast, flanks and vent plain white. Greater wing coverts, flight feathers and tail grey above and white below, with broad, black tips. **Female** dark brown overall with white rump. Above, feathers tipped, and flight feathers and tail broadly barred, with rufous. Pale eyebrow; head and throat streaked with white. Chest and neck pale rufous, streaked with dark brown. Breast, thighs and vent pale rufous streaked with brown. Both sexes: bill black; eyes, cere and long, bare legs yellow. Female slightly larger.

R. Thorstrom

C.m. macrosceles adult female

JUVENILE Both sexes similar to adult female, but with pale rump washed with brown, broader, paler tips to feathers above, and neck and breast more brown. Eyes brown.

DISTRIBUTION, HABITAT AND STATUS Endemic to Madagascar and the islands of the Comoros and Réunion. Most common in damp, marshy areas, whether freshwater or saline, but hunts far from these habitats – prefers rather to hunt (especially on Réunion and the Comoros) over mountainsides and forest patches. Common on Réunion (subspecies *maillardi*) but local and uncommon on the Comoros islands and Madagascar (subspecies *macrosceles*).

SIMILAR SPECIES The only harrier on these islands. On Madagascar, only the brown adult female and juvenile might be confused with **Madagascar Buzzard** (p. 112) (also hunts low over grasslands and canopy; but no white rump; large, rounded head; short, broad wings and tail; heavy flight; pale grey cere and yellow legs) or **Black Kite** (p. 186) (shorter, forked tail; dark eyes; no white rump).

ALTERNATIVE NAMES: Malagasy Harrier, Malagasy Marsh Harrier, Réunion Marsh Harrier, Réunion Harrier (race *maillardi*)

Large, long, slender harrier. Adult male black above, with pale grey secondaries and tail. Below white streaked with black. Appears silvery black and white in bright sunlight. Adult female and juvenile dark brown, with pale streaky head and paler rufous underparts. Medium-sized (about 50 cm tall, 130 cm wingspan). Found on Madagascar, Comoros and Réunion.

Madagascar Buzzard, Black Kite

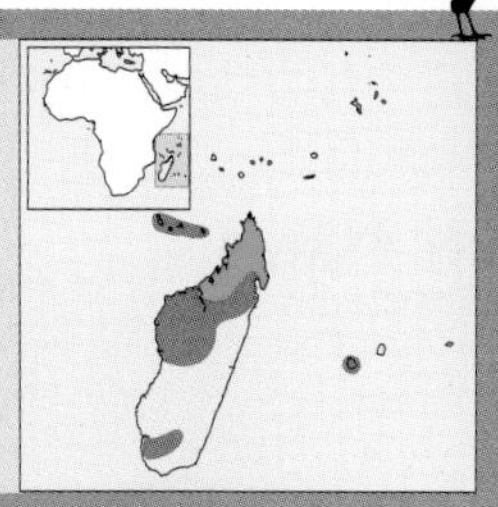

Black Harrier

Circus maurus

AT PERCH

Stands tall on long legs, the long wing tips often crossed over the long tail but well above its tip. Appears all-black with obvious yellow eye, cere and legs. Long, sooty-grey tail with pale bars notable and, if wings drooped, white rump obvious. Juvenile appears similar but browner from behind. From in front, the rufous or cream underparts, buff eyebrow, ears and throat and dark gorget become obvious.

IN FLIGHT

Relatively broad, short wings for a harrier, but tail long. Courses relatively slowly, but capable of bursts of speed and agile turns. Broad, white rump and clearly barred tail obvious at all ages. Black body and wing coverts of adult contrast above with pale grey patches at base of primaries; together with black trailing edge contrast below with white underwings. Dark chest and underwing notable on juveniles and at close range yellow eyes also notable.

DISTINCTIVE BEHAVIOUR

Spends much time perched on the ground, less often on a mound or post. Hunts with low, coursing flight, especially when windy, with frequent fanning and twisting of the long tail. Small birds, rodents and large insects form the main prey. Builds an untidy platform of weeds and reeds on the ground as its nest, often (but not always) in a marshy area.

H. von Hörsten/ABPL

Adult showing tail pattern

ADULT Black overall, including underwing coverts, but with broad, white rump. Some birds have a few white spots on the underparts, fine, white tips to the belly and thigh feathers, or slight white bars on the underwing coverts. Flight feathers and tail silvery-grey above with dark tips; outer primaries with white bases, inner primaries and secondaries with dark bars. Below white, outer primaries with grey tips, inner primaries and secondaries with broad, black tips, and tail with broad, black bars and a narrow, white tip. Bill black; eyes, cere and long, bare legs bright yellow. Sexes similar in plumage, female about 8% larger (male 383 g, female 574 g).

JUVENILE Dark grey-brown above, except for broad, white rump. Nape feathers with white bases for mottled effect. Wing coverts tipped buff for scaled effect. Head and underparts pale rufous, broadly streaked with dark brown on crown, ear coverts, chest, underwing coverts and, to a lesser extent, flanks. Flight feathers sooty-brown above with greyer secondaries, faint black bars and black trailing edge. Below white, with similar markings but extra line of dark grey on greater underwing coverts. Tail grey above and white below with broad black bars. Eyes yellow.

DISTRIBUTION, HABITAT AND STATUS Breeds on the southern tip of Africa during the southern spring, after the winter rains. Main habitat is fynbos scrub- and bushlands, interspersed with marshes and sedgelands or, more recently, cereal croplands. Not dependent on wetlands. Ranges further north during the southern autumn and winter, especially into Namibia (where a small resident population may occur in the northwest) and southern Botswana. Winter habitat includes dry Karoo steppes, semi-desert and dry grasslands.

SIMILAR SPECIES Only really confusable during southern summer with the non-breeding migrant harriers which are rare in its breeding habitat. Adult resembles only the rare dark forms of **Montagu's** and **Western Marsh harriers** (pp. 140 and 134) (both with dark rumps; inconspicuously barred tails; grey-barred wings). Juvenile resembles adult females and juveniles of **Montagu's** and **Pallid harriers** (pp. 140 and 142) (paler brown; generally more rufous below; finely streaked on chest and underwing coverts; brown eye; notable dark ear-patches; small, U-shaped, white rump). Occurs alongside resident **African Marsh Harrier** (p. 132) in several areas (larger; browner; dark rump; barred primaries and secondaries; heavier in flight).

All-black, with broad, white rump. White bars across flight and tail feathers. Sexes identical. Juvenile dark grey-brown above and pale rufous below, the chest heavily streaked with brown. Yellow eye at all ages. Medium-sized (about 35 cm tall, 100 cm wingspan). Breeds only in southern South Africa.

Females, juveniles and dark forms of Pallid Harrier, Montagu's Harrier, Western Marsh Harrier

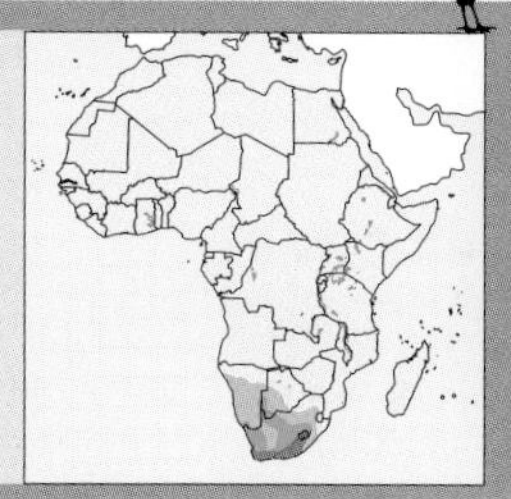

R. du Toit/ABPL

Montagu's Harrier

Circus pygargus

AT PERCH

Stands rather horizontally on relatively short legs. Long wings reach tip of long tail. Adult male grey, with black wing-bar and tips, and prominent yellow eyes. Chestnut streaks on white breast only visible in good light. Adult female and juvenile appear brown above and buff (female) or rufous (juvenile) below, variously streaked and blotched; face pattern of white-rimmed eye and dark ear-patch notable. Dark forms appear black, with pale tail-bars on female.

IN FLIGHT

Very long, rather broad, pointed wings, often held slightly bent and raised in shallow V-shape, and long tail. Flight very buoyant and agile, with regular glides between short bouts of flapping. Adult male grey above with black ends to the wings and black bar across centre. Breast and underwing coverts streaked, with 2 black bars and black tips to secondaries below. Tail with dark bars showing on sides. Adult female and juveniles dark brown above with white rump, buff and streaky (female) or plain rufous (juveniles) below with barred wings and tail, or secondaries sometimes all dark. Dark ear-patches and eyes prominent on pale face.

DISTINCTIVE BEHAVIOUR

Usually seen on the wing, flapping and gliding low over bushes and scrub. Makes fast twists to the ground after its main prey (in Africa) of large insects, birds and rodents. Stops to perch on low vantage points, less often on flat ground. Roosts on the ground, sometimes in pairs or small groups, often in tall, marshy vegetation or among bushes.

ADULT Male blue-grey overall, including head and chest. Breast and underwing coverts white with obvious chestnut streaks. Primary coverts black. Rump mottled with white. Outer primaries black; inner primaries and secondaries plain grey above, with black median upperwing coverts above and 2 dark bars and dark tip below. Tail grey to white and barred with rufous-brown on outer feathers. **Female** dark brown above, with fine rufous or buff feather tips. Head pale rufous, white feather bases on nape. Face marked with white lines around eye, dark brown ear bracket. Below buff or pale rufous, with dark brown streaks. Underwing coverts even paler, with heavier chestnut-brown streaks. Flight feathers and tail dark brown above and grey-brown below, primaries with paler bars and dark tips, secondaries similarly barred or all-dark. Rare all-black form occurs, male washed grey above and on tail with silvery patches below at base of primaries: female more sooty-brown with narrower pale grey bands across tail. **Both sexes**: bill black; eyes, cere and long, bare legs yellow. Female about 2% larger (male 227–305 g, female 310–445 g).

JUVENILE Both sexes very similar in plumage to adult female. Darker overall with more rufous tips above. Head, neck and underparts (including underwing coverts) rufous. Secondaries often all-dark. Only crown, neck and primary underwing coverts streaked with dark brown. White eyebrow and cheeks and sooty ear-patch even more obvious. Rump white with rufous tips. Eye dark brown. Moults directly into adult plumage in second year.

DISTRIBUTION, HABITAT AND STATUS Rare breeder in northwestern Africa. Most are non-breeding migrants from Europe to sub-Saharan Africa during the northern winter. Migrates on a broad front and may be encountered almost anywhere on the continent. Most common over areas of scattered bushes and tall grass, but occurs on more open grasslands and agricultural lands. Now generally uncommon, even in good habitat. Vagrant to Madeira and Canary Islands.

SIMILAR SPECIES Adult female and juvenile very similar to female and juvenile **Pallid Harrier** (p. 142) (more obvious pale collar; dark stripe through eye; narrower white eyebrow; paler abdomen; narrower wings; secondaries always appear barred; longer legs; juvenile less rufous) and, to a lesser extent, **Hen Harrier** (p. 144) (wings even broader; dark trailing edge to secondaries and tail), both of which are also migrants to Africa. Also very similar to juvenile of resident **Black Harrier** (p. 138) (very dark above with buff scallops; pale rufous or cream below; pale ear-patch and rim; darkly streaked throat gorget and underwing; yellow eye). Adult male **Pallid** and **Hen harriers** (pp. 142 and 144) are also grey, but differ in markings of underparts and wings (Pallid very pale grey, plain white underparts, plain grey wings, black wedge at wing tips; Hen has dark trailing edge to secondaries, longer tail, wider wings, smaller black wedges at wing tips). Brown **marsh harriers** (pp. 132–134) are larger and heavier and lack white on the rump.

Adult male blue-grey. Black line across centre of wings and black outer primaries. Chest grey; breast and underwing coverts white streaked with chestnut. Adult female dark brown above, buff or rufous below with dark brown streaks; face marked with white lines around eyes and dark ear-patches. Both sexes have yellow eyes. All birds have relatively short legs. Juvenile like adult female with brown eye. Medium-sized (about 35 cm tall, 97–115 cm wingspan).

Pallid Harrier, Hen Harrier, Black Harrier (juv.)

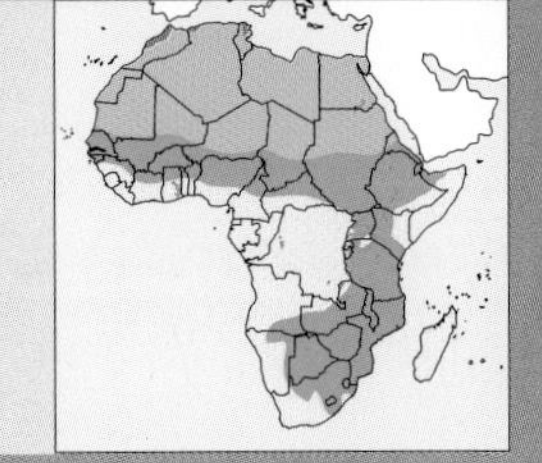

F. Cahez/BIOS

Pallid Harrier

Circus macrourus

AT PERCH

Stands quite upright for a harrier, on relatively long legs. Long wings do not reach tip of long tail. Male appears very pale with black wing tips. Adult female and juvenile appear dark brown with pale shoulders, the head notable with its distinctive pattern of dark and light markings and pale collar. Eye yellow in adult, brown in juvenile.

IN FLIGHT

Appears to have narrower, more pointed wings and a longer tail than other harriers. Flies with great buoyancy and agility, almost like a small gull, with the wings held in a high V-shape. The pale grey male with black wedges at the end of the wings is distinctive. Female and juvenile, with obvious white rumps, facial pattern and dark secondaries, almost indistinguishable from Montagu's Harrier.

DISTINCTIVE BEHAVIOUR

Spends even more time on the wing than other harriers and glides even less. Courses low over open vegetation, rarely soars, and lands more often on the ground than on a perch. Captures small animals and especially insects (in Africa), after a sharp twist down to the ground or vegetation. Roost singly or communally on the ground, usually in a patch of tall grass.

ADULT Male pale grey above, barred white on rump. Below white. Four outer primaries with black ends, outer tail feathers with white bars, rest of flight and tail feathers pale grey above and white below. **Female** dark brown above with grey cast, feathers tipped buff especially on lesser wing coverts. Rump white. Dark brown streaks on crown, patch across eye and edge to facial disc, with buff in between to form distinctive face pattern; pale collar behind. Head, neck and underparts paler and more rufous-brown. Breast streaked with chestnut, fading to pale buff vent. Flight feathers and tail white below, and dark brown above, with buff tips, and grey-brown bars especially wide on primaries. **Both sexes:** bill black; eyes, cere, and relatively long, bare legs yellow. Female about 8% larger (male 235–325 g, female 425–454 g).

JUVENILE Both sexes very similar in plumage to adult female. Buff feather tips more obvious on slightly darker upperparts and flight feathers, forming pale patch on lesser wing coverts. Secondaries often darker. Darker sides to neck, more rufous and less marked below, with paler head to make dark facial markings and pale collar more obvious. Eyes dark brown. Moults direct into adult plumage in second year.

DISTRIBUTION, HABITAT AND STATUS Non-breeding migrant from eastern Eurasia to sub-Saharan Africa during the northern winter. Main crossing, where a few birds remain, is the eastern Mediterranean basin and Arabia. May be encountered almost anywhere on the continent. Most common on open floodplains, grass- and steppelands, and open agricultural fields. Less common over wooded savanna. Now generally uncommon to rare, even in good habitat.

Dr E. Sauer/Bruce Coleman Ltd

Adult male

SIMILAR SPECIES Adult female and juvenile very similar to female and juvenile **Montagu's Harrier** (p. 140) (pale collar less marked; white lines around eye less prominent; abdomen as dark as breast; wings broader; no barring visible in dark secondaries; shorter legs; juvenile more rufous) and, to a lesser extent, **Hen Harrier** (p. 144) (lacks dark stripe through eye; wings even broader; abdomen heavily streaked; dark trailing edge to secondaries and tail), both of which are also migrants to Africa, and also to juvenile of resident **Black Harrier** (p. 138) (very dark above with buff scallops; pale rufous or cream below; pale ear-patch and rim; darkly streaked throat gorget and underwing; yellow eye). Adult male **Montagu's** and **Hen harriers** (pp. 140 and 144) are also grey, but darker and differ in markings of underparts and wings (Montagu's has grey chest, chestnut streaks on white lower breast and underwing coverts, black bar across centre of upperwing, larger black wedge at ends; Hen has grey chest, white breast, and dark trailing edge to secondaries and tail). Brown **marsh harriers** (pp. 132–134) are larger and heavier and lack white on the rump.

Adult male pale grey; black wedge at ends of long, narrow wings; rump barred with white. Adult female dark brown above, below paler and more rufous; rump white; dark facial disc markings and pale collar; flight feathers broadly barred with white. Both sexes have yellow eyes. Relatively long legs. Juvenile like adult female but with brown eye and darker ear coverts. Medium-sized (about 35 cm tall, 95–120 cm wingspan).

Montagu's Harrier, Hen Harrier, Black Harrier (juv.)

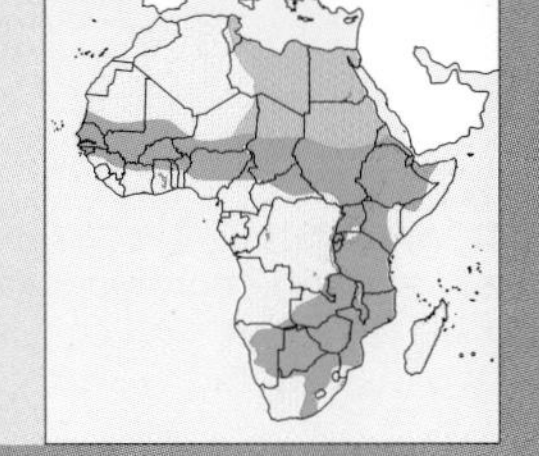

J. Scott/Planet Earth Pictures

Hen Harrier

Circus cyaneus

AT PERCH

Appears rather stocky for a harrier, with relatively long, stout legs. Long wings do not reach tip of long tail. Adult male appears solid grey above, with grey head and chest and white breast. Note black wing tips and grey tail faintly barred on sides. Adult female and juvenile appear dark brown above, paler below with dark streaks. Dark ear-patch the only facial mark.

IN FLIGHT

Relative to other small harriers, wings rather broad; flies with deeper wingbeats, possibly somewhat slower, but very agile and buoyant with large tail. Glides with wings relatively flat. All ages have broad, white rump. Adult male grey above and on head, white below, with black wedge at end of wing and dark trailing edge. Adult female and juveniles dark brown above, streaky below with rufous wash on upperwing coverts, barred underwings and tail clearly barred above and below. No obvious facial markings or pale collar.

DISTINCTIVE BEHAVIOUR

Hunts relatively slowly but with great agility over scrub and heath. Main food small rodents. Appears larger, heavier and slower than other small, ring-tailed harriers. Roosts communally on the ground.

L. Campbell/NHPA

Adult female at the nest with chicks

ADULT **Male** grey above, including head and throat, with broad, white rump. Breast white. Outermost primaries black, rest of primaries and secondaries grey above, below white with broad, grey tips. Tail grey with faint darker grey bars on outer feathers. **Female** dark brown above, with paler feather tips and slight brown marking on white rump. Dark ear coverts are the only facial markings. Cheeks and underparts buff, heavily streaked with dark brown on breast and underwing coverts. Flight feathers and tail brown with darker bars and broad, dark tips. Bill black; eyes deep yellow; cere and long, bare legs yellow. Female about 10% larger (male 300–400 g, female 410–708 g).

JUVENILE Both sexes very similar in plumage to adult female. More rufous overall, with broader pale tips above. Darker below, including underwing coverts. Sooty ear-patches the only obvious facial marks. Rump white with rufous feather tips. Eyes dark brown in female, yellow in male. Moults directly into adult plumage in second year.

DISTRIBUTION, HABITAT AND STATUS Non-breeding migrant from northern Europe to the Mediterranean basin during the northern winter. Uncommon but regular visitor, known to hunt mainly over agricultural crops, marshes and bush-clad hillsides. Very rare vagrant to sub-Saharan Africa.

SIMILAR SPECIES Adult female and juvenile very similar to female and juvenile **Montagu's Harrier** (p. 140) (wings slightly longer and narrower; legs shorter; dark ear-patch and crown with white eyebrow) and, rarer in range, **Pallid Harrier** (p. 142) (more obvious pale collar; dark stripe through eye and dark ear-patch; pale collar; narrower wings; secondaries often appear barred; juvenile less rufous), both of which are also migrants to Africa. Male **Montagu's** and **Pallid harriers** (pp. 140 and 142) also grey but differ in markings of underparts and wings (Montagu's darker grey, and with black line(s) across wings; Pallid very pale grey; plain grey wings; black wedge at wing tips). **Western Marsh Harrier** (p. 134) is larger and heavier, and differs in always having unbarred secondaries and lacking white on rump.

ALTERNATIVE NAME: Northern Harrier

Robust harrier with relatively short, broad wings and long tail. Broad, white rump at all ages. Adult male otherwise grey with white breast, black primaries and grey trailing edge to wing. Adult female dark brown above; below buff to rufous, heavily streaked with brown; flight feathers clearly barred; no obvious facial pattern. Yellow eyes in both sexes. Juvenile like adult female with brown (female) or yellow (male) eyes. Medium-sized (about 35 cm tall, 99–121 cm wingspan). Found only around Mediterranean.

Montagu's Harrier, Pallid Harrier

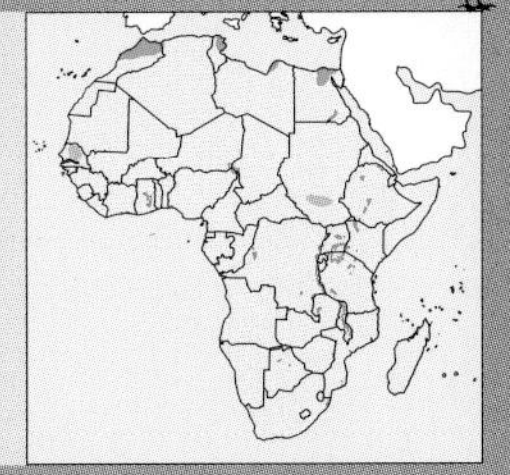

Long-tailed Hawk

Urotriorchis macrourus

AT PERCH

Rather vertical stance, long tail obvious. Dark body, grey from behind or chestnut from front, contrasts with white tail coverts, white tail-bars, and yellow eyes, cere and legs.

IN FLIGHT

White-barred flight feathers (especially on long tail) and white tail coverts prominent. Fast, agile, dashing flight with strong wingbeats. Chestnut underparts, including wing coverts, contrast with dark flight feathers and grey upperparts as its twists and turns in direct and rapid flight.

DISTINCTIVE BEHAVIOUR

Perches for long periods between flights through branches and foliage. Wings and tail often fanned open and twisted from side to side to facilitate dextrous flight. Rarely flies above forest canopy. Apparently hunts mainly on trunks and in foliage for small-mammal and some bird prey. Utters a drawn-out cry; sometimes quite vocal – often the first indication of its presence.

ADULT Above, including head and neck, dark grey. Throat white to pale grey. Below deep chestnut, or dark grey on rare dark form. Sometimes has fine, white bars on flanks – possibly subadult plumage. Rump and vent pure white. Flight feathers black with white bars, long, graduated tail with fine, white bars and white tip. Bill black; eye deep yellow; cere and long, bare legs with stout feet and short toes yellow. Sexes similar in plumage, female about 6% larger (male 492 g, female about 675–775 g).

JUVENILE Rufous-brown, with grey wash over head and neck and white streaks, and white feather bases on nape. Upperwing coverts have faint, pale brown bars; rump has white bars and feather tips. Below white with large, dark brown blotches, forming dark bars on flanks and thighs. Vent white with a few black spots. Flight and tail feathers have narrow, pale and dark bars and white tips. Possibly eye dark brown, cere and legs pale yellow.

DISTRIBUTION, HABITAT AND STATUS Specialist rainforest species of western and central Africa. Little-known and not often recorded, mainly owing to secretive habits and preference for forest canopy. Widespread and may be common in undisturbed forest, less often recorded from large stands of adjacent riverine forest. Most often seen near small villages, where sometimes quite bold and reported to raid poultry.

SIMILAR SPECIES Same size and build as long-tailed **Congo Serpent-eagle** (p. 68) (tail not graduated; dark brown above with dark rump; white below; large, dark spots on breast and bars on flanks; dark 'moustache' and central throat stripes; eye yellow (male) or brown (female); dark crown; large, floppy nape feathers; wings and tail brown with sooty-brown bars; juvenile most similar in colour and markings but with spotted underparts). Also resembles **Long-tailed (White-crested) Hornbill** (*Tockus albocristatus*), with its long, graduated tail and dextrous flight. Spotted underparts of juvenile resemble those of much smaller juvenile **African Cuckoo-hawk** (p. 190) or **African Goshawk** (p. 172).

Large, long, slim hawk. Dark grey overall. Long, graduated, white-tipped tail distinctive. White rump and vent. Rufous below, except in rare all-grey form. Wings and tail black with white bars and tips. Eyes, cere and long, bare legs yellow. Juvenile brown with spotted white breast. Large (about 55 cm tall with long tail, 90 cm wingspan).

Congo Serpent-eagle, Long-tailed Hornbill

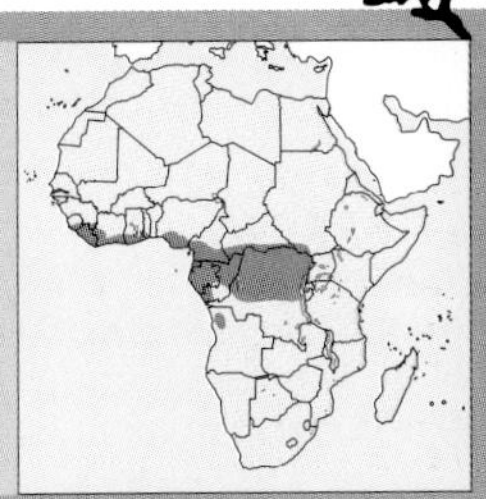

Dark Chanting-goshawk

Melierax metabates

AT PERCH

Upright stance with long, red legs and overall dark grey appearance. Red cere obvious, dark eye notable at close range. Pale grey secondary coverts unmarked over most of range, but with fine, white vermiculation in northeastern Africa. Brown juvenile with streaked chest and yellow eye. Barred rump in adult or juvenile rarely visible and appears dark rather than white.

IN FLIGHT

Flies with steady, rather shallow wingbeats. Rarely shows good turn of speed and agile pursuit of prey. Note dark wing tips. Tail dark above or with broad, white bars below. Secondaries pale grey above, barred with dark grey below. Pale grey base to black primaries also distinctive. Rump appears paler grey than body, with barring rarely visible. Juvenile more narrowly barred with brown on flight feathers, especially on secondaries.

DISTINCTIVE BEHAVIOUR

Perches for long periods on a prominent perch, on the lookout for lizards, insects and rodents on the ground. Most prey taken on the ground after a steep, fast dive from a perch, less often after an aerial pursuit. Calls often from a perch or in flight prior to breeding (but not as vociferous as other chanting-goshawks), with an accelerating series of piping notes, almost a song. Builds a small stick-nest on an inside fork of a woodland tree.

L. Hes

Juvenile hunting on foot

ADULT Above dark grey, including chest and head. Secondary wing coverts slightly paler. Rump and underparts, including underwing coverts, white finely barred with grey. Outer primaries black with pale grey bases below; inner primaries and secondaries pale grey above with fine pale bars below. Tail black, outer feathers barred with white. Bill black; eye dark red-brown; cere, long, stout, bare legs and feet red. **Male** has slightly paler wing coverts; female about 2% larger (male 645–695 g, female 841–852 g).

JUVENILE Above brown, with paler feather tips. Rump and underparts white; chest streaked and breast and rump barred with brown. Flight feathers and tail dark brown above, and white with narrow, brown bars below, especially on secondaries and outer tail feathers. Cere and legs orange, eye yellow.

DISTRIBUTION, HABITAT AND STATUS Well-wooded African savanna, extending to Morocco and the Arabian Peninsula. Favours broad-leafed woodland, barely overlapping with the drier habitats of other chanting-goshawks. Generally common, but not as conspicuous as its open-country congeners.

SIMILAR SPECIES Just overlaps in southern Africa with larger **Pale Chanting-goshawk** (p. 150) (rump plain white; paler grey; secondaries mainly white; fine, white, wavy bars on secondary coverts) and in eastern Africa with smaller **Eastern Chanting-goshawk** (p. 152) (rump plain white; browner plumage; paler yellow cere and orange legs). Juvenile darker and more grey-brown than other chanting-goshawks. Also common in same habitat is **Lizard Hawk** (p. 154) (smaller; dumpier; dark line down white throat; white rump; white band across black tail). Much smaller **Gabar Goshawk** (p. 156) similar in coloration but with white rump. Juvenile resembles some large-headed, short-legged ***Buteo*** **buzzards** (pp. 108–120).

ALTERNATIVE NAME: Chanting-goshawk

Dark grey upperparts and chest. Finely barred underparts. Red cere and legs. Perches upright. White rump finely barred with grey. Dark eye at close range. Juvenile has brown upperparts, streaked chest and rump, and yellow eye. Medium-sized (about 40 cm tall, 100 cm wingspan).

Pale Chanting-goshawk, Eastern Chanting-goshawk, Lizard Hawk, Gabar Goshawk

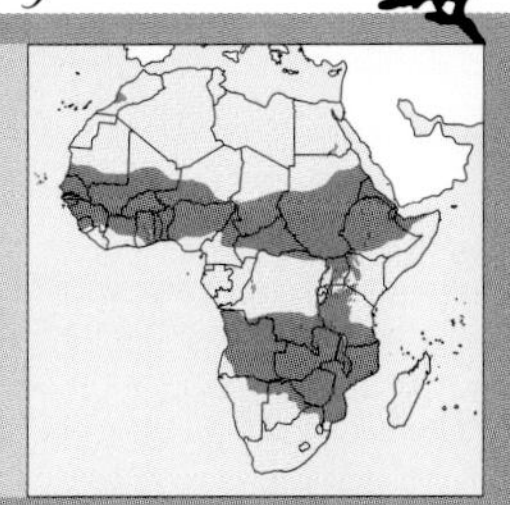

M. Goetz

Pale Chanting-goshawk

Melierax canorus

AT PERCH

Upright, pale grey raptor with long, red legs, red cere and dark eyes. White rump and vermiculation on secondary coverts visible at close range. Juvenile brown with blotched chest and yellow eye.

IN FLIGHT

Usually flies with steady, shallow wingbeats, but can change to fast, agile pursuit. Black outer primaries and white, unbarred underwing, with white secondaries above and below, are distinctive against the grey head and body. Tail appears dark above and broadly barred with white below. Juvenile narrowly barred with brown on all flight feathers. White rump obvious at all ages.

DISTINCTIVE BEHAVIOUR

Perches prominently, most often on a utility pole or bush, but also on the ground, especially in more arid areas. Dives or runs after its main prey of rodents, lizards and insects. Sometimes makes agile aerial chases after birds or hares, including quite large species. Calls often, usually from a perch but also in a circling display flight with slow wingbeats. The main chant is a melodious series of piping notes. Stick-nest built in upper fork of tree or on utility pole.

R. du Toit

Juvenile standing at rest

ADULT Above pale grey, including chest and head. Secondary coverts have fine, wavy, white bars. Rump white. Below white with fine, grey bars; underwing coverts unmarked. Outer primaries black with white bases below. Inner primaries and secondaries mainly white. Tail dark grey, outer feathers barred with white. Bill black; eye dark red-brown; cere, long, stout, bare legs and feet red. Sexes similar in plumage, female about 7% larger (range, male 413–750 g, female 750–1 000 g).
JUVENILE Above brown. Rump and underparts white; chest blotched and breast barred with brown. Flight feathers and tail brown above, and white with narrow, brown bars below, especially on secondaries and outer tail feathers. Cere and legs orange, eye yellow.

DISTRIBUTION, HABITAT AND STATUS Open savanna, arid steppe and desert margins of southern Africa. Most common in areas of limited groundcover among scattered bushes and low thorn trees, and along watercourses in more arid areas. Widespread and generally common, most obvious where utility poles provide perches.
SIMILAR SPECIES Overlaps along northeastern edge of range with smaller **Dark Chanting-goshawk** (p. 148) (darker grey overall; grey barring on white rump, making it appear dark; plain grey upperwing coverts; grey secondaries). Much smaller **Gabar Goshawk** (p. 156) (darker grey; plain grey coverts and secondaries) also common in most of range. Juvenile resembles some large-headed, short-legged ***Buteo* buzzards** (pp. 108–120).

♀ Pale grey plumage overall. Red cere and long, red legs. Dark eye. Rump white. Secondaries and inner primaries white. White vermiculation on secondary coverts. Perches upright. Often on the ground. In flight, black primary tips contrast with white underwing. Juvenile pale brown with yellow eye. Medium-sized (about 45 cm tall, 110 cm wingspan).

Dark Chanting-goshawk

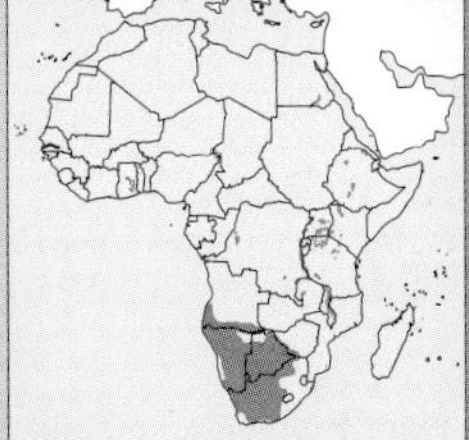

Eastern Chanting-goshawk

Melierax poliopterus

AT PERCH

Upright stance, long, orange legs, pale orange-yellow cere and overall grey-brown appearance. Dark eye and white vermiculation on secondary coverts visible at close range. Juvenile brown with streaked chest, banded breast and yellow eye. Unmarked white rump often visible at all ages.

IN FLIGHT

Usually flies with steady, shallow wingbeats or rarely in fast, agile pursuit. Black outer primaries and pale grey underwing and secondaries. Tail black with white bars on outer feathers. All juvenile flight feathers barred with brown and white. White rump obvious at all ages.

DISTINCTIVE BEHAVIOUR

Often perches in the open. Main prey lizards, plus some small birds, rodents and insects. Hunts mainly on the ground. Calls often from a perch or in flight, including during communal nocturnal aerial display flights. The main song is a melodious series of piping notes.

ADULT Above dull grey with brown wash, including chest and head. Fine, wavy, white bars on upperwing coverts. Rump white. Below, including underwing coverts, white finely barred with grey. Outer primaries black, with pale grey bases below. Inner primaries and secondaries pale grey. Tail black, outer feathers barred with white. Bill black; eye dark red-brown; cere dull yellow; long, stout, bare legs and feet dull orange. Sexes similar in plumage, female about 4% larger (male 514–581 g, female 673–802 g).

JUVENILE Above pale brown with white rump. Below white, but chest streaked and breast barred with brown. Flight feathers and tail brown above, and white with narrow, brown bars below, especially on secondaries and outer tail feathers. Cere and legs even paler orange than adult, eye yellow.

DISTRIBUTION, HABITAT AND STATUS Inhabits savanna, steppe and desert regions of eastern Africa. Most common in dry savanna grassland with scattered bushes and thorn trees. Widespread, but only common in undisturbed habitats; sparse in desert regions. Ecological replacement for Dark Chanting-goshawk of more wooded habitats.

SIMILAR SPECIES Range abuts that of larger **Dark Chanting-goshawk** (p. 148) (darker grey; white rump barred with dark grey, to appear dark rather than white; secondaries darker than inner primaries; cere and legs red rather than orange; juvenile darker grey-brown). Also common in same habitat is **Lizard Hawk** (p. 154) (smaller; dumpier; red cere and legs; dark line down white throat; white band across black tail). Juvenile resembles some large-headed, short-legged ***Buteo*** **buzzards** (pp. 108–120).

I. Sinclair

Adult

ALTERNATIVE NAMES: Eastern Pale Chanting-goshawk, Somali Chanting-goshawk

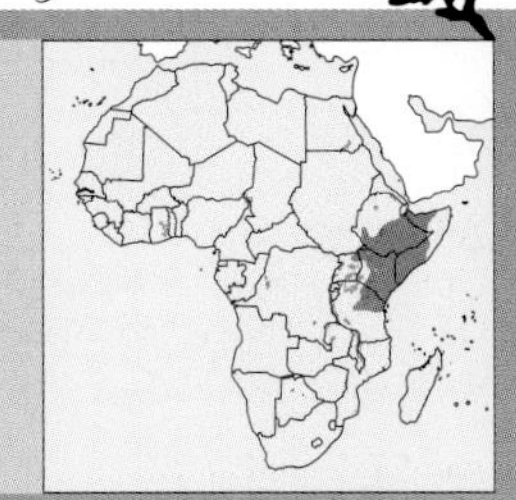

Dull grey-brown overall. Yellow cere and orange legs. Upright stance when perched. Eye dark. Below barred with white. White rump. Pale grey inner primaries and secondaries. Juvenile brown, chest streaked and breast barred with white, eye yellow. Medium-sized (about 40 cm tall, 95 cm wingspan).

Dark Chanting-goshawk, Lizard Hawk

I. Davidson

Lizard Hawk

Kaupifalco monogrammicus

AT PERCH

Appears dumpy and large-headed. Grey overall, with black flight and tail feathers and white band across tail, and with red cere and legs. Barred belly, dark eye and white throat with median stripe obvious at close range. Sits rather vertically. Juvenile best distinguished by eye colour.

IN FLIGHT

Grey-black wings and tail contrast above with grey body, white rump and white tail band(s). Below appears white or pale grey with black-banded tail and wings. Dark eye and red cere and legs obvious at close range. Flies with fast, stiff beats of the short, pointed wings, interspersed with short glides.

DISTINCTIVE BEHAVIOUR

Perches alone for long periods, often staring fixedly at the ground below. When moving between perches, it often drops low from one and then swings up to the next. Regularly calls with a series of melodious whistles, almost a song, uttered with the bill raised and open. Sometimes soars and calls in flight. Makes hard, fast dives at its main prey of large insects, lizards and other small animals, caught mainly on the ground and even in long grass. Builds a small stick-nest in fork of woodland tree, rarely in exotic plantations.

ADULT Above grey, palest on face and wing coverts, darkest on rump. Below grey, plain on chest, and with fine, white bars on breast and underwing. Throat white with obvious black line down centre. Rump and vent white, forming white base to black tail with a broad, white bar across centre, rarely 2, and white tip. Flight feathers dark grey above, white below with narrow, dark bars and black tips to primaries. Bill black; eye dark red-brown; cere, short stout legs and stubby toes red. Sexes similar in plumage, female about 6% larger (male 220–275 g, female 248–410 g).

M. Goetz

Adult

JUVENILE Very similar to adult, but with fine, brown tips to feathers of back and upperwing coverts, faint brown wash to underparts, and obvious pale tips to flight feathers and tail when fresh. Eye brown, becoming pale yellow, within a year like adult; cere and legs more orange.

DISTRIBUTION, HABITAT AND STATUS Occurs throughout the well-wooded savannas of sub-Saharan Africa. Favours deciduous broad-leafed woodland, extends along riparian woodland into drier savanna, and enters clearings and secondary growth in forests. Resident and common in central and western Africa, less predictable in drier areas of eastern and southern Africa.

SIMILAR SPECIES In appearance, behaviour, and relationships, very like the much larger **Dark**, **Eastern** and **Pale chanting-goshawks** (pp. 148, 152, and 150) and smaller **Gabar Goshawk** (p. 156) (all are more slender and longer-legged, and lack the white tail band(s) and the white throat with its black stripe; juveniles brown with streaked chest, barred breast and yellow eye). Resembles and overlaps in habitat with some smaller grey accipiters, such as **Shikra** (p. 166) (pale yellow cere and legs; plain grey tail and rump; red eye) and **Ovambo Sparrowhawk** (p. 158) (orange cere and legs; rump grey with white-tipped feathers; white tail markings; dark red eye), which also differ in being more slender and longer-legged.

ALTERNATIVE NAME: Lizard Buzzard

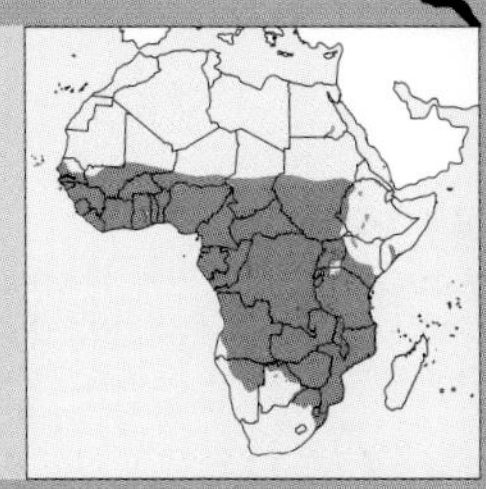

A dumpy, grey hawk. White throat with black stripe down centre. Rump white; tail black with 1 white bar across centre, rarely 2. Chest grey, breast white with fine, grey bars. Note red cere and legs. Eyes very dark red. Medium-sized (about 30 cm tall, 79 cm wingspan).

Dark, Pale and Eastern chanting-goshawks, Gabar Goshawk

Gabar Goshawk

Micronisus gabar

AT PERCH

Appears grey, with white-barred belly. White rump not always visible. Pale tail-bars, dark eye and red cere and legs obvious. Juvenile appears brown with streaked chest. Perches upright on long legs and often in the open.

IN FLIGHT

White rump and clearly barred flight feathers most prominent, together with plain (adult) or streaked (juvenile) chest. Dark eye and red legs of adult evident at close range. Flies fast and low, flapping and then gliding, on broad, rather pointed wings.

DISTINCTIVE BEHAVIOUR

Varied hunting techniques for main prey of small birds – mainly strikes from cover but also makes fast, searching flights, follows prey deep into bushes, or tears open and robs nests. Often tame; hunts in pairs and perches on sides of bushes or trees, waiting for prey to emerge; also sometimes seen clambering through branches. Sometimes shares hunting with Red-necked Falcon. Small stick-nest in crown of indigenous, often thorny, tree is usually covered in spider web.

N. Dennis

Adult

Juvenile

ADULT Plain grey above and on chest. Prominent white rump. Below white with grey bars. Vent plain white. Flight feathers sooty-grey with broad, pale grey bars (almost white) on tail. Bill black; eyes deep dark red; cere, long bare legs and stocky toes bright red. Sexes similar in plumage, female about 16% larger (male 110–173 g, female 180–221 g). Melanistic form (6–25% of population) black with barring across flight feathers (pale grey above and white below) but no white rump. Red leg scales flecked with black.

JUVENILE Individual variation in shades of brown, from pale rufous to dark fuscous. Above brown, with pale streaking on head, pale tips to feathers. Below white, broadly streaked brown on chest and barred brown on breast. Eye pale yellow and cere and legs more orange-red than on adults. Melanistic form dark from hatching.

DISTRIBUTION, HABITAT AND STATUS Widespread in savanna and steppe habitats of sub-Saharan Africa, extending to Arabian Peninsula. Only absent from the densest forest and driest deserts. Most common in drier thorn savanna, including along watercourses; more local in dense woodland and around forest patches. Avoids plantations of exotic trees.

SIMILAR SPECIES Most resembles **Ovambo Sparrowhawk** (p. 158) (bars below extend to throat; small snaky head; only white flecks on rump; white tail-feather shafts; orange legs and cere; juvenile has prominent pale eye-stripe and dark crown and ear-patches). Adult and juvenile coloration almost identical to that of much larger chanting-goshawks, and dumpier **Lizard Hawk** (p. 154) (stout legs; 1 or 2 white tail-bands; black stripe on white throat). Same size as **Shikra** (p. 166) (pale dove-grey; yellow cere and legs; bright red eye; plain grey rump) and uncommon migrant male **Levant Sparrowhawk** (p. 170) (pale grey upperparts; black tips to white underwing; grey rump; yellow cere and legs), which male Gabar Goshawk resembles.

Prominent white rump and pale bars on tail feathers at all ages. Adult plain grey above and on chest. Breast white with grey bars. Juvenile brown, streaked on chest and barred on breast. Note red or orange cere and legs at all ages. Regular melanistic form all black, including rump, with pale grey bars on wings and outer tail feathers. Small (about 25 cm tall, 60 cm wingspan).

Ovambo Sparrowhawk, chanting-goshawks, Lizard Hawk

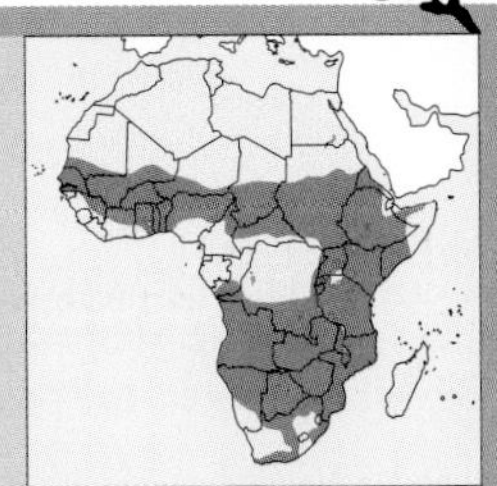

C. Paterson-Jones

Ovambo Sparrowhawk

Accipiter ovampensis

AT PERCH

Plain grey upperparts and entirely barred underparts the most obvious feature, together with white tail shaft spots. Dark eye and orange cere and legs distinctive. Small head often twisted about, snake-like. Often perches rather horizontally in comparison with other accipiters.

IN FLIGHT

Overall grey in appearance, with bars on flight feathers not very obvious and only white flecks on grey rump. Dark eye often visible. Fast, direct flight with long, rather pointed wings and long tail.

DISTINCTIVE BEHAVIOUR

Most often seen in fast, dashing, dextrous flight, hunting its main prey of small birds or harassing larger raptors. At such times perches in the open and appears very alert. Also soars over open habitats and stoops at prey, but for much of the time difficult to locate as it perches quietly within wooded cover. Call a high, nasal 'kiep' note repeated. Builds a small stick-nest in the crown of a large tree.

ADULT Above plain grey, darker on wing coverts and flight feathers. Rump has white bars. Below white finely barred with grey from throat to belly and thighs. Thighs only lightly barred and vent almost plain white. Flight and tail feathers dark grey with only slightly paler broad bars. Tail with 4 or 5 bars and white shaft marks. Bill black; eyes deep dark red; cere, rather short legs and long, slender toes orange. Female about 14% larger (male 105–190 g, female 180–305 g). A rare (1–2%) melanistic form is dark sooty-brown (may appear black) with very little white on the tail or rump; bars on the flight feathers faint above but obvious below.

JUVENILE Dark brown upperparts with pale rufous feather tips and broad, pale rufous eyebrow-stripe. Crown and ear coverts dark grey-brown to give capped and masked effect. Two colour forms, with underparts either plain rufous or streaked and barred with white. Eyes dark brown, cere and legs pale orange.

Distribution, habitat and status Recorded widely in savannas of sub-Saharan Africa but nowhere common, and status in several areas uncertain. Most records from tall, deciduous woodlands, recently including plantations of exotic trees. Rare north of the equator, breeds sporadically in eastern Africa but apparently only a breeding migrant to western Africa in the June–Oct. wet season. More common in southern Africa, especially in teak and miombo woodlands and, in South Africa, in exotic plantations.

SIMILAR SPECIES Most similar to **Gabar Goshawk** (p. 156) (larger head; plain grey chest; cere and legs bright red; prominent white rump; tail with clear, pale grey and white bars below, black and dark grey above, with dark tip; juvenile has streaked chest and barred breast; melanistic form is blacker and lacks white rump, also has more prominent flight-feather barring). Juvenile, especially rufous form, also similar to juvenile **Rufous-breasted Sparrowhawk** (p. 160) (less prominent eyebrow-stripe; yellow eyes; yellow cere and legs; lacks dark crown and ear-patches).

M. Goetz

Juvenile

ALTERNATIVE NAME: Ovampo Sparrowhawk

Plain grey above. Below white, finely barred with grey including on the throat and thighs. Dark red eyes. Orange cere and legs. Small, snaky head. White shaft spots on tail. Note rather long, pointed wings and short legs. Juvenile brown above and rufous below, either plain or streaked and broadly barred with white; eye on average dark brown and pale eyebrow-stripe always conspicuous. Rare melanistic form. Small (about 30 cm tall, 67 cm wingspan).

Gabar Goshawk, Rufous-breasted Sparrowhawk (juv.)

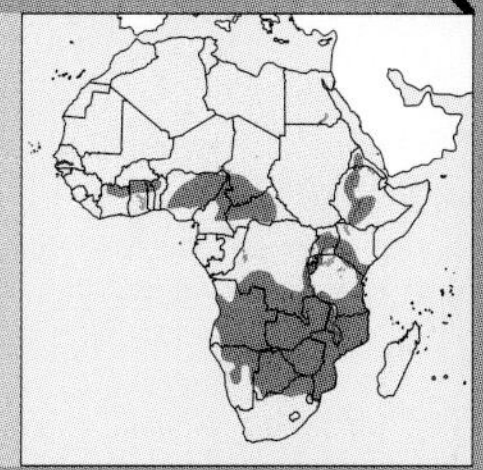

N. Dennis/SIL

Rufous-breasted Sparrowhawk

Accipiter rufiventris

AT PERCH

Dark upperparts and rufous underparts distinctive, with vent forming white patch below. Yellow eye notable at all ages. Pale eyebrow obvious in juvenile, but streaking and barring gives mottled effect unless seen close up.

IN FLIGHT

Rufous underparts, including underwing coverts, contrast with dark upperparts and dark flight feathers with broad, pale bars. Travels with slow, deep wingbeats and wavy flight path, but makes fast, direct dashes after prey.

DISTINCTIVE BEHAVIOUR

Usually seen perched in the open at edge of forest, flying fast over open country nearby, or soaring high overhead. Hunts mainly on the wing for small birds, especially doves, from fast, coursing flights or even stooping from a soar. Less often seen to strike from a perch or descend to the ground for insects or small reptiles and mammals. Noisy in display flight above forest, often with white vent fluffed out on either side of tail. Call a series of high, fast 'kew' notes. The small stick-nest is built high in a tree, often at the edge of a forest patch.

N. Myburgh

Juvenile perched on rim of nest

ADULT Male dark grey-brown above, with white feather bases showing on nape, shoulders and rump. Below plain rufous, paler (sometimes even white) on throat. **Female** browner above and deeper rufous below. Vent white, sometimes with fine, rufous bars. Flight feathers and tail grey above and white below. Tail white-tipped. Bill black; eyes, cere and long, bare legs with slender toes yellow. Female about 10% larger (both sexes: 180–210 g).
JUVENILE Dark brown above with paler rufous feather tips and slight, pale cream eye-stripe. Below rufous, finely streaked on breast and barred on belly with white.
DISTRIBUTION, HABITAT AND STATUS Occupies patches of montane forest within grasslands, from Cape Town to the Ethiopian highlands and including the Rift Valley. Locally common but with a disjunct range that includes many small forest patches. Even occupies plantations of exotic pine and poplar trees within open steppe and grassland habitats of southern Africa, extending to adjacent open country that includes hill slopes and cliff faces.
SIMILAR SPECIES All ages most similar to juvenile **Ovambo Sparrowhawk** (p. 158), especially form with rufous underparts which also has a prominent pale eyebrow-stripe (head rufous; dark crown and ear-patches; eyes dark brown; rufous underparts less mottled). Juvenile also similar in northeast Africa to juvenile of migrant **Eurasian Sparrowhawk** (p. 164) (below more obviously streaked and barred with dark brown; eye yellow at all ages).

ALTERNATIVE NAMES: Red-breasted Sparrowhawk, Rufous-chested Sparrowhawk, Rufous Sparrowhawk

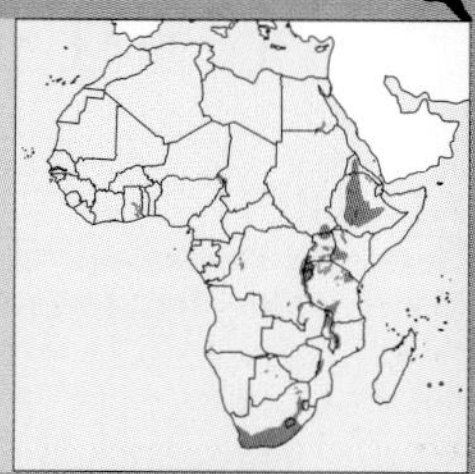

Dark grey-brown above, including sides of head. Below plain rufous. Vent plain white. Eyes yellow at all ages. Juvenile has rufous underparts finely streaked and barred with white. Small (about 30 cm tall, 72 cm wingspan).

Ovambo Sparrowhawk (juv.), Eurasian Sparrowhawk

Madagascar Sparrowhawk

Accipiter madagascariensis

AT PERCH

Dark grey-brown above and clear barring below highlights the plain white vent. Juvenile appears mottled above and clearly streaked below. Eye yellow at all ages.

IN FLIGHT

Broad, rounded wings and long tail clearly barred, including underwings, which appear dark against the paler, more widely barred flight feathers. Appears plain dark brown above but with barred tail.

DISTINCTIVE BEHAVIOUR

Hunts mainly on the wing after small birds and insects, but also takes frogs and rodents on the ground. Not recorded to perform aerial display flights over the forest but does occasionally soar overhead. Builds a stick-nest high in a forest tree. Calls with high, fast 'kee' notes.

D. Halleux/BIOS

Adult perched on nest rim

ADULT Above dark grey-brown, with white bases to nape feathers and fine, rufous tips to rump. Below white, sharply barred with dark grey. Throat and ear coverts streaked with dark grey. Vent plain white. Flight feathers and tail dark grey-brown above and pale grey-brown below, with slightly paler broad brown bars below, showing substantial contrast; only slight contrast above. Bill black; eyes, cere, long bare legs and slender toes yellow. Sexes similar in plumage, female about 18% larger.

JUVENILE Dark grey-brown above, with white feather bases and narrow rufous tips giving generally mottled effect. Below white, broadly streaked with dark brown to form centre stripe and 'moustache' stripes on throat and barred on wings.

DISTRIBUTION, HABITAT AND STATUS Occurs throughout Madagascar in primary forest, including dry and thorn forests, but rarely enters secondary habitats. Uncommon and rarely seen in degraded habitats.

SIMILAR SPECIES Slightly larger than **Frances's Sparrowhawk** (p. 168) (paler and greyer overall; slight rufous wash over the barred underparts; finely barred vent; juvenile with barred rather than streaked underparts; eyes yellow in both species). Build quite different to that of **Banded Kestrel** (p. 230) (dumpy build; short, stout legs; heavy bill; prominent yellow cere and bare eye-ring; long, pointed wings).

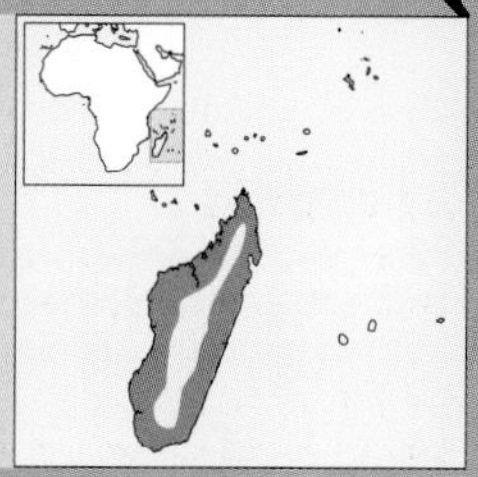

Dark grey-brown above. Below white with clearly defined, fine, dark grey-brown bars. Vent plain white. Juvenile streaked and barred with grey-brown below. Yellow eye at all ages. Small (about 30 cm tall, 70 cm wingspan).

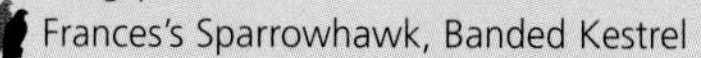

Frances's Sparrowhawk, Banded Kestrel

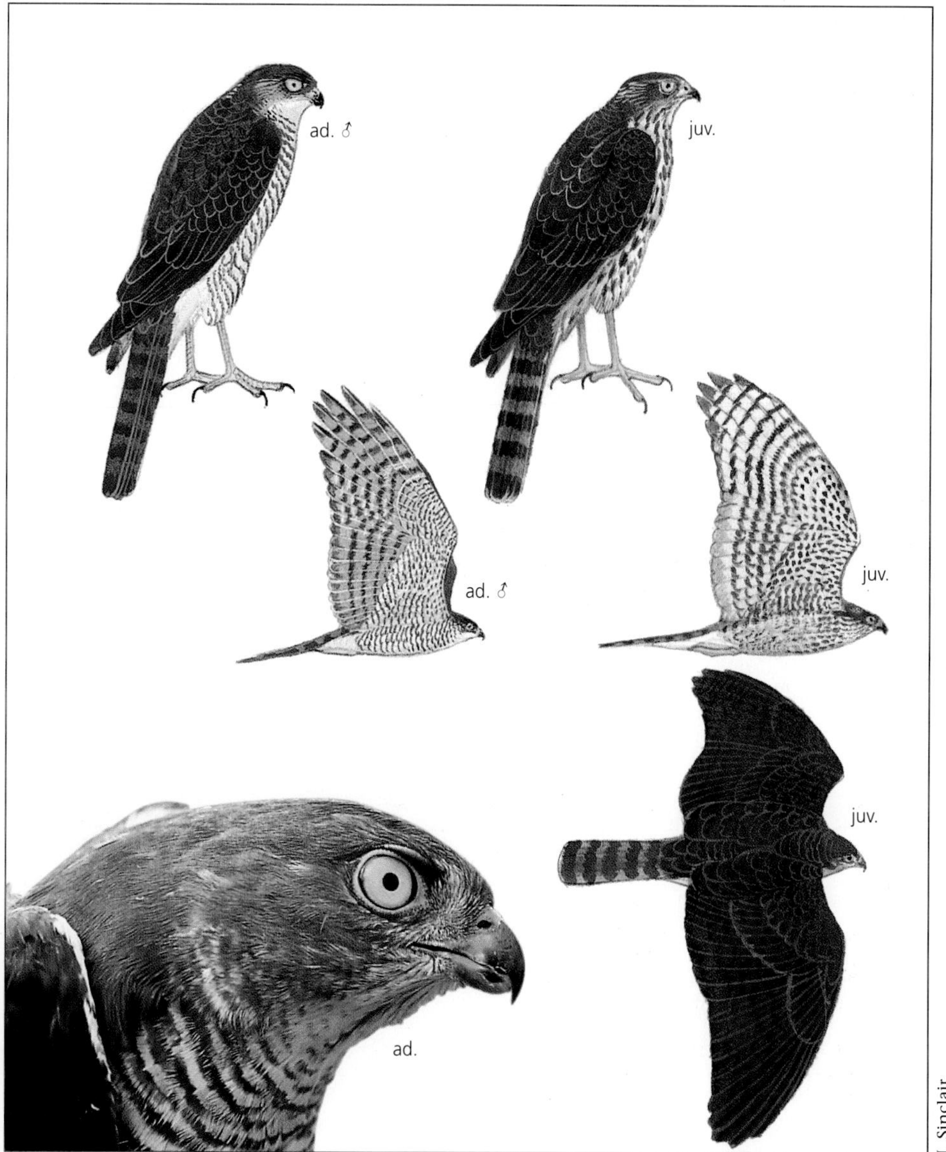

I. Sinclair

Eurasian Sparrowhawk

Accipiter nisus

AT PERCH

Plain grey (male) or brown (female) above and barred below, with a few white patches on nape and rump visible from behind. Throat streaked; rufous wash (male) or pale eyebrow-stripe (female) and white vent visible from front. Note orange or yellow eye at close range.

IN FLIGHT

Appears dark grey (male) or brown (female) above, and paler cream and rufous on belly and underwing coverts; coloration, together with broad bars on the flight feathers (below), makes plain white vent obvious. Wings broad and rounded and tail long. Flies with deep, strong wingbeats and direct flight.

DISTINCTIVE BEHAVIOUR

Usually perches deep within wooded cover, emerging in aerial pursuit of the small birds that are its main prey. Most often seen coursing fast over vegetation or soaring high overhead to try to surprise prey. Vocal when breeding and flies above woods with white vent fluffed out. Call a high series of 'kew' notes. Small stick-nest placed high in tree fork.

ADULT Male plain dark grey above, including on sides of head; below white with rufous wash over fine, rufous streaks on throat and bars on breast and flanks. **Female** browner above, with thin, rufous tips to feathers, thin, white eyebrow-stripe and no rufous wash below. Both have plain white vent. Flight and tail feathers dark grey (male) or dark brown (female) above and pale below, with broad, dark bars including broad subterminal bar on tail. Bill black; eyes orange (male) or yellow (female); cere, long, bare legs and slender toes yellow. Female 15% larger (male 110–196 g, female 185–342 g).

JUVENILE Brown above with rufous tips to feathers. Below white, washed and streaked with rufous on throat and barred with brown and rufous below. Eyes yellow.

DISTRIBUTION, HABITAT AND STATUS Resident population to the north of the Atlas Mountains in northern Africa, in patches of woodland, and on Canary Islands. Non-breeding migrants from Europe also visit the Libyan coast and the Nile Valley, extending as far west as Gambia and Mali or southwards to northern Tanzania. Residents generally uncommon and migrants rare, the latter being found mainly in drier thorn savanna within Africa. Old specimen from South Africa.

SIMILAR SPECIES Overlaps in northern Africa with small race of **Northern Goshawk** (p. 184) (black crown and neck; broad, white eyebrow-stripe in adult; rufous below with dark brown streaks in juvenile). In sub-Saharan Africa, usually found in drier habitat than **Rufous-breasted Sparrowhawk** (p. 160) (plain rufous breast) but shares habitat with juvenile **Ovambo Sparrowhawk** (p. 158) (dark eye; broad, pale eyebrow-stripe; dark crown and ear-patches; lightly marked below; orange cere and legs), juvenile **Gabar Goshawk** (p. 156) (broad, white rump; streaked chest and barred breast; yellow eye; orange cere and legs) and smaller **Shikra** (p. 166) (pale dove-grey; lightly marked; eyes deep red; plain grey central tail feathers).

M. Lane/NHPA

Juvenile with pigeon prey

ALTERNATIVE NAME: European Sparrowhawk

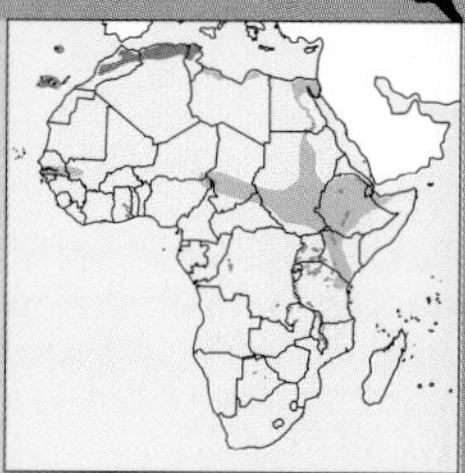

Plain dark grey (male) or brown (female) above. Thin, pale eyebrow-stripe. Below white, washed and barred with rufous (male) or finely barred with dark grey (female). Vent all-white. Eye yellow (female) to orange (male). Juvenile browner above, washed rufous below with finer brown streaks on throat and bars on breast. Small (about 30 cm tall, 60–75 cm wingspan). Rare in north-western and eastern Africa.

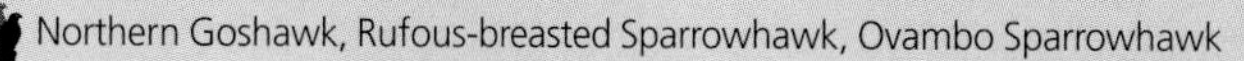

Northern Goshawk, Rufous-breasted Sparrowhawk, Ovambo Sparrowhawk

D. Karp/NHPA

Shikra

Accipiter badius

AT PERCH

Rather large head and plump body distinctive, as well as pale dove-grey colour, faint barring below and obvious pale yellow cere. At close range red eye conspicuous. Tail and rump plain grey from behind, narrowly barred dark grey and white from below. Juvenile has notable dark throat-stripe, blobs on side of breast and yellow eye. Slight pale eyebrow sometimes useful for identification.

IN FLIGHT

Note lack of marking on tail and rump, and lack of contrast in barring of flight feathers. Barring of outer tail feathers distinctive against plain centre when spread. Tail faintly barred with dark and light brown in juvenile. Eye appears dark (adult) or pale (juvenile) and cere pale yellow at close range. Not as fast as some accipiters, often dropping low from a perch and then swooping up to the next one.

DISTINCTIVE BEHAVIOUR

Perches more in the open than many other accipiters, searching for the lizards, insects and small birds that are its main diet. Takes most prey on or near the ground in a fast dive, or plucks it from tree trunks and foliage. Small stick-nest built in fork of woodland tree. Noisy at nest, with high, shrill 'kwik' notes.

Adult Above pale grey. Below white finely barred with pale grey (male) or rufous-brown (female). Flight feathers dark grey above with broad, pale grey bars, and paler below with white bars. Tail plain grey above when closed; from below, dark grey with broad, white bars. Bill black; eyes red (male) or deep orange (female); cere pale yellow and bulbous; bare legs and stubby toes a darker yellow. **Female** slightly darker, more heavily barred and about 10% larger (both sexes, 75–158 g).

Juvenile Dark brown above with paler feather tips. Below white with broad, dark rufous bars, overlaid with large spots on sides of chest. Throat has a dark central stripe. Flight feathers dark brown with broad, slightly paler bars; below rufous-brown with white bars. Tail grey-brown above with broad, paler bars. Eyes brown at fledging, turning yellow, then orange.

Distribution, habitat and status Widespread and common in the drier woodlands and savanna of much of sub-Saharan Africa, including clearings within forest. Only absent from dense forest and the driest deserts. Range extends to India and southeast Asia, where also common.

Similar species Overlaps with several small grey or brown hawks in same range and habitat. These include the larger **Ovambo Sparrowhawk** (p. 158) (small snaky head; orange cere and legs; slender build with longer wings; dark eye; partly white rump; juvenile brown or rufous above with dark eyes and pale eyebrow) and **Gabar Goshawk** (p. 156) (broad white rump; chest plain grey; only breast barred; dark eyes; red cere and legs; juvenile brown with yellow eyes but chest streaked and breast barred with brown, legs orange) or **Lizard Hawk** (p. 154) (white rump; white band (or 2) across tail; prominent black stripe on white throat) and the smaller **African Little Sparrowhawk** (p. 176) (2 white tail-spots; white rump-bar; yellow eye; spotted below in juvenile).

Juvenile

Alternative name: Little Banded Goshawk

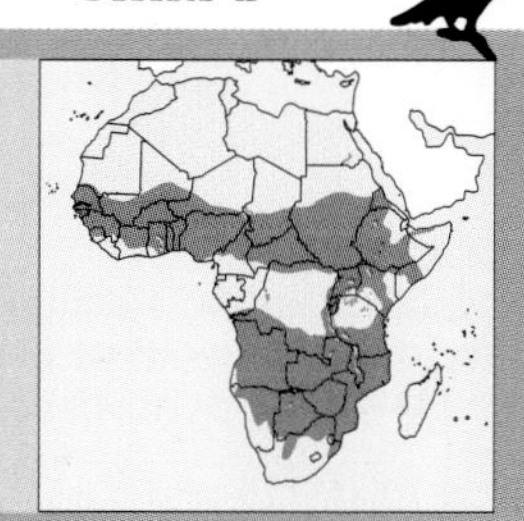

Pale dove-grey above. Below white finely barred with grey (male) or rufous-brown (female). No distinctive rump or upper tail markings. Eyes deep red (male) or orange (female). Prominent pale yellow cere. Juvenile brown, with streaked chest and barred breast, a dark central throat streak and yellow eyes. Small (about 22 cm tall, 58 cm wingspan).

Ovambo Sparrowhawk, Lizard Hawk, African Little Sparrowhawk, Gabar Goshawk

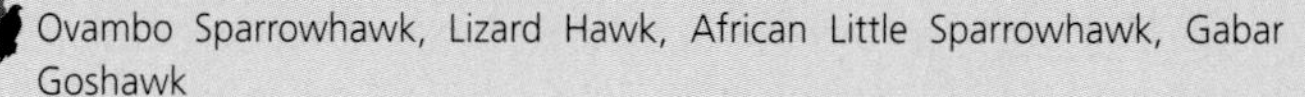

Frances's Sparrowhawk

Accipiter francesiae

AT PERCH

Multi-barred tail notable from front or behind. Dark upperparts and pale underparts – almost plain (male) or clearly barred (female) – distinctive, especially when washed rufous below as on Comoros. Juvenile resembles adult female, except when spotted below as on Mayotte (Maore).

IN FLIGHT

Wings broad and rounded, tail relatively long. Flight feathers pale with darker bars extending to tips. Male underwing white with only pale grey bars across flight feathers, darker in female. Tail dark above and much paler below, with narrow, darker bars.

DISTINCTIVE BEHAVIOUR

Often perches alone in the open. Flies with fast, shallow beats, often close to the ground then swinging up to a perch. Diet mainly lizards, frogs and large insects, with some small birds and rodents. Hunts after a short dash from a perch. Builds a substantial stick-nest in the upper fork of a large tree, not necessarily within forest. Calls with high, fast 'kee' notes.

I. Davidson

Adult female

ADULT Male plain dark grey above; white with faint, rufous barring below. Variable rufous wash below, extending to chest in some on Comoros. **Female** dark brown above, clearly barred brown below. Similar to male, with rufous wash below on Grande Comore (Ngadzija). Flight feathers white below with broad, light grey bars. Tail dark grey or brown above with narrow, paler bars. Bill black; eyes pale yellow; cere and long, bare legs and stubby toes yellow. Female about 11% larger; each of 3 Comoros subspecies about 10% smaller than Madagascan subspecies.

JUVENILE Resembles brown, barred adult female, but often with rufous wash on breast and head. Feather tips rufous above. Juveniles on Mayotte (Maore) have dark brown blobs on underparts.

DISTRIBUTION, HABITAT AND STATUS Madagascar and the main islands of the Comoros group (Grande Comore, Anjouan (Ndzuani) and Mayotte). Generally common and widespread in primary and secondary forests on Madagascar. Most obvious accipiter, by use of forest margins and some of the more degraded secondary habitats. Now uncommon and local on Comoros, rare on Anjouan.

SIMILAR SPECIES Only sparrowhawk on the Comoros. On Madagascar, smaller and more slender than **Madagascar Sparrowhawk** (p. 162) (grey-brown above, with crisp, dark grey barring below; longer-winged; without rufous wash below). Larger than **Madagascar Kestrel** (p. 240) (deep rufous above; below white or rufous with dark streaks; broad, dark terminal tail-band). Heavily barred female and juvenile resemble **Banded Kestrel** (p. 230) (tail barred white; heavy bill; prominent yellow cere; bare yellow eye-ring).

ALTERNATIVE NAMES: Madagascar Goshawk, Malagasy Goshawk, Frances's Goshawk

Above uniformly dark grey (male) or brown (female and juvenile). Below white, with faint stripe down throat and fine rufous bars (male) or broad brown bars (female and juvenile) on breast and vent. Tail has narrow, dark bars. Small (about 25 cm tall, 65 cm wingspan). Consistent differences between Madagascan population and 3 populations of Comoros.

Madagascar Sparrowhawk, Madagascar Kestrel

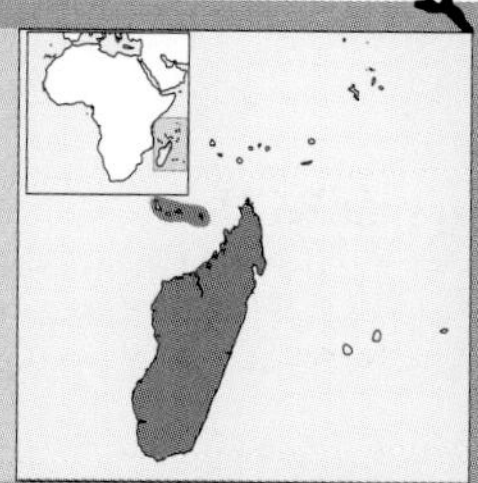

I. Davidson

Levant Sparrowhawk

Accipiter brevipes

AT PERCH

Adult male pale grey above, barred white and rufous below, female brown above, barred white and dark brown below, juvenile brown above and spotted below; all distinctive with good view. Dark tail-bars most distinctive feature at all ages. Dark eyes of adults also notable.

IN FLIGHT

Wings rather long, pointed and narrow for a sparrowhawk. Adult male underwing appears white with black tips; fine, dense, rufous barring from throat to belly. Female similar, with more brown wash on underwings. In both sexes, multi-barred tail appears rather dark. Above, male wing tips also appear dark and tail has unbarred centre; female is overall darker and brown in colour but with head and tail slightly paler. Juvenile appears dark, with clear brown spots all over underparts; barred tail again distinctive.

DISTINCTIVE BEHAVIOUR

Often migrates into Africa in flocks; even soars and hunts together. Flies low over the ground to hunt, often along the edges of woodland and late into the dusk. Insects or birds reported as main African prey.

ADULT Male dove-grey above. Below (including underwing) white with faint, fine rufous bars on breast. Flight feathers faintly barred with dark grey and tipped with black, contrasting with white underwing coverts. Tail plain grey in centre with dark bars on outer feathers. **Female** browner overall, with dark centre stripe on throat. More heavily and broadly barred on underparts, less so on underwing coverts. Flight feathers and tail grey-brown with broad, dark grey bars. Bill black; eyes dark red; cere and long, bare legs and stocky feet yellow. Female 6% larger (male 155–223 g, female 220–275 g).

JUVENILE Similar to female. Dark brown above, with paler feather tips and white bases to nape feathers. Below white, with dark streak down centre of throat and large, dark spots on breast and underwing coverts.

DISTRIBUTION, HABITAT AND STATUS A non-breeding migrant to northeastern parts of Africa during the northern winter, from September to April. Crosses into Africa in thousands, sometimes as large flocks, but rarely sighted within Africa and main non-breeding range uncertain. Recorded from Niger eastwards to Ethiopia and southwards to Kenya; vagrant to Tunisia. Rarely observed south of the Sahara but probably overlooked.

SIMILAR SPECIES Very similar to **Shikra** (p. 166) (fewer dark tail-bars; lacks black wing tips; more dumpy and about 20% smaller; wings much shorter and more rounded; below more obviously barred; eye also red, but not as dark; cere pale yellow). Adult female resembles female **Eurasian Sparrowhawk** (p. 164) (broader, more rounded wings; obvious white eyebrow-stripe; orange eye). Juvenile more heavily spotted than most African accipiters, except for **African Goshawk** (p. 172) (barred rather than spotted underwing).

Dove-grey above. Below white, very finely barred with pale rufous. Tail has narrow, pale bars, but appears dark overall. Red eyes. Cere yellow. In flight, black wing tips contrast with white underwing coverts in male. Juvenile and female browner, with heavy bars or spots below and dark stripe on throat. Small (about 25 cm tall, 70 cm wingspan).

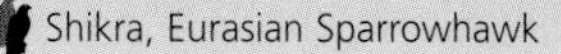
Shikra, Eurasian Sparrowhawk

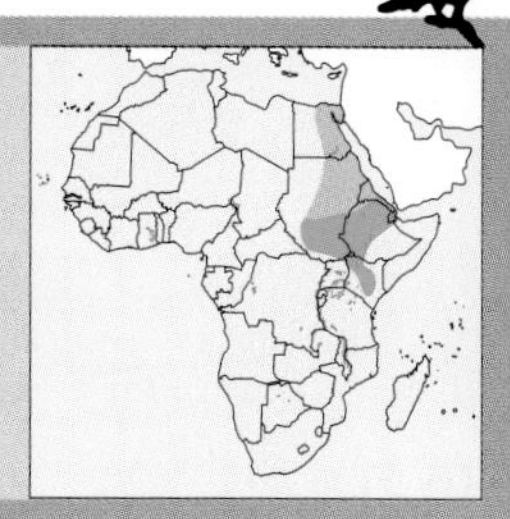

African Goshawk

Accipiter tachiro

AT PERCH

Appears dark above and pale below from a distance. At closer range, note deep yellow eye, barred or plain rufous (adult) or spotted (juvenile) underparts and pale tail-patches (above) or bars (below). Grey (adult) or greenish (juvenile) cere distinctive. Sits vertically for much of the time.

IN FLIGHT

Broad, rounded wings and long tail distinctive. Relatively broad, pale bars on underside of flight feathers contrast with the brown body and pale plain or finely barred underwing coverts. Most often seen from below when soaring during the morning display flight but, when flushed from or dashing through dense bush, the lack of a pale rump is distinctive.

DISTINCTIVE BEHAVIOUR

Most obvious through clicking call note uttered at regular intervals for long periods, either while perched or, more often, during early-morning display flights. The circling display flights are performed with slow wingbeats and bouts of gliding, often so high that it is difficult to locate the bird. Spends much of the rest of the day perched in wooded habitat, often overlooking a track or clearing, and occasionally in dashing flight after its prey of small mammals, reptiles and birds and large insects. Often hunts late into dusk, including after bats. Juvenile very vocal after fledging, with a monotonous begging cry. Small stick-nest usually placed deep in foliage, often among creepers.

ADULT Variable across wide range; smaller rainforest and Pemba birds darker above, more rufous below with few or no breast-bars. In any given area, male darker grey above and more rufous below than female. Above dark grey to grey-brown (male) or grey to brown (female), including rump. Below white washed or barred finely with rufous (male) or barred broadly with brown (female), especially on flanks. Throat off-white to pale grey (rufous on Bioko). Underwing pale rufous to white. Flight and tail feathers sooty-brown to grey with faint paler brown bars above, white with grey bars below. Tail-bars paler and forming 2 or 3 white patches on central rectrix pair in male. Bill black; cere grey; eyes, long, bare legs and stubby feet yellow. Female about 14% larger (male 150–340 g, female 270–510 g). A rare melanistic sooty-grey form, with paler grey tail-bars, occurs in tropical Africa.

JUVENILE Above dark brown with paler feather edges giving mottled effect. Obvious pale eyebrow. Below white or buff, plain (rainforest birds) or spotted with large, sooty-brown blobs (savanna birds). Dark stripe down centre of throat. Flight feathers dark brown with broad, slightly paler brown bars. Eyes brown; cere pale grey-green, legs and feet greenish-yellow. Sexes differ only in size.

DISTRIBUTION, HABITAT AND STATUS Throughout forest and woodland habitats of sub-Saharan Africa, including Bioko, Pemba and Zanzibar islands. Widespread and often common on forest edge, in dense woodland, at forest patches, along riverine strips and on wooded hill slopes. Darker, more rufous rainforest forms are sometimes considered a separate species, Red-chested Goshawk (*A.t. toussenelli*).

SIMILAR SPECIES Most similar to much smaller **African Little Sparrowhawk** (p.176) (yellow cere; tail with only 2 white spots in both sexes; unmarked throat in juvenile; longer and more slender toes) and **Red-thighed Sparrowhawk** (p. 178) (darker – almost black – above; red eyes and cere; orange legs; pale bar across rump and broken white tail bars; unmarked throat in juvenile; longer and more slender toes). Overlaps in lowland rainforest of central and western Africa with very similar **Chestnut-flanked Sparrowhawk** (p. 174) (darker – almost black – above; fine, grey barring or rufous streaks on throat and underwing coverts; no aerial display calling; more rufous, less barred underparts; dark-tipped tail).

A.t. tachiro adult female

ALTERNATIVE NAMES: Red-chested Goshawk, West African Goshawk (*A.t. toussenelli*)

Above dark grey (rainforest) or brown (savanna). Below white with variable rufous wash, or with a few bars on flanks (rainforest), or with whole breast barred brown (savanna). Dark rump. Pale bars form 2 or 3 pale patches on folded tail of male. Yellow eyes; grey cere. Juvenile brown above, white below with variable number of dark spots and dark throat-stripe. Note regular clicking call, uttered at perch or in high, soaring flight. Small (male) to medium-sized (female) (about 25–30 cm tall, 70 cm wingspan).

African Little Sparrowhawk, Red-thighed Sparrowhawk, Chestnut-flanked Sparrowhawk

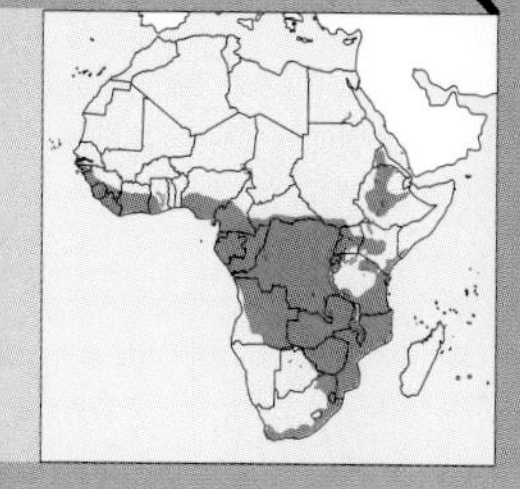

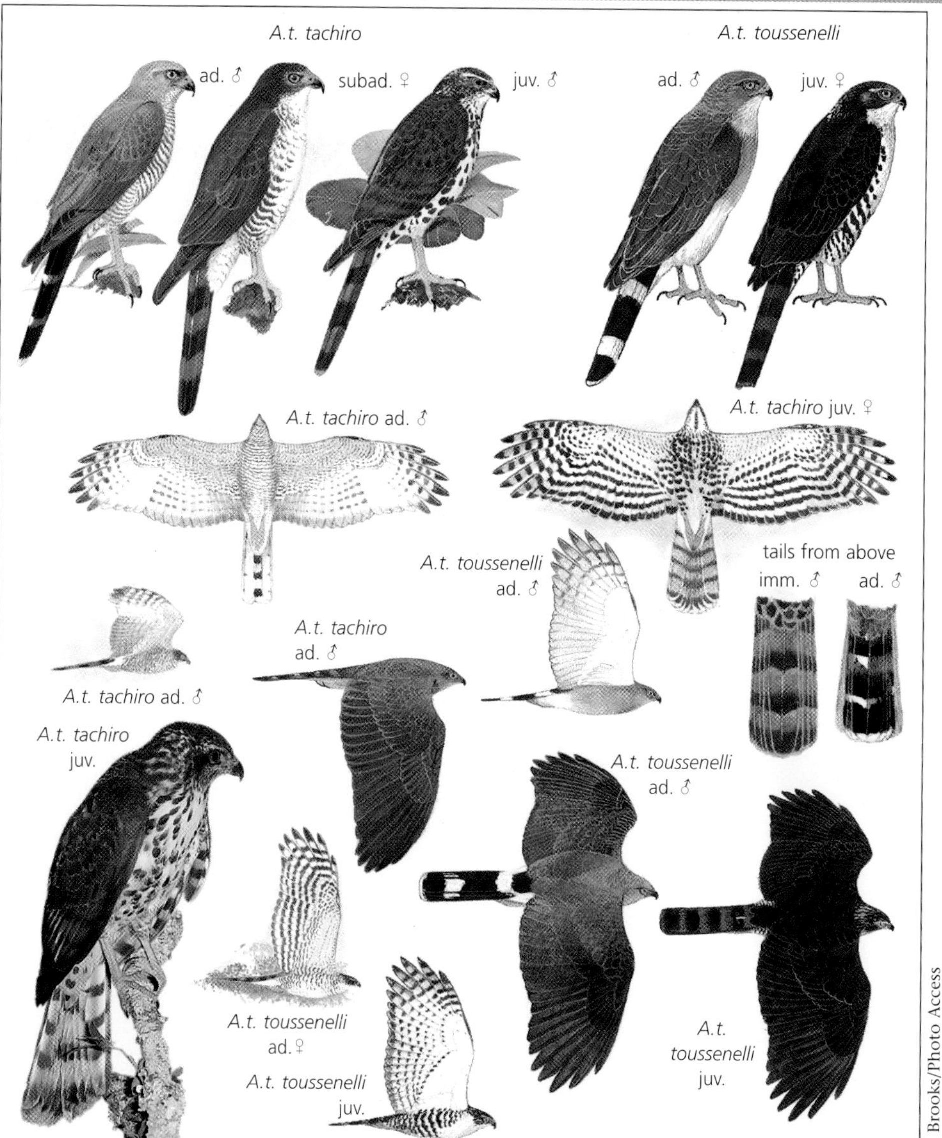

J.J. Brooks/Photo Access

Chestnut-flanked Sparrowhawk

Accipiter castanilius

AT PERCH

Black above and rufous below, with obvious white vent below and tail-spots above. Typical upright sparrowhawk posture for much of time, less often horizontal when active. Long, slender toes sometimes evident across perch.

IN FLIGHT

Short, rounded wings and long tail typical of a forest sparrowhawk. Dark upperparts, including rump, and rufous underparts, contrast with white vent and paler bars across flight feathers and pale underwing coverts. Spends most of time below the forest canopy, occasionally soaring or flying in the open.

DISTINCTIVE BEHAVIOUR

Perches unobtrusively amongst foliage for long periods. Less often seen making short dashes after its main prey of small birds. Not recorded to call much or fly above the forest often, so generally poorly known.

ADULT Above (including cheeks) dark grey, almost black, giving capped effect. Below deep rufous, with white barring across breast. Vent white. Throat and underwing coverts white barred with grey or faintly speckled with rufous. Flight feathers dark grey with faint, broad bars above, below appears white to grey with dark grey bars. Tail with pale bars almost white, forming 3 obvious bands above. Bill black; eyes, cere, long, bare legs and long, slender toes yellow. **Male** more rufous, female about 13% larger (male 135–150 g, female 152–165 g).

JUVENILE Dark grey-brown above with paler feather tips and broad, pale bars across flight feathers and especially across tail. Below white with a few large, dark brown spots on the chest and wing coverts and, in some birds, with a rufous wash on the flanks and thighs; wings white to grey below with dark grey bars on flight feathers. Paler eyebrow. Eye grey-brown; cere, legs and feet yellow.

DISTRIBUTION, HABITAT AND STATUS Occupies lowland rainforest habitat, especially primary forest, and virtually confined to the Congo River basin. Never recorded as common, but easily overlooked and also difficult to differentiate with certainty from other forest accipiters.

SIMILAR SPECIES Most resembles much smaller **Red-thighed Sparrowhawk** (p. 178) (also black above and rufous below or pale grey with rufous sides; but red or orange eyes and cere; orange legs; long, slender toes; white rump; broken white tail-bars). Similar in size to widespread and common **African Goshawk** (p. 172) (paler grey above; often more obviously barred below; unbarred white throat; pale rufous underwing coverts only faintly marked).

ALTERNATIVE NAMES: Chestnut-flanked Goshawk, Chestnut-bellied Sparrowhawk

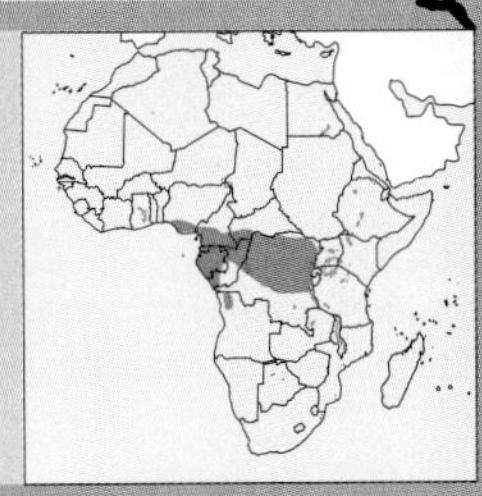

Black above. Deep rufous below. White vent. 3 white bars across centre tail feathers. Note long, slender toes at close range. Small (male) to medium-sized (female) (about 25 cm tall, 60 cm wingspan).

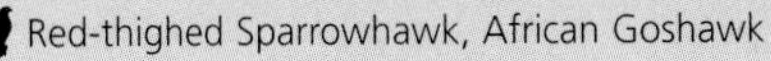

Red-thighed Sparrowhawk, African Goshawk

African Little Sparrowhawk

Accipiter minullus

AT PERCH

White rump-bar and 2 tail-spots distinctive. Small size and dumpy body notable. Dark grey above; finely barred white undersides. Rufous wash on flanks and darker back often distinguish male. Obvious large, round spots on juvenile breast. All ages, note yellow eye and cere at close range.

IN FLIGHT

White rump-bar and 2 white tail-spots contrast with dark upperparts. Fine barring below, including on underwing coverts, appears as rufous wash. Flies with fast, choppy wingbeats and swerving flight path.

DISTINCTIVE BEHAVIOUR

Spends much time perched but more active than larger sparrowhawks, often jumping through branches and making short flights between trees. Hunts main prey of small birds on the wing in short dashes from cover, also hawks insects and, as it is a crepuscular bird, bats. The tiny stick-nest is placed in a high fork and is often well-concealed. Noisy at nest with high, sharp 'kwit' notes.

W. Tarboton

Adult

I. Carlyon

Subadult with prey

ADULT Above dark grey, almost black (male), or dark brown (female), including cheeks which contrast with white throat. Below white with fine, rufous bars (male) or broader, brown bars (female). Individually variable rufous wash on flanks, especially in male. Bill black; eyes deep yellow; cere, long bare legs and slender toes yellow. Female about 12% larger (male 75–85 g, female 82–105 g). **JUVENILE** Dark brown above with pale tips to feathers and with white feather bases on nape and scapulars. Below white or cream with large, dark brown spots on breast and dark streaks on chest; throat plain white. Rump feathers only white-tipped to appear dark overall. Tail has 3 or 4 pale brown bars. Eyes brown. Cere greenish-yellow. **DISTRIBUTION, HABITAT AND STATUS** Widespread in patches of dense bush and woodland, from Ethiopia southwards to Cape Town, and extending westwards to southern Congo and northern Namibia. Secretive in behaviour but often common and reveals presence by high, shrill calls.

SIMILAR SPECIES Most similar in coloration and habitat to larger **African Goshawk** (p. 172) (dark rump; grey cere; male has 2 or 3 white tail-spots; juvenile has pale eyebrow and dark throat-stripe). However, range overlaps with various larger hawks, all greyer, with dark red eyes and no tail-spots, including **Shikra** (p. 166) (plump, pale yellow cere; dark rump; adult pale dove-grey; juvenile brown with yellow eye but sooty-brown throat stripe, chest streaks and breast-bars; stubby toes), **Ovambo Sparrowhawk** (p. 158) (small, snaky head; fine, grey throat- and breast-bars; partly white rump; long wings; juvenile brown or rufous with dark eye and pale eyebrow) and **Gabar Goshawk** (p. 156) (broad, white rump; chest plain grey; only breast barred, with grey; red cere and legs; juvenile brown with yellow eye but chest streaked and breast barred with brown). A lighter version of similar-sized **Red-thighed Sparrowhawk** (p. 178), which replaces it in rainforest.

ALTERNATIVE NAME: Little Sparrowhawk

Tiny hawk. Grey above and white below; finely barred with brown across breast and washed with rufous on flanks. White rump-band and 2 white spots on upper tail feathers distinctive. Juvenile browner, with large brown spots on underparts. Very small (about 18 cm tall, 50 cm wingspan).

African Goshawk, Shikra

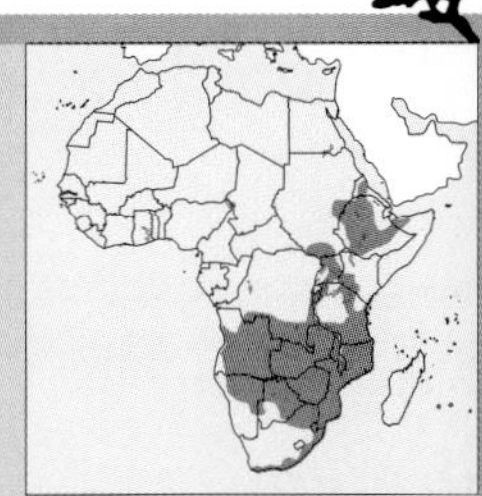

J. Knobel

Red-thighed Sparrowhawk

Accipiter erythropus

AT PERCH

Black upperparts contrast with white rump and broken tail-bars. Below more rufous and grey (adult) or white (juvenile), with dark barring obvious on juvenile only. Red or orange eye obvious within dark cap.

IN FLIGHT

Dark above, showing grey wing- and tail-bars. White rump and broken tail-bars notable. Below appears pale rufous and pale grey; underwing coverts white with only faint rufous bars. Very short, rounded wings and long tail distinctive. Flies with fast, deep wingbeats and swerving flight.

DISTINCTIVE BEHAVIOUR

Spends much time perched, but also hops about within branches or makes short, dashing flights between trees. Rarely if ever flies above forest canopy; most often found in understorey. Main prey is small birds and some insects, taken in fast, swerving flights. Habitually follows and harasses bird parties, often as a pair working together. Hunts in forest canopy, understorey or undergrowth, and often attacked by similar-sized Velvet-mantled and Shining drongos (*Dicrurus modestus* and *D. atripennis*). Tiny stick-nest is built in a thin upper fork of a forest tree.

V. Salewski

Adult male

ADULT Above sooty-grey, almost black, including cheeks; rump white. Below white with deep rufous wash on flanks and thighs. Faint rufous barring across centre of breast in some birds. Flight feathers black with faint bars. Tail has white bars, broken in the centre. Bill black; eyes and cere red; long, thin bare legs and slender toes orange. **Female** browner than male and about 12% larger (male 78–82 g, female 132–136 g). Markings of females and subadults very variable and easily confused.

JUVENILE Dusky-brown above with white feather bases showing on nape and scapulars. Below variable – some white but for few brown spots on breast, most with rufous bars especially on thighs and flanks. Flight feathers lightly barred; pale tail-bands below, faint above. Eyes yellow.

DISTRIBUTION, HABITAT AND STATUS Only tiny sparrowhawk in western Africa, extending to central Africa in lowland primary rainforest and adjoining forest edge and secondary forest. Nowhere common; also secretive and difficult to detect except when it hunts.

SIMILAR SPECIES Coloration and habitat is most similar to the much larger dark-rumped **Chestnut-flanked Sparrowhawk** (p. 174) (grey-barred white throat and underwing coverts; yellow eyes, cere and legs) and larger **African Goshawk** (p. 172) (dark rump; 2 or 3 white tail-spots; stubbier toes; yellow eyes and legs, grey cere). A darker version of the similar-sized **African Little Sparrowhawk** (p. 176), which replaces it in adjoining drier eastern habitats.

ALTERNATIVE NAMES: Red-thighed Goshawk, Western Little Sparrowhawk

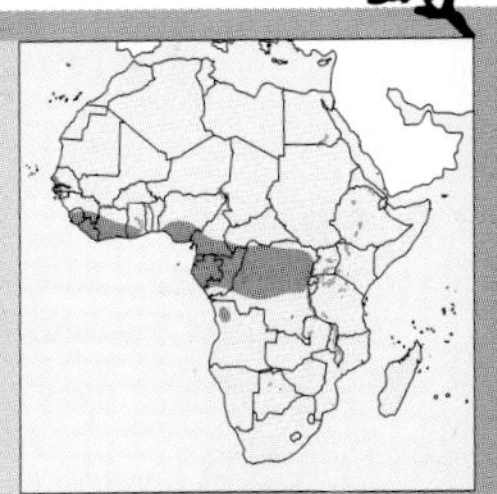

Very small accipiter. Black above and rufous below. White rump and 3 broken white tail-bars distinctive. Darkly capped head. Red (male) or orange (female) eyes and cere conspicuous. Orange legs. Juvenile white below with dark spots on chest or fine rufous bars on breast and flanks. Very small (about 18 cm tall).

Chestnut-flanked Sparrowhawk, African Goshawk, African Little Sparrowhawk

Black Sparrowhawk

Accipiter melanoleucus

AT PERCH

Typical accipiter poses, standing vertical with foot up when at rest or horizontal with legs exposed when active. Unstreaked white underparts with black 'skirt' often most obvious. Juvenile rufous below, darker brown above.

IN FLIGHT

Fast, strong wingbeats characteristic. Flies straight and fast across country, or twists and turns through vegetation. The overall appearance is of a dark hawk with white underparts, except for melanistic form. Juvenile mainly brown except for a pale 'window' at base of upperwing primaries and light and dark grey barred flight feathers below.

DISTINCTIVE BEHAVIOUR

Spends much time perched within the tree canopy, where difficult to detect, but sometimes makes flights high above surroundings. Hunts passing prey – mainly doves, pigeons, ground and other birds – from a perch, or makes fast, low flights to surprise prey. Often descends to bathe in streams. Noisy at nest, with drawn-out whistling and shrill, yelping calls. Builds a large stick-structure high in a forest or plantation tree.

N. Myburgh

Adult pale form standing on nest

Adult, melanistic form

ADULT Above black, except for white feather bases on nape and wing coverts. Below white, including throat and underwing, except for black blotches on flanks and armpit. Flight and tail feathers black, barred with dark grey above and white below. Rare melanistic form all-black, except sometimes for white on throat. Bill black; eyes wine-red; cere, eyebrow and bare legs generally yellow. Sexes similar in plumage, female about 14% larger (male 430–490 g, female 650–790 g).

JUVENILE Above brown, head and rump streaked with black and white, rest darker brown with broad, rufous feather edges. Below variable, from deep rufous to white, and with dark brown streaks varying from fine to broad. Female often paler and less streaked. Eye brown at fledging, soon turning grey and then yellow. Cere dull olive.

DISTRIBUTION, HABITAT AND STATUS Forested habitats across sub-Saharan Africa, from extensive tropical lowland forest to small patches of montane or riverine forest and, more recently, exotic plantations. Widespread but nowhere common, although most easily located in and around smaller forest patches. Melanistic form unknown in western Africa, most frequent in South Africa.

SIMILAR SPECIES Colour pattern most resembles **African** and **Ayres's hawk-eagles** (pp. 96 and 100) (feathered legs; black streaks on white below; obvious black bars or markings on wings and tail; yellow eyes; juvenile body colours are similar, but not wing or tail markings). Ayres's Hawk-eagle also has a slight crest and white 'headlights' at base of forewing.

ALTERNATIVE NAMES: Great Sparrowhawk, Black Goshawk, Pied Goshawk, Black-and-white Goshawk, Black-and-white Sparrowhawk

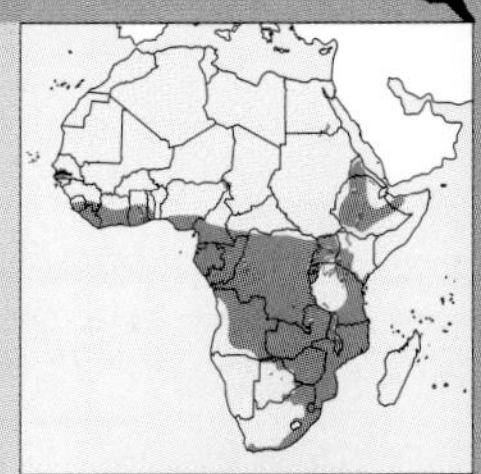

Pied raptor, black above and white below. Rare black form has white throat. Note black thighs and bare yellow legs. Red eyes. Cere yellow. Juvenile brown above with pale scallops, below chestnut or white with dark brown streaks; eyes yellow. Medium-sized (about 40 cm tall, 102 cm wingspan).

African Hawk-eagle, Ayres's Hawk-eagle

Henst's Goshawk

Accipiter henstii

AT PERCH

Spends long periods at rest in rather vertical stance, but most often seen more horizontal and ready to depart. Note grey-brown upperparts and fine bars below. Juvenile appears more heavily mottled below. Pale eyebrow-stripe obvious. Yellow eye at all ages is an important feature.

IN FLIGHT

Short, rounded wings and rather long tail are dark above; bars are faint on flight feathers, but more obvious on tail. Below appears uniform pale brown, with barring only visible at close range. Yellow cere and legs can be seen at close range.

DISTINCTIVE BEHAVIOUR

Secretive and not often seen outside wooded habitat. Spends much time perched within trees. Main prey is medium-sized birds and small mammals, which it probably hunts from a perch or during a low, searching flight. Utters high, cackling calls and loud, piercing screams in the nest area. Builds a relatively large stick-nest in a high tree fork.

A. Hawkins

Adult calling from perch

ADULT Dark grey-brown above, with white feather bases showing on the nape and faint barring across the white eyebrow-stripe. Below (including throat and underwing coverts) white finely barred with dark grey, and feather shafts dark grey. Vent white with only very slight barring. Flight and tail feathers grey-brown with broad, brown bars only slightly paler above but very pale brown below. Bill black; eyes, cere and heavy, bare legs and feet yellow. Sexes similar, female about 14% larger.

JUVENILE Above dark brown with rufous feather edges. Feather bases barred with paler grey-brown on flight feathers, upperwing coverts, shoulders and upper tail coverts. Below white to pale yellowy-rufous with brown blobs at feather tips and bars at bases creating heavily blotched effect. Flight feather bars more obvious than in adult.

DISTRIBUTION, HABITAT AND STATUS Forest habitat on Madagascar that has not been too degraded by felling; occurs mainly in primary rainforest but also in dry deciduous forests.

SIMILAR SPECIES Most similar in size and colour at all ages to rare **Madagascar Serpent-eagle** (p. 70) (very long, broad, multi-barred tail – even longer and more barred than Henst's; darker upperparts; more coarsely barred below; no white eye-stripe; large, loose nape feathers; longer, more slender legs; darker chest and throat) and common **Madagascar Buzzard** (p. 112) (lacks pale eyebrow-stripe; broad bars and blobs below; juvenile differs more in shape than in details of plumage or colour; stockier; larger head; heavier legs and feet; shorter tail; longer wings). Same basic colour pattern as **Madagascar Harrier-hawk** (p. 130) (unbarred chest; small, bare, yellow face; small, dark eye; tail black with white band across centre).

ALTERNATIVE NAME: Madagascar Goshawk

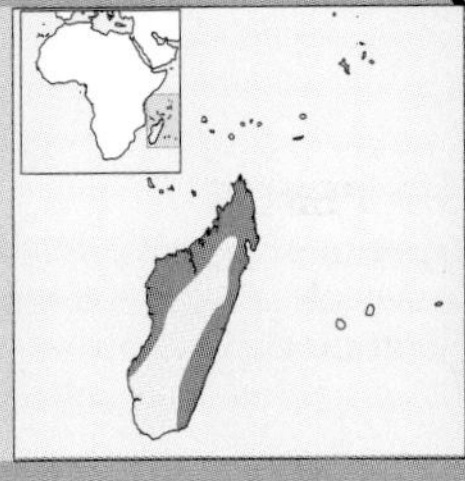

Dark grey-brown above. Below finely barred grey and white. White eyebrow-stripe. Note yellow eye, cere and bare legs. Juvenile is mainly brown, with underparts white or washed rufous broadly blobbed and streaked with brown. Medium-sized (about 50 cm tall, 110 cm wingspan).

Madagascar Serpent-eagle, Madagascar Buzzard, Madagascar Harrier-hawk

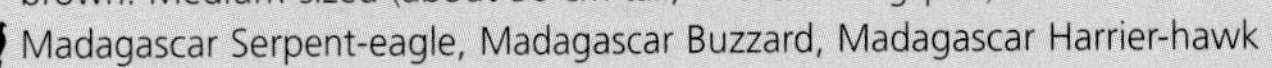

Northern Goshawk

Accipiter gentilis

AT PERCH

Often perches upright with long, barred tail hanging down and pale underparts fluffed out. Black and white (male) or brown and white (female) head pattern distinctive, highlights orange eye. Juvenile appears brown above and cream to pale rufous below, mottled above and streaked with rufous below; has yellow eye and pale eye-stripe.

IN FLIGHT

Wings relatively long and pointed for an accipiter. Flies with strong, steady wingbeats, soars and glides with wings horizontal, often slightly bent at the wrist. Appears grey or brown above and pale grey below. Bars evident on flight feathers, and especially on the tail.

DISTINCTIVE BEHAVIOUR

Spends long periods perched at the edge of cover on the lookout for its main prey of medium-sized mammals and birds. Hunts with fast dash at prey, sometimes after a prospecting flight or from soaring over open habitats near its wooded base. Builds a large stick-nest in a high fork and often defends it vociferously.

D. Karp/NHPA

Adult female

A. Rouse/NHPA

Juvenile on pheasant prey

ADULT Charcoal grey (male) or brown (female) above, except for black crown, ear coverts and sides of neck. Eyebrow and bases to nape feathers white. Below white finely barred with grey, but throat and vent plain white. Flight feathers dark grey or brown with broad, pale grey bars. Bill black; eyes deep orange , becoming red with age; cere and stout, short, bare legs and heavy feet yellow. **Female** often has darker bars and fine, dark shaft streaks on throat and breast; about 12% larger (average: male 690 g, female 1 206 g).

JUVENILE Brown above with paler feather tips, more cream to rufous below with dark brown streaks. Crown dark brown. Scapulars and rump finely barred with pale brown. Eyes yellow.

DISTRIBUTION, HABITAT AND STATUS Restricted to the wooded hills of northwestern Africa. Uncommon but known to breed here; an extension of the Iberian population that may be augmented with a few non-breeding European birds during the northern winter.

SIMILAR SPECIES No other large accipiters occur in northern Morocco. The only other common raptor of similar size is the **Eurasian Buzzard** (p. 108) (mainly brown; only confusable with juvenile Northern Goshawk; longer, rounded wings; shorter tail; blotched and broadly barred below; yellow to dark brown eye).

ALTERNATIVE NAMES: Eurasian Goshawk, European Goshawk

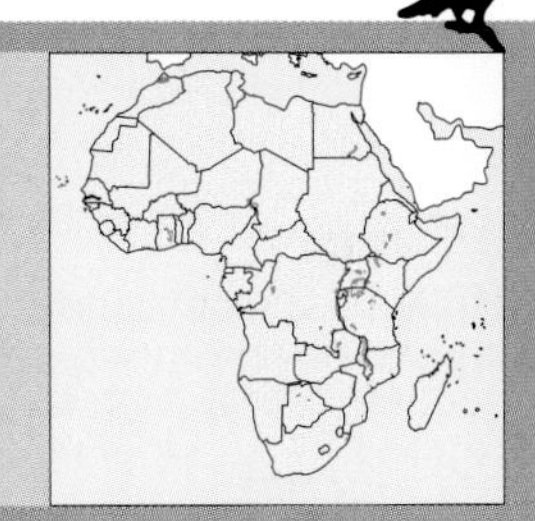

Large, thickset accipiter. Pale grey above, white below with fine, grey barring. Broad, dark bars across grey wings and tail. White eyebrow-stripe. Dappled nape. Black crown and ear coverts. Voluminous white vent. Orange eyes. Bare, heavy, yellow legs and feet. Medium-sized (about 45 cm tall, 105 cm wingspan). Restricted to northwestern Africa.

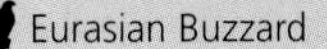
Eurasian Buzzard

P. Steyn

Black Kite

Milvus migrans

AT PERCH

Appears dark brown, with contrast of brown upperparts and rufous underparts discernable. Grey head of European subspecies (*M. m. migrans*) and yellow bill of African subspecies (*M. m. aegyptius*) notable. Forked tail obvious. Often stands with rather horizontal posture on short legs.

IN FLIGHT

Dark brown flight feathers contrast with paler brown coverts and body. Long wings, often bent at wrist, and forked tail most obvious. Flaps methodically and glides, with frequent changes of direction, during which the long tail is opened, closed and twisted. Also soars high on outstretched wings and open tail. Tail usually more deeply forked in African subspecies.

DISTINCTIVE BEHAVIOUR

Spends long periods in low, searching flight, seeking small-animal prey or offal. Takes prey in flight or from the ground. Often flies down roads or tracks in search of carrion, swooping down to snatch food with agile twists and turns, even among people and traffic. Utters a shrill squeal in aggression and courtship. Gregarious, often gathers at food sources and roosts communally when not breeding. Chases other birds to rob them of prey.

I. Davidson

M.m. migrans adult soaring

M.m. aegyptius adult gliding

ADULT Dark brown above, with paler rufous-brown edges to coverts and back feathers. Head pale brown to grey with fine, dark brown streaks in migrant European subspecies, but same colour as upperparts in African races. Below rufous-brown (*migrans*) to rufous (African subspecies), with dark brown streaks on breast. Tail red-brown above with dark brown bars in African subspecies, and grey-brown below. Bill black in *migrans*, but all-yellow in African subspecies; eyes red-brown; cere, short bare legs and toes yellow. Sexes similar in plumage, female about 4% larger (both sexes 567–941 g).

JUVENILE Very similar to adult, except for less forked tail, narrow, pale brown tips to upperpart feathers, and more streaked underparts. Head brown and bill black in all subspecies.

DISTRIBUTION, HABITAT AND STATUS Occurs throughout Africa as migrant and breeding populations. European subspecies (*migrans*) also breeds in northwestern Africa, but most widespread and common as non-breeding migrant to all of sub-Saharan Africa. African subspecies *aegyptius* mainly resident and breeding in northeastern Africa, with some post-breeding movement south to Kenya and Tanzania. Other African race *parasitus* widespread in sub-Saharan Africa, plus islands of Cape Verde, Gulf of Guinea, Comoros and Madagascar, but only a breeding migrant to southern Africa and vagrant to Aldabra and Seychelles. Occupies wide range of habitats, from urban areas in deserts to openings in rainforest. Favours woodland for roosting and nesting but will roost on the ground when on migration. Widespread and generally common, especially at urban, aquatic or other concentrations of food.

SIMILAR SPECIES Most similar to **Red Kite** (p. 188) (breeding resident of Morocco and Cape Verde Islands, only a rare vagrant elsewhere in Africa; larger; paler grey head; longer, more deeply forked, plain rufous tail; white patches at base of primaries).

ALTERNATIVE NAME: Yellow-billed Kite *(M.m. aegyptius/parasitus)*

Large, dark brown kite, more rufous below. Long, forked tail. Long wings often held bent at wrist in flight. Grey head in migrant subspecies *migrans* from Europe, brown in African subspecies *aegyptius* and *parasitus*. Eyes red-brown. Bill black in *migrans*, all-yellow in adults of African subspecies. Medium-sized (about 45 cm tall, wingspan 145–180 cm).

Red Kite

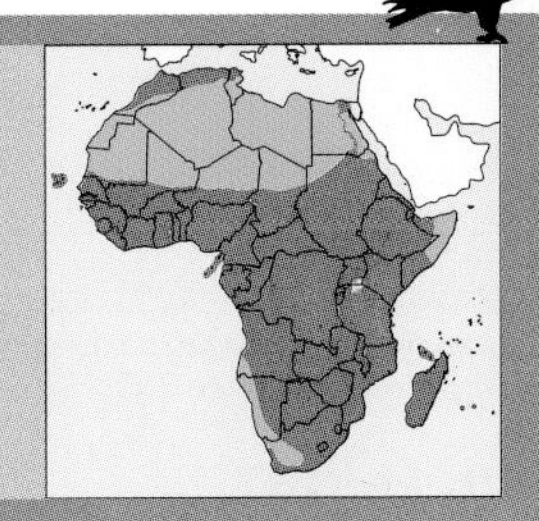

D. Balfour

Red Kite

Milvus milvus

AT PERCH
Pale grey head and long, forked, rufous tail contrast with brown, streaky body. Often stands rather horizontally, in typical posture of short-legged kites. Pale yellow eye obvious at close range. Pale yellow base of bill notable.

IN FLIGHT
Typical *Milvus* silhouette with bent wrists, strong wingbeats and, especially, flexible twists of the notably long and deeply forked tail. Dark flight feathers contrast with red-brown coverts and, from below, with white patches at base of primaries. From above, plain rufous tail obvious. Pale head not always obvious.

DISTINCTIVE BEHAVIOUR
Hunts mainly in flight and spends long periods flying back and forth at low altitude, or soaring high overhead in search of small animals and carrion. Notably agile and buoyant on the wing, with dramatic twists and swoops when necessary. Congregates at abundant food, such as termite swarms or rubbish dumps, and joins communal roosts when not breeding or migrating.

R. Tidman/NHPA

Juvenile

Juvenile

ADULT Rufous-brown overall. Above dark brown, with rufous edges to feathers giving broadly streaked effect. Head and neck pale grey with fine, black shaft streaks. Below paler rufous-brown with dark brown streaks. Flight feathers dark brown with paler inner primaries, especially below. Long, deeply forked tail plain rufous with bright central and browner outer feathers. Bill yellow with black tip; eyes pale yellow; cere, bare legs and feet yellow. Sexes similar in plumage; female about 3% larger (both sexes 757–1 221 g). Subspecies *fasciicauda* of the Cape Verde Islands is noticeably smaller and darker.

JUVENILE Similar to adult but tail more brown and with faint bars above, feathers of upperparts more obviously edged with pale brown and underparts more heavily streaked, head often browner and eye pale yellow-brown.

DISTRIBUTION, HABITAT AND STATUS Resident, with small breeding populations, in Morocco and Cape Verde Islands (extinct on Canary Islands), but only a palaearctic summer migrant elsewhere in northern Africa. Rare vagrant south of the Sahara (questionable record south to South Africa). Favours forest edge and patches of woodland, up to 2 500 m a.s.l. in Morocco. Crosses desert on migration and generally overwinters in more open habitats.

SIMILAR SPECIES Most resembles **Black Kite** (p. 186) (with which it hybridizes in Cape Verde Islands; browner and less rufous; grey head less obvious; tail dark brown, shorter and less forked; eye brown; bill black (migrants) or all-yellow (African subspecies)).

Large, rufous kite. Pale grey head. Long, deeply forked, tail plain rufous above and faintly barred below. Agile, twisting flight. Pale yellow eye. Large (about 50 cm tall, wingspan 185 cm).

Black Kite

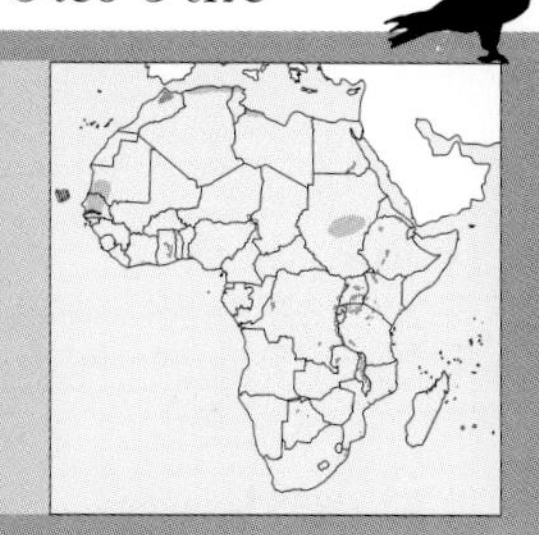

D. Kjaer/Planet Earth Pictures

African Cuckoo-hawk

Aviceda cuculoides

AT PERCH

Stands horizontally on rather short legs, often with the wings slightly drooped to expose the tail and rump. Usually perches within cover or below canopy, rather than on exposed sites. Primaries come close to but not past tip of tail.

IN FLIGHT

Long, broad, pointed wings and rather slow wingbeats recall *Milvus* kite, but long, barred tail, usually held closed, is distinctive. Seen from in front, makes deep wingbeats with tips almost touching below. Head appears small but is often held rather high. Flight usually direct and quite fast; flaps and glides, only rarely soars. Underwing patterns and colours obvious in good light, especially if soaring.

DISTINCTIVE BEHAVIOUR

Spends much time perched looking for insects and small vertebrates, especially lizards. Makes slow descent to pounce on prey, hop after it on the ground or snatch it from foliage. Noisy, particularly before breeding, with loud, whistling calls. Pairs also perform twisting display flights, especially at dawn and dusk. Untidy nest recognised by leaves still attached to twigs.

P. Ginn

Adult

ADULT Above dark grey with paler face. Rump black with white bars. Below, chest pale grey; breast and underwing coverts white with broad, dark rufous bars. Small crest usually folded, with pale centre and rufous nape-patch. Flight feathers dark grey with narrow, pale grey bars. Tail has 2 narrow, pale bars and pale tip. Bill black with double-notched cutting edge; eyes deep red-brown (male) or yellow (female); cere, feet and short, stocky, bare legs deep yellow. Sexes similar in size (220–296 g).

JUVENILE Brown above with pale rufous feather tips. White streaks on crown, forehead and eyebrow, and white throat with faint brown streaks, leave dark face-mask. Below white, with heavy black streaks becoming blobs on flanks. Tail has 3 lighter bars, not 2. Crest less prominent. Eye dark brown. Moults into subadult plumage at end of first year.

SUBADULT Similar to adult, but with rufous wash over chest and finer grey barring on underparts. Eye pale yellow. Probably moults into adult plumage at end of second year.

DISTRIBUTION, HABITAT AND STATUS Throughout sub-Saharan Africa in suitable habitat of woodland, light forest and riverine forest and along forest edges; avoids desert and semi-desert. Generally uncommon, but a most unobtrusive species except before breeding. Easily overlooked, even in suburban habitats.

SIMILAR SPECIES Adult, and especially immature, resembles smaller **African Goshawk** (p. 172) and even smaller **African Little Sparrowhawk** (p. 176) (no crest; wings short and rounded; long tail, also with 2 or 3 pale bars and/or spots; underwing pale and finely barred or spotted; adult breast, chest and throat has fine, brown bars; juvenile breast has narrow, dark stripes and flank spots).

ALTERNATIVE NAMES: African Baza, (African) Cuckoo-falcon

Dark grey hawk with short, pointed crest. Broad, rufous bars across white breast. Head and chest pale grey with rufous nape patches. Bulging brown (male) or pale orange (female) eye without prominent eyebrow ridge. Horizontal posture on short, stocky legs. Long, barred tail. Rump black with white or pale grey bars. Juvenile brown above, white below with spotted breast and dark face-mask. Medium-sized (about 30 cm tall, wingspan 90 cm).

African Goshawk

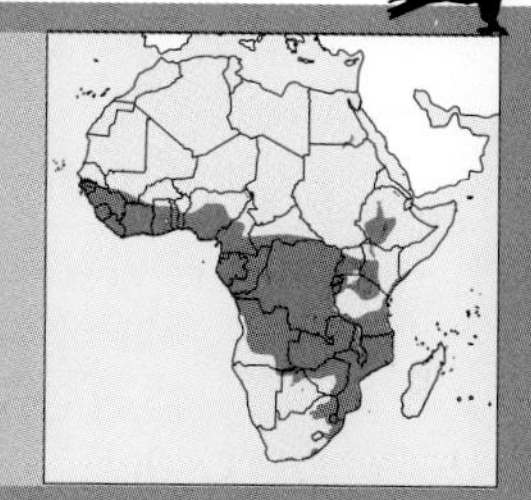

Madagascar Cuckoo-hawk

Aviceda madagascariensis

AT PERCH

Perches close to perch on very short legs, often rather horizontally with the wings slightly drooped to expose the tail and rump. Usually perches within cover or below canopy, but most often seen on exposed sites on forest edge. Primaries do not extend past tip of tail.

IN FLIGHT

Long, broad, pointed wings and long, barred tail, usually held closed. Head held rather high and prominent. Flight usually direct and quite fast; mainly deep flaps and short glides, but also soars regularly.

DISTINCTIVE BEHAVIOUR

Spends much time perched looking for insects and small vertebrates, especially lizards (chameleons, geckos). Makes slow descent to snatch prey, often grabbing it off foliage and tree trunks but also from the ground. Often perches at edge of clearings and active into dusk. Call undescribed.

O. Langrand

Adult

ADULT Above dark brown with paler head. Rump off-white. Crest usually folded. Below white, chest and sides of breast heavily blotched with brown. Flanks finely streaked with brown. Tail has 2 narrow, pale bars and pale tip. Underwing coverts densely barred with brown, and flight feathers with broad, dark brown to grey bars. Bill brown, with double-notched cutting edge; eyes and cere grey or pale yellow; bare legs and feet grey. Sexes similar in plumage and size.

JUVENILE Similar to adult in coloration but with narrow, pale tips to feathers of upperparts, paler underwing coverts, and lacking white on the throat. Tail has 3 lighter bars, not 2, and a white-tipped rump. Crest less prominent than that of adult. Eye brown. Subadult plumage as yet undescribed.

DISTRIBUTION, HABITAT AND STATUS Throughout Madagascar in suitable habitat. Favours evergreen or dry deciduous forest with clearings, from sea level to 1 600 m a.s.l. Avoids dense forest, but least common in arid south and deforested central plateau of the island. Generally uncommon and unobtrusive.

SIMILAR SPECIES Most resembles plumage of **Madagascar Buzzard** (p. 112) (dark band across lower breast rather than chest; lacks crest; short, rounded wings; cere pale grey; eye and longer legs pale yellow; usually perches more vertically; large rounded head).

ALTERNATIVE NAMES: Madagascar Baza, Madagascar Cuckoo-falcon

Dark brown hawk with obvious crest. Brown above, with pale rump. Below white with brown blotches. Rather horizontal posture on short, stocky legs. Tail has 2 pale bars and pale tip. Bulging yellow eye without prominent eye-brow ridge. Medium-sized (about 30 cm tall, wingspan 85 cm).

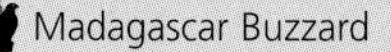
Madagascar Buzzard

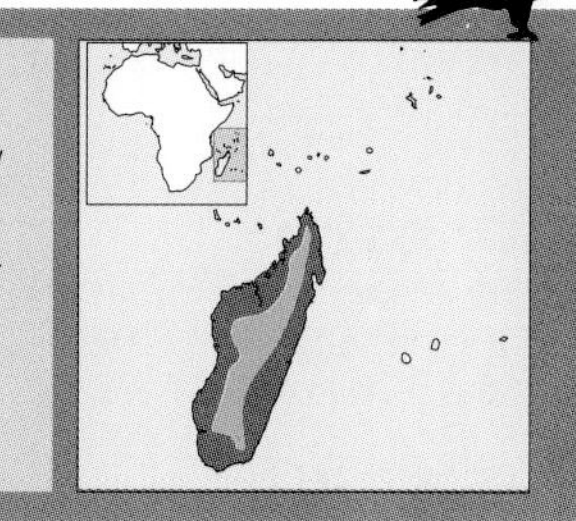

Black-shouldered Kite

Elanus caeruleus

AT PERCH
Pale colour and black shoulder distinctive. Stocky appearance, often with rather horizontal stance on short legs, and long, dark wing tips extending past short tail. Broad, rounded head with red eye obvious at close range.

IN FLIGHT
Black shoulder on upperwing and black primaries below are obvious against pale grey and white plumage. Plain white tail. Buoyant, gull-like flight, usually with bent wrists. Wings often held up at marked dihedral when gliding; also soars with wings extended horizontally and tail fanned.

DISTINCTIVE BEHAVIOUR
Favours exposed perch sites, from which it hunts its main prey of small rodents, with some birds, reptiles or insects. Also hunts on the wing, with extended bouts of hovering, especially at dusk. Hunt often ends in a controlled descent on prey with wings held high and legs extended. When excited, pumps tail up and down. Flies with shallow, fluttering flight during courtship while uttering a high, rasping call or a drawn-out whine; often follows this by gliding to a prospective nest site in a small tree. Roosts in trees, often communally with up to 500 birds together.

ADULT Above pale dove-grey, including central tail feathers. Palest on crown and darkest on tertials. Black shoulder-patch (actually a wrist-patch) formed by lesser upperwing coverts. Eyebrow, forehead, face, underparts, underwing coverts and outer tail feathers white. Outer primaries dark grey above and black below; inner primaries and secondaries grey with white tips above and white below. Bill black; eye red; cere, short stocky bare legs and feet butter yellow. Sexes similar in plumage and size (197–343 g).

JUVENILE Similar to adult, but with sooty-grey shoulder, and with brown wash and pale brown tips to feathers above, especially on scapulars and upperwing coverts. Below white, with pale brown wash on breast and fine, brown streaks down centre of breast feathers and dark grey to black primary flight feathers. Eye pale brown, later turning yellow then red.

DISTRIBUTION, HABITAT AND STATUS North African coast, Nile Valley and sub-Saharan Africa. Favours savanna and grasslands. Also uses clearings within forest or woodland, desert margins after good rains, agricultural areas, and road verges with utility poles. Widespread and locally common, but widely nomadic to track rodent populations, including extensive movements such as from South Africa to Uganda. May breed in any season when prey is abundant.

SIMILAR SPECIES Overlaps in western and eastern Africa with very similar **African Swallow-tailed Kite** (p. 196) (smaller, with long, deeply forked tail; darker grey above; lacks black shoulder-patch). Dove-grey colour and red eye like **Shikra** (p. 166) (long, thin legs; short, rounded wings; long, barred tail; finely barred below; grey head;. no shoulder-patch), but unlike other small, grey hawks.

Adult gliding

ALTERNATIVE NAMES: Black-winged Kite, Common Black-shouldered Kite

Pale grey above and white below, with black shoulder patch. Ruby-red eye. In flight, all white except for black primaries. Often stops to hover in flight. Sometimes pumps tail up and down when perched. Small (about 20 cm tall, 84 cm wingspan).

African Swallow-tailed Kite, Shikra

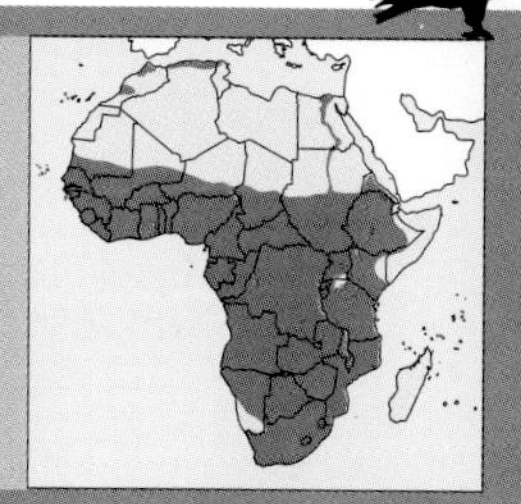

African Swallow-tailed Kite

Chelictinia riocourii

AT PERCH

Note small size and elongate form. Grey above and white below. Red eye visible at close range. Often perches within cover during heat of the day.

IN FLIGHT

Appears dark grey in flight, with slender, pointed wings and long, forked tail. White underwings with black carpal patch obvious only when it hovers, soars overhead or banks to one side. Flight buoyant and tern-like, with bouts of hovering, swooping on prey or soaring.

DISTINCTIVE BEHAVIOUR

Hunts staple prey of arthropods mainly from on the wing, often after hovering briefly. Prey either caught on the wing or snatched from the ground in dextrous flight, often then also consumed in flight. Usually forms small flocks when hunting and when nesting in loose colonies. Roosts colonially. Gathers in largest numbers to hunt around game or livestock, during veld fires and at locust or termite emergences. Pumps long tail up and down when excited.

ADULT Dark grey upperparts and tail, except for narrow, white tips to inner primaries and secondaries. Eyebrow, forehead, face and underparts white. Underwing coverts white, except for black primary underwing coverts forming carpal patch. Bill black; eye deep red; cere, short, bare legs and toes yellow. Sexes similar in plumage and size (about 90–110 g).

JUVENILE Similar to adult but with pale brown tips to back feathers and pale brown wash on breast. Outer tail feathers shorter. Eye probably brown.

DISTRIBUTION, HABITAT AND STATUS Confined mainly to semi-arid interface of Sahara and savanna, favouring Sahel steppe and scrub savanna and extending into East Africa along arid corridor of western Rift Valley. Locally common, but highly nomadic as it tracks availability of insect prey and nest sites. Performs regular annual movements, southwards to forest margins of western Africa at start of southern winter dry season and northwards into the Sahel to breed when rains start in summer. In eastern Africa, moves south with rains to breed in March, and small numbers may remain during dry season.

SIMILAR SPECIES Most resembles widespread **Black-shouldered Kite** (p. 194) (larger; paler grey; black upperwing coverts forming the shoulder; plain white underwing coverts; short, square, white tail).

ALTERNATIVE NAMES: Swallow-tailed Kite, Fork-tailed Kite, Scissor-tailed Kite

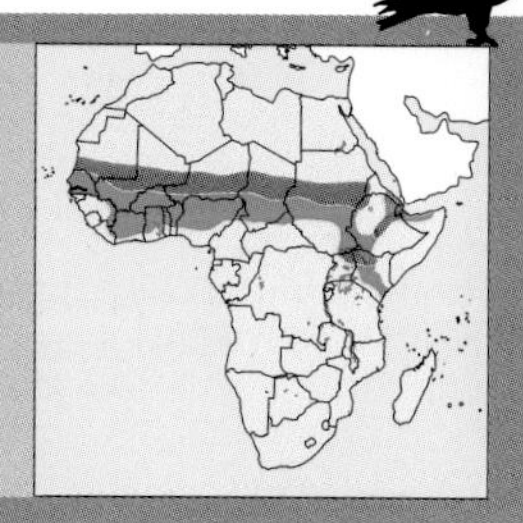

Dark grey above and white below with white wingtips and small black carpal patch. Slender build. Long, pointed wings. Elongate outer tail feathers. Often in loose flocks. Small (about 25 cm tall, 90 cm wingspan).

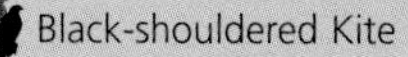
Black-shouldered Kite

Bat Hawk

Macheiramphus alcinus

AT PERCH

All-dark hawk, with pale eyelids and false eye spots on nape, a slight crest and white belly sometimes visible. Long, pointed wings extend past tip of short, square tail. Usually perches upright in falcon-like pose. Pale feet often notable.

IN FLIGHT

Long, pointed wings with wide base; short, square tail. Appears broad-headed from below; pale legs and feet and yellow eyes often visible, especially at close range. Looks like a large, dark falcon, except for pronounced bend at wrist and, usually, more deliberate and deeper wingbeats. Usually languid, leisurely flight broken by fast, twisting pursuits.

DISTINCTIVE BEHAVIOUR

Spends the day perched, usually in a tall tree, but may emerge to chase other birds, to soar around midday, or occasionally to hunt. Often found in or near the nest tree, which is usually large and pale-barked, and stands taller than surrounding trees. At dusk emerges to hunt; main prey insectivorous bats but also small birds; may continue through moonlit nights and into the dawn. Usually captures, kills and swallows prey on the wing; only larger prey taken to a perch. Often seen hunting at dusk around areas of open water or floodlights. Nest obvious for being draped over an open, horizontal branch. Often noisy around nest, with high 'kwik' notes.

ADULT Sooty-brown overall, except for white eyelids, white 'eye' spots on nape, off-white throat with black streaks down the centre and white belly and flank feathers. Flight feathers and tail have narrow, pale grey bars. Bill black; eyes deep yellow; cere, wide gape and bare legs with long slender toes pale blue-grey. Sexes similar in plumage and size (600–650 g).

JUVENILE Similar to adult, but with more white on underparts, especially on throat, chest and belly, leaving only a dark band across breast. Fine, pale grey bars on flight feathers and tail give a spotted appearance.

DISTRIBUTION, HABITAT AND STATUS Widespread in sub-Saharan Africa, except for the most arid regions. Most common in woodland and on forest edge. Often found around rivers, dams and gorges, with numerous bats. Enters cities, with their enhanced light levels, and also found around exotic eucalyptus plantations with their tall, pale-barked trees. Nowhere common but difficult to assess status.

SIMILAR SPECIES Several other raptor species hunt the same type of prey at dusk, when their colours and markings are difficult to see. Most often confused with slightly smaller **Lanner** and **Peregrine falcons** (pp. 200 and 202) (pointed, broad-based wings held straight out; underparts paler; dark eyes; yellow legs; never swallow prey in flight; tails longer and more rounded; faster wingbeats and more driving flight). Various other small falcons and accipiters also hunt bats at dusk, but their smaller size and different wing shapes are usually obvious.

W. Tarboton

Juvenile

 ALTERNATIVE NAMES: Bat Falcon, Bat Pern, Bat Kite, Bat-eating Buzzard, Bat-eating Hawk

All-dark falcon-like hawk. Note large, yellow eyes, white eyelids and false white 'eye' spots on nape. Pale blue-grey legs. White belly may be either exposed or concealed. White throat with dark centre stripe not always obvious. Small bill but very wide blue-grey gape. Slight crest. In flight, wings bent at wrist, and shows pale grey bars on dark wings and tail. Medium-sized (about 40 cm tall, 110 cm wingspan).

Lanner Falcon, Peregrine Falcon

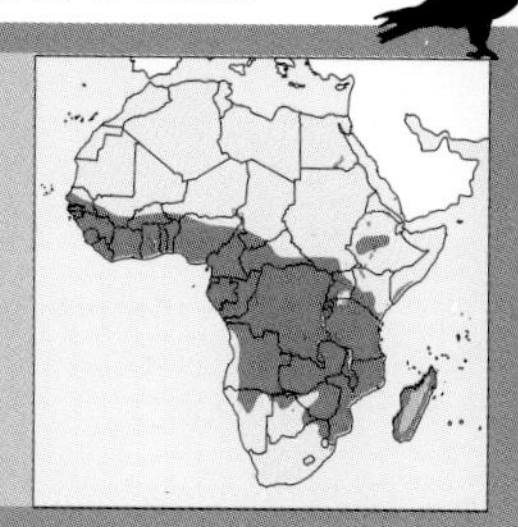

Lanner Falcon

Falco biarmicus

AT PERCH

Usually stands upright, with long wings reaching tip of rather long tail. Dark upperparts and pale underparts. Head pattern (with dark eye and 'moustache' stripes) and yellow cere and eye-ring usually obvious.

IN FLIGHT

Long, narrow tail and wings, with rather blunt tips. Flies with steady, flat wingbeats, or soars with wings held well forward and tail fanned. Bars obvious across flight feathers and tail, contrasting with pale, plain or spotted flanks (adults) or heavily streaked (juvenile) body and underwing coverts. Face-stripes and rufous crown evident at close range.

DISTINCTIVE BEHAVIOUR

Often perches inconspicuously on a rockface, tree or utility pole. Usually noticed in flight, soaring, sometimes to great heights, or flying fast and low. Often on the wing early and late, or soaring in the middle of the day. Hunts mainly small birds and, in more desert areas, mammals and reptiles, using fast pursuit or hard stoops as main technique, often in pairs and near water-holes. Eggs are laid in old stick-nests of other birds, in trees, in pylons, or on rock or building ledges, even on the ground among rocks in open desert. Utters harsh, grating calls 'kek kek kek', especially around nest area.

ADULT Head rufous with black stripes behind, below and between the eyes. Above grey to grey-brown with paler bars; rump and tail palest. Below cream, often with pink wash; black, arrow-shaped flecks across breast and flanks (northwestern, northern and western/northeastern African subspecies *erlangeri*, *tanypterus* and *abyssinicus* respectively), or restricted to flanks (central/southern African subspecies *biarmicus*). Desert forms palest. Flight feathers and tail sooty-brown with evenly spaced paler bars. Bill black with pale grey base; eye dark brown; cere, eye-ring and short, bare legs with long toes deep yellow. Sexes similar in plumage, female about 9% larger (male 430–600 g, female 700–910 g).

JUVENILE Individual variation from dark to pale and from heavily to lightly marked. Browner than adult, above less barred and with buff feather edges. Head and throat pale rufous to white, but with same black facial markings as adult. Below pale rufous to white with dark brown streaks on breast and underwing coverts; rest plain. Flight and outer tail feathers dark brown with narrow, pale rufous bars; central tail feathers often unmarked, and fresh tail has broad, cream tip. Cere and eye-ring blue-grey at fledging, turning yellow after 3–4 months. Legs pale yellow. Moults directly into adult plumage in second year.

DISTRIBUTION, HABITAT AND STATUS Throughout Africa wherever prey and nest sites exist. Most common around small gorges or patches of large trees in open savanna. Occurs from forested mountain cliffs to treeless desert, only absent from continuous forest. Reported vagrant to Socotra.

SIMILAR SPECIES Similar, including in size, to **Peregrine Falcon** (p. 202) (black to dark grey above; finely barred (adult) or narrowly streaked (juvenile) below; all-black head with white cheeks and throat; broad 'moustache' bar; wings more pointed and faster, shallower wingbeats), especially pale desert subspecies ***pelegrinoides*** (only nape rufous; brown above; buff below with barring on flanks). Most similar in plumage to smaller **Red-necked Falcon** (p. 212) (rufous head and neck; only slight 'moustache' stripe; densely barred below; broad, black subterminal band on tail). At dusk, same general form and flight as **Bat Hawk** (p. 198) (pointed 'wrist' joint in flight; all-dark with faint, pale tail bars; yellow eye; white eyelids, false 'eye' spots and legs); juvenile also resembles more slender **Eleanora's Falcon** (p. 208) (black crown; sharp, dark streaks below; blue cere and eye-ring in adult female and juvenile).

F.b. biarmicus adult

Large for a falcon. Note rufous crown. Dark brown 'moustache', forehead and eye-stripes. Adult grey-brown above. Below salmon pink, plain or lightly spotted with brown. Juvenile sooty-brown above, buff below with heavy, dark brown streaks. Small (about 25 cm tall, 90–115 cm wingspan).

Peregrine Falcon, Red-necked Falcon

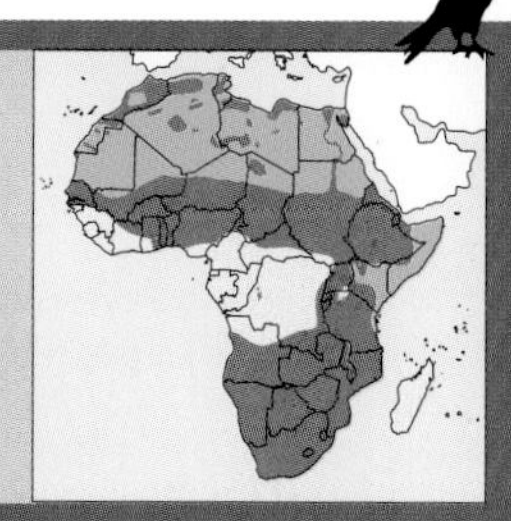

J. Carlyon

Peregrine Falcon

Falco peregrinus

AT PERCH

Thickset, with long, pointed wings that reach the tip of the tail. Often sits upright, facing into ledges with tail and wings pointing downwards. Note rounded head with black or dark grey 'helmet', broad 'moustache' stripe and contrasting white cheeks and throat. From a distance, looks black above and white below. Adult has plain or faintly spotted chest, finely barred below, juvenile all finely streaked below. Large, yellow feet obvious at close range.

IN FLIGHT

Long, broad-based pointed wings, like elongated triangles, that make the long tail look relatively short. Flies with powerful, choppy wingbeats, rarely glides, but soars on extended wings and tail. Dark head and pale chest contrast with darkly barred (adult) or streaked (juvenile) underparts. Paler barred rump notable from above.

DISTINCTIVE BEHAVIOUR

Fast, dashing flier, travelling everywhere at speed in level, driving flight or in spectacular dives. Perches for long periods on ledges or trees. Feeds on small to medium-sized birds, mostly taken on the wing and then carried to a perch to be eaten. Often hunts in the early morning and late evening in dim light. Nests on bare ledges on high cliffs, often noisy in nest area with harsh screams 'kak kak kak' and whines 'keer-ik keer-ik'.

ADULT Above dark grey to black, darkest on side of head and malar stripe. Fine, grey tips to wing coverts, rump pale grey finely barred with black. Nape often has rufous feather tips. Cheeks and underparts white to pale rufous; tiny black streaks on chest; breast has black bars, heaviest on flanks and undertail coverts. Flight feathers and tail slate-grey with narrow, white bars below, broad, dark ends and fine, white tips, especially on tail. Bill black with pale grey base; eye dark brown; cere, eye-ring and short, bare legs with long toes deep yellow. Sexes similar in plumage, female about 10% larger (male 500–740 g, female 610–1 300 g). Five subspecies in Africa. Smallest is sub-Saharan resident *minor* and largest is migrant *calidus*; also northwestern African *peregrinus*, Mediterranean *brookei* and Saharan *pelegrinoides* (often regarded as a full species, Barbary Falcon; paler, with rufous nape and pale rufous breast with less barring). Small Madagascan *F.p. radama* most resembles *F.p. minor*. **JUVENILE** Sooty-brown above with fine rufous edges to feathers, especially on forehead and nape. White to buff below with narrow, pointed streaks, heaviest on flanks. Cere and eye-ring blue-grey. **DISTRIBUTION, HABITAT AND STATUS** Widespread resident throughout Africa, Madagascar, Comoros, Bioko and Cape Verde islands, in areas with cliffs and plentiful avian prey. Augmented throughout by Eurasian migrants in northern winter. Residents uncommon and local, except for a few areas with ideal habitat. Vagrant on Canary, Mauritius, Seychelles, Aldabra and Socotra islands. **SIMILAR SPECIES** Resident Peregrine Falcon subspecies smaller and stockier than widespread **Lanner Falcon** (p. 200) (rufous crown; plain, pale grey rump; pink breast either plain or finely spotted; plain, pale underwing coverts; juvenile heavily blotched below), also resembles smaller, more slender and black-capped **Eleanora's Falcon** (p. 208) (long, thin wings and tail; relaxed deep wingbeats; blue cere and eye-ring in adult female as well as juvenile) and even smaller **Eurasian Hobby** (p. 218) (pale eyebrow; grey upperparts; heavily streaked breast; deep rufous thighs and undertail coverts). Similar in size and shape to crepuscular **Bat Hawk** (p. 198) (pointed 'wrist' joint in flight; all-black with light tail bars; yellow eye; white eyelids, false eye spots and legs).

D. Balfour

F.p. minor adult

ALTERNATIVE NAMES: Barbary Falcon or Shaheen (*pelegrinoides*)

Peregrine Falcon

Large for a falcon and thickset. Head black or dark grey and back dark grey. Cheeks and throat white. Broad, black 'moustache' stripe. Below pale rufous or white with black bars (adult) or streaks (juvenile). Fast, strong flier. Broad-based, pointed wings and tail narrowly barred black and pale grey. Saharan Barbary Falcon (*F.p. pelegrinoides*) paler, browner, with rufous nape; below rufous with reduced barring). Small (about 20–30 cm tall, 80–117 cm wingspan).

Lanner Falcon, Eleonora's Falcon

Peregrine Falcon *Falco peregrinus*

F.p. peregrinus ad.
F.p. peregrinus juv.
F.p. peregrinus ad.
F.p. peregrinus juv.
F.p. radama ad.
F.p. radama ad.
F.p. radama juv.

Saker Falcon

Falco cherrug

AT PERCH

A lanky, small-headed falcon with a broad chest, long, pointed wings and long tail. Note the pale head and white streaky breast, brown upperparts and darkly marked flanks and thighs. Dark 'moustache' and pale eye-stripe often not obvious. Dull yellow (adult) or blue-grey (juvenile) bare skin differentiates age classes at close range.

IN FLIGHT

Long, pointed wings with slightly rounded tips (used in flat, powerful wingbeats for fast, driving flight) and long, square-ended tail obvious. Note paler head. Dark, streaky body and larger under-wing coverts obvious against adult's plain underside, pale tail and flight feathers with their relatively broad, dark trailing edges.

DISTINCTIVE BEHAVIOUR

An open-country 'desert' falcon, usually seen flying fast and low, interspersed with glides. Sometimes seen perched on a tree or on the ground. Often flies down prey over a long distance and takes it on the ground. Main prey is medium-sized to large birds. Uncommonly soars and occasionally hovers briefly.

A. Rouse/NHPA

Adult

N. Downer/Planet Earth Pictures

Juvenile

ADULT Individual variations in colour and density of markings. Above brown with broad, rufous feather edges, giving overall russet-brown colour. Head white to pale rufous, with fine, brown streaks on crown and nape and pale forehead, eyebrow and nape patches. Dark brown line behind and below eye. Underparts white; plain on throat and cheeks; finely streaked dark brown on chest; more heavily streaked on breast, with large spots on flanks and dark streaked thighs. Flight feathers sooty-brown above with buff tips, below grey-brown with darkly streaked coverts. Tail plain dark fawn, with darker spots on more rufous outer feathers forming pale stripes, dark subterminal band and white tip. Bill black with pale yellow base; eye dark brown; cere, eye-ring, stocky bare legs and long toes pale yellow. Sexes similar in plumage, female about 9% larger (male 730–890 g, female 970–1 330 g).

JUVENILE Very similar to adult, usually more finely streaked on breast, more streaked and less spotted on flanks. Note blue-grey cere, eye-ring and legs.

DISTRIBUTION, HABITAT AND STATUS Rare, non-breeding migrant to the eastern Mediterranean and northeastern Africa, in Libya, Egypt, Ethiopia, Sudan and northern Kenya and Chad, vagrant to Morocco, Senegal, Mali, Cameroon and Burundi. Present during the southern winter, mainly around flat, open marshes and desert margins.

SIMILAR SPECIES Largest falcon in Africa. Most resembles more slender **Lanner Falcon** (p. 200) (pale grey above; obviously barred flight and tail feathers; pinkish below with fine spotting; rufous crown and dark, facial stripes; bright yellow cere, eye-ring and feet; juvenile darker brown above, more heavily blotched below, with yellow feet) and the stockier **Peregrine Falcon** (p. 202) (dark grey above with black head; grey, barred rump; broad 'moustache' stripe; bright yellow cere and feet; barred underparts; juvenile sooty-grey, dark head and 'moustache', yellow feet).

Large for a falcon. Red- to sandy-brown colour. Pale, streaky head. Heavily streaked white underparts. Facial skin dull yellow (adult) or blue (juvenile). Medium-sized (about 40 cm tall, 102–129 cm wingspan). Uncommon migrant to northeastern Africa.

Lanner Falcon, Peregrine Falcon

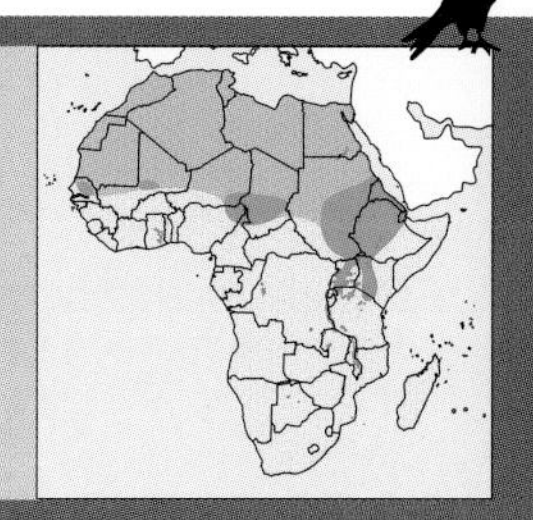

Eleonora's Falcon

Falco eleonorae

AT PERCH

Often stands rather horizontally, which exaggerates the short legs and long wings and tail. Slim, elongate build always notable. Appears dark overall. White cheeks and throat contrast with heavy 'moustache'. Rufous underparts vary in shade and streaking.

IN FLIGHT

Long, broad-based, pointed wings and long, rounded tail. Very fast, agile flight, but with rather slow, shallow wingbeats in more casual flight. Soars on flat, straight wings and often glides on air currents for long periods. Areas of pale barring on underside of flight feathers and tail contrast with dark coverts and tips, the whole wing darker than the body and tail in pale forms. White throat and cheeks and dark 'moustache' often obvious.

DISTINCTIVE BEHAVIOUR

Usually seen on the wing, less often perched quietly on a tree or rock. Notable for very fast, driving flight with deep wingbeats, accentuated by slim body and long, slender wings and tail. Feeds mainly on small, flying migrating birds when breeding and includes many large, aerial insects during migration. Hunts mainly in early morning and late evening. Nests in colonies on coastal and island cliff-faces and often seen in groups when migrating. Not very vocal.

P. Pickford/SIL

Juvenile

ADULT Above plain dark sooty-brown. Throat and cheeks white, separated by broad malar stripe. Grades into deep rufous breast, with broad, black streaks on breast and underwing coverts, or all-dark form with only faint trace of paler markings. Flight feathers and tail all-black above, and with narrow, pale grey bars and broad, black tips below. Bill black with grey base; cere and eye-ring blue in female but greenish-yellow in male; eye dark brown; short legs and long toes dull yellow. Sexes similar in plumage, female about 5% larger (male 350 g, female 340–850 g).

JUVENILE Similar to adult, but with rufous edges to dark feathers of upperparts, pale rufous throat, cheeks and underparts, and rufous bars across underwing and undertail feathers. Dark form like adult. Cere and eye-ring blue in both sexes, legs pale yellow.

DISTRIBUTION, HABITAT AND STATUS Breeds on islands and coastal cliffs of northern Africa from Canary Islands eastwards through Mediterranean to Cyprus. Migrates to Madagascar during the northern winter, crossing northern and eastern Africa. Vagrant to Mauritania, Comoros (including Aldabra), Mauritius, Seychelles, Mozambique and Zimbabwe. Occupies arid shores when breeding, but occurs over savanna, woodland and forest during migration.

SIMILAR SPECIES Pale form most similar in slender form and colouring to much smaller **Eurasian Hobby** (p. 218) (underwings appear rufous and barred; only thighs and undertail coverts deep rufous; chest white with dark streaks, not washed rufous, grey upperparts contrast with black head; thin 'moustache' stripe) and juvenile **Sooty Falcon** (p. 228) (grey upperparts; dark grey blobs below on buff background; yellow cere and eye-ring), and in colouring to similar-sized but stocky **Peregrine Falcon** (p. 202) (shorter wings and tail; off-white below with dark bars; barred underwings; pale grey barred rump). Dark form resembles much smaller adult **Sooty Falcon** (p. 228) (uniformly grey with darker flight feathers and head; pale throat; bright yellow cere and eye-ring) and adult male **Red-footed** (p. 222) or **Amur falcons** (p. 220) (grey body and dark grey head; deep rufous undertail coverts; grey (Red-footed) or white (Amur) underwing coverts; red cere, eye-ring and legs).

Large but slender black falcon; 2 forms. Dark head and broad 'moustache' stripe. White throat and cheeks. Breast deep rufous with dark streaks or (25%) all-dark. Cere and eye-ring blue (adult female and juvenile) or greenish-yellow (adult male). In flight, note long, slender wings with dark underwing, long, rounded tail and agile flight. Juvenile paler than adult. Medium-sized (about 30 cm tall, 90–105 cm wingspan).

Eurasian Hobby, Sooty Falcon, Peregrine Falcon

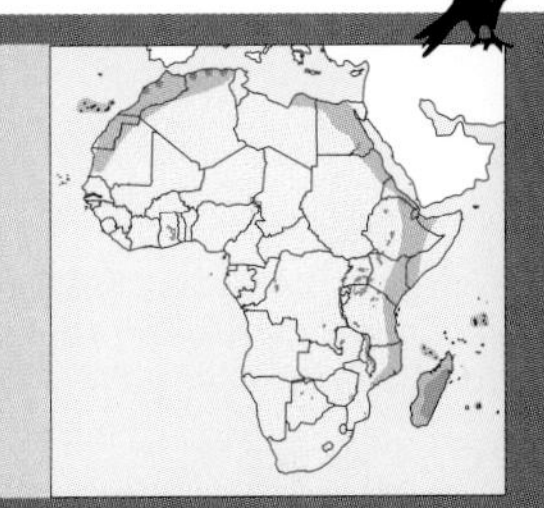

G. Robbrecht/Wildlife Pictures

Taita Falcon

Falco fasciinucha

AT PERCH

Compact build, accentuated by short tail and wide shoulders, with relatively large, dark head. Pale nape, cheek and throat patches obvious, and plain rufous breast also notable. Pale grey rump and large, yellow feet sometimes evident. Juvenile with dark rump and more spotted flanks.

IN FLIGHT

Wide-based pointed wings, heavy body and short tail, like a flying 'bullet'. Flies with fast, stiff wingbeats; rarely glides but dives at great speed. Above, black flight feathers contrast with paler coverts and with pale grey rump and tail. Below, rufous underparts and underwing coverts contrast with darkly barred flight feathers and tail, and with a white throat.

DISTINCTIVE BEHAVIOUR

Spends much time perched inconspicuously on a ledge or cliff-side tree; emerges to fly with fast, parrot-like wingbeats in pursuit of small avian prey, less often to soar overhead. Catches prey on the wing, usually after a fast dive, then carries it back to the cliff to eat. Lays eggs on bare rock ledge or in pothole. Shrill, harsh 'kek kek kek' call.

ADULT Above slate grey, darkest on head, palest on rump. Head with 2 rufous nape patches and narrow collar, pale rufous forehead and cheek patches. White ear-patches, throat and neck grade into rufous underparts, with fine, black shaft streaks and small spots on flanks and underwing coverts. Flight feathers black with narrow, paler bars above, more rufous below, matching greater underwing coverts. Tail plain grey with faint, darker outer bars, broader subterminal bar and white tip. Bill black with grey base; eye dark brown; cere, eye-ring, stout bare legs and long toes yellow. Sexes similar in plumage, female about 12% larger (male 212–233 g, female 297–346 g).
JUVENILE Similar to adult but browner, with fine, buff edges to feathers of upperparts, broadest on dark rump. Nape patches paler rufous. More heavily streaked below and spotted with black on flanks, tail tip buff when not worn. Cere and eye-ring probably pale blue and legs pale yellow. Throat and ears washed rufous.
DISTRIBUTION, HABITAT AND STATUS Patchily distributed along major cliffs and river gorges from Ethiopia south along the Rift Valley to northern South Africa. Rare throughout its range, most widespread in Zimbabwe, Zambia, Malawi and Uganda.
SIMILAR SPECIES Stocky build, especially in flight, like much larger **Peregrine Falcon** (p. 202) (all-dark head, rump barred with black, white cheek patches, underparts barred (adult) or streaked (juvenile) including on underwing coverts), similar size and coloration to more slender **African Hobby** (p. 216) (pale eyebrow; deep rufous underparts with dark brown streaks, rufous throat, long, slender wings and tail).

P. Ginn

Juvenile

Stocky, short-tailed falcon. Pale grey rump. Plain, pale rufous below with fine, black streaks on flanks. Black head with rufous nape patches and white cheeks and throat. In flight, wings and tail darkly barred against rufous underparts and grey upperparts. Small (about 20 cm tall, 70 cm wingspan).

Peregrine Falcon, African Hobby

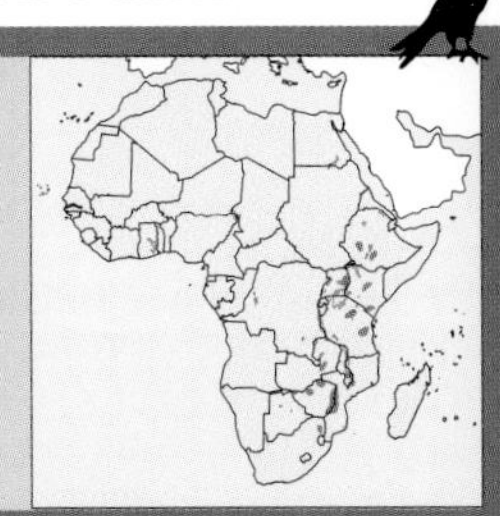

P. Steyn/Photo Access

Red-necked Falcon

Falco chicquera

AT PERCH

Upright on rather long legs. Short, pointed wings not near end of the long tail, with its broad, black subterminal band and white tip. Often appears slim and large-headed. Rufous crown and hindneck, together with plain, pale white cheeks and pink chest, are diagnostic. Malar stripe less obvious and barring more so in adult than in juvenile.

IN FLIGHT

Relatively short, broad wings and long tail. Flaps interspersed with glides, more like a sparrowhawk, but wings more pointed especially when soaring or stooping. Very fast with rapid, thrusting wingbeats. Note grey tail with broad, black subterminal band and white tip, together with densely barred underwing and body. Rufous crown and pale chest and face evident in good light.

DISTINCTIVE BEHAVIOUR

Dashes after its main prey of small birds, often from an open perch below the canopy of a tree or palm. Less often soars overhead and dives on prey. Catches most birds in flight, sometimes after a long, fast, agile, ringing pursuit. Often hunts as a pair and sometimes in association with Gabar goshawks. Nests on old stick structures of other birds, usually in the crown of an isolated tree. Calls infrequently with high, repeated, sharp notes.

ADULT Above pale blue-grey with narrow, dark grey to black bars. Crown, nape and hindneck deep rufous with black eyebrow and malar stripes. Throat, cheeks and chest pale pinky-rufous or white. Rest of underparts, including underwing coverts, white to pale grey with narrow black bars. Flight feathers black and tail grey above, white with narrow, black bars below, with black tips to primaries and wide, black subterminal band to white-tipped tail. Bill black with yellow base; eye dark brown; cere, eye-ring and long, slender bare legs and toes deep yellow. Sexes similar in plumage, female about 10% larger (male 139–178 g, female 190–305 g).

A. Wilson/Photo Access

Adult

JUVENILE Similar to adult bird, but crown, hindneck and forehead dull red-brown. Also, eyebrow and malar stripe more prominent. Below pale rufous, almost unmarked throughout except for barred flanks.

DISTRIBUTION, HABITAT AND STATUS Widespread across sub-Saharan Africa; also occurs on the Indian subcontinent. Occupies open, sparsely vegetated habitats with scattered large trees, such as watercourses in desert, flood and coastal plains, and edges of woodland. Most common along dry watercourses or in palm savanna, elsewhere uncommon and local.

SIMILAR SPECIES Juvenile, and to some extent adult, most resembles the larger and heavier **Lanner Falcon** (p. 200) (dark forehead band; only crown and nape rufous; hindneck grey; underparts pale rufous and almost unmarked (adults, southern and central Africa), spotted (adults, eastern and western Africa) or heavily streaked (juvenile); tail evenly barred throughout). Same size and basic colours as female **Red-footed Falcon** (p. 222) (back faintly barred rufous; underparts unbarred; tail evenly banded; cheeks white; cere, eye-ring and feet orange to red in colour).

ALTERNATIVE NAMES: Red-headed Merlin, Red-headed Falcon, Turumti

Small, short-winged falcon. Pale grey above with fine, black bars, below white with fine, dark grey bars on breast and flanks. Head and hindneck deep rufous. Face and chest unmarked pale rufous. Eyebrow and 'moustache' dark brown. In flight, note finely barred underwing and broad, black subterminal band to white-tipped tail. Juvenile has brown crown and hindneck, only barred below on flanks. Small (about 25 cm tall, 69 cm wingspan).

Lanner Falcon, Red-footed Falcon (female)

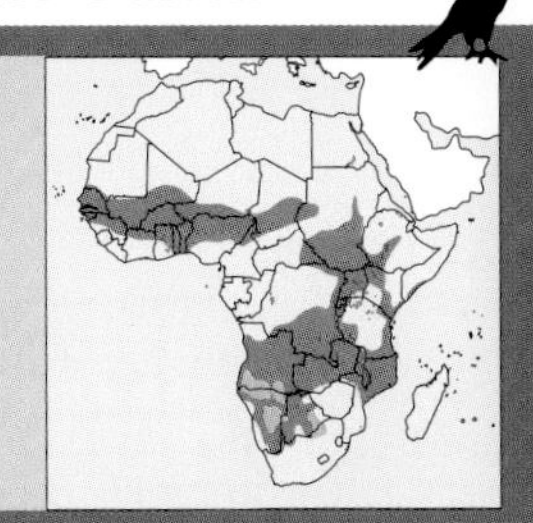

N. Dennis

Merlin

Falco columbarius

AT PERCH

Appears stocky, and often stands rather horizontally. Dark streaking all over evident on grey male and brown female. Rather short wings only reach dark end-bar on tail.

IN FLIGHT

Very fast with quick, direct, driving wingbeats. Soars, often with tail fanned to show barring. Never hovers. Note dark, barred wings, underwing coverts and tail, from above and below, with pale tips to wing and tail feathers.

DISTINCTIVE BEHAVIOUR

Hunts on the wing, mainly for small birds, with fast, level flight interspersed with glides. Chases down prey, often after agile, swerving flight. Stops to land on the ground or on low, open perches. Call a fast 'kik kik kik'.

A. Rouse/NHPA

Adult male

Juvenile taking off

ADULT Male dark grey above with fine, black streaks. Whitish eyebrow. Below pale rufous with fine, dark brown streaks, including neck and cheeks. Faint dark malar stripe. **Female** browner with heavier streaking, grey or brown rump and more prominent pale forehead and eyebrow. Flight feathers and tail dark grey (male) or brown (female) above and below with narrow, pale grey bars, and with broad, dark end (male) to white-tipped tail. Bill black with grey base; eyes dark brown; cere, eye-ring and slender, bare legs and long toes dull yellow. Female about 9% larger (male 150–210 g, female 189–255 g).

JUVENILE Both sexes similar to adult female, but with browner rump. Facial skin blue.

DISTRIBUTION, HABITAT AND STATUS Rare, non-breeding migrant to coastal lowlands of northwestern Africa and Nile Delta during northern winter. Old specimen record from Durban in South Africa. Rare vagrant to Senegal and Sudan.

SIMILAR SPECIES Only overlaps, and female might be confused, with larger, more slender **Common Kestrel** (p. 232) (deep rufous back spotted with black; below buff streaked with black; grey tail unbarred (male) or narrowly barred with rufous (female); broad, dark subterminal band) and female or juvenile **Lesser Kestrel** (p. 238) (heavily streaked buff underparts; rufous, evenly barred tail; dark 'moustache'; weak feet with white claws), both of which hover regularly.

ALTERNATIVE NAME: Pigeon Hawk

Small, stocky falcon, sexes different. Male dark grey above and pale rufous below, streaked with black. Female browner, especially above, with heavier streaks below. In both sexes, note whitish eyebrow and faint 'moustache' stripe. In flight, dark barred underwing and tail with pale tips. Juvenile similar to adult female. Very small (about 20 cm tall, 50–67 cm wingspan).

Common Kestrel, female Lesser Kestrel

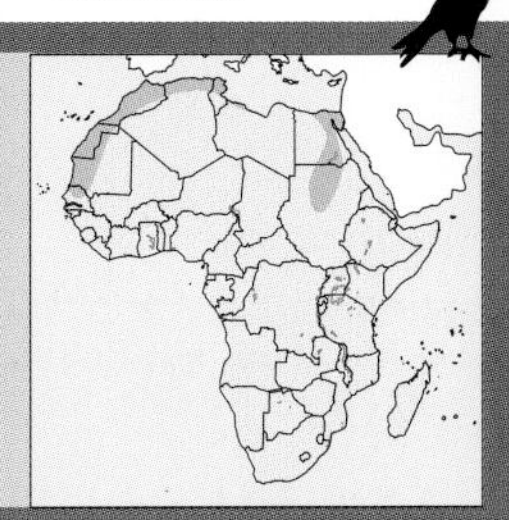

S. Dalton/NHPA

African Hobby

Falco cuvierii

AT PERCH

Appears upright, slim and neat, with the tail extending just past the tips of the long, pointed wings. Dark overall, but double-notched rufous facial markings and underparts obvious in good light. Adult appears unstreaked below, juvenile streaked. Note yellow facial skin and legs at close range.

IN FLIGHT

Note slender, pointed wings and long tail. Flies fast with shallow wingbeats and great agility. Appears black above, below dark rufous with black edges to wings and tail. Fine, black streaks on flanks and coverts, and bars on wings and tail, obvious only in good light. Markings heavier and more obvious on paler juvenile.

DISTINCTIVE BEHAVIOUR

Hunts its main prey of small birds and insects on the wing, often at dawn and dusk. Flies back and forth over good hunting areas, or makes fast sorties from a prominent perch. Usually found singly, in pairs or in small family groups. Nests high up in old stick-nests of other birds or at the base of palm fronds. Calls with repeated high-pitched chipping notes 'chek chek chek'.

Seitre/BIOS

Adult

ADULT Black above, greyer on back and wings. Eyebrow and lower nape tipped rufous. Throat, cheeks and outer nape patches rufous, divided by black malar stripe and partial ear-stripe. Below deep rufous, finely streaked with black, especially on flanks. Flight feathers black and tail slaty-black above; narrowly barred with rufous below and with broad, black edges. Bill black with grey base; eye dark brown; cere, eye-ring and short, bare, slender legs with long toes yellow. Sexes similar in plumage, female about 7% larger (male 125–178 g, female 186–224 g).

JUVENILE Similar to adult, browner above with fine rufous edges to feathers, especially on crown and nape. Below paler rufous with heavier black streaks, including on the underwing coverts. Cere and eye-ring pale blue-green at fledging, but soon turn yellow.

DISTRIBUTION, HABITAT AND STATUS Resident on the edge of moist woodlands and forests of sub-Saharan Africa. Most common in palm savanna and gallery forest of West Africa and western East Africa, including in several cities. Uncommon in northeastern and central Africa and only a rare vagrant to southern Africa.

SIMILAR SPECIES In size and plumage most resembles female **Red-footed Falcon** (p. 222) (grey above; thin 'moustache' streak; plain pale rufous below; cere, eye-ring and legs orange-red; usually in flocks with plain dark grey males; often hovers) and stocky, short-tailed **Taita Falcon** (p. 210) (pale grey rump; paler rufous below with very fine black streaks; sides of nape dark; upper nape with 2 rufous patches; white throat and cheeks; plain rufous underwing coverts; flight feathers and tail more grey), especially juvenile Taita Falcon (light underpart streaking; contrasting underwing coverts and flight feathers). **Eurasian Hobby** (p. 218) has similar habits but is longer-winged and shorter-tailed (pale eyebrow; underparts white with heavy black streaks on breast; deep rufous only on thighs and vent).

Small, slim falcon. Black above, deep rufous below, including underwings. Rufous throat, cheek and side-of-nape patches. At close range note fine, black streaks on breast and flanks. Yellow facial skin and legs. Fast and agile flyer. Juvenile browner above, more heavily streaked with black below, paler throat, cheek and nape patches. Small (about 20 cm tall, 70 cm wingspan).

Red-footed Falcon (female), Taita Falcon

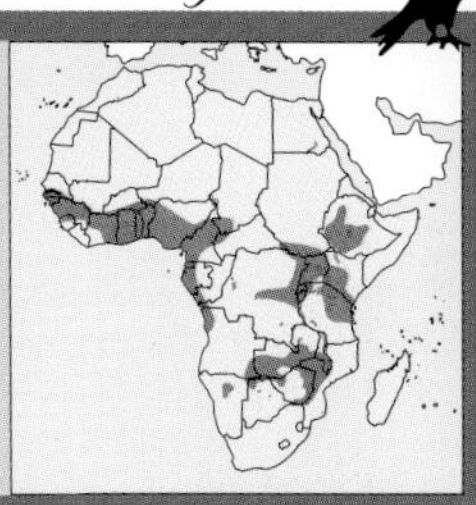

P. Steyn

Eurasian Hobby

Falco subbuteo

AT PERCH

Usually stands erect with long, pointed wings reaching tip of tail. Note slim body and small head. Head pattern consists of white nape, cheek and throat patches divided by black 'moustache' and ear-stripes, often connected by pale collar across nape, and is distinctive, as are deep rufous thighs and vent. Appears black, heavily streaked below with white or buff. Yellow (adult) or blue (juvenile) cere and eye-ring obvious at close range.

IN FLIGHT

Long, slender wings and fairly short tail, plain above, barred below. Distinctive fast, agile flight with slow, shallow, driving wingbeats, like a giant swift. Rufous vent and dark head with distinctive white pattern often obvious.

DISTINCTIVE BEHAVIOUR

Most active at dawn and dusk, catching its main prey of small birds and insects on the wing. Usually seen singly or in pairs, less often several birds together at good hunting areas or in small groups on migration. Often perches on prominent dry branches, unless resting in the shade. Note fast, swerving flight, covering large areas, and consumption of small prey on the wing. Nests in old stick-nests of other birds in tall woodland trees.

ADULT Above sooty-brown. Head black, with pale cream forehead, eyebrow and feather edges on lower nape. Throat, cheeks and sides of nape white, divided by heavy, black malar stripes and partial ear-stripes. Below, including underwing coverts, buff with heavy, black streaks. Thighs and vent deep rufous with fine, black streaks. Flight feathers and tail slaty-black, plain above, below narrowly barred buff with broad, dark edges. Bill black with blue-grey base; eye dark brown; cere, eye-ring and short, bare legs with long, slender toes yellow. Sexes similar in plumage, female about 4% larger (male 131–222 g, female 141–340 g).

JUVENILE Similar to adult, but browner above and with more obvious buff edges to feathers, paler crown and rufous bars on scapulars. White below with broader black streaks and pale rufous thighs and vent. Cere and eye-ring blue-grey, legs pale yellow.

DISTRIBUTION, HABITAT AND STATUS Small breeding populations in northwestern Africa (Morocco, Tunisia), which join main European population on migration to sub-Saharan Africa in southern winter. May occur anywhere and in any habitat on migration, but favours woodland and wooded savanna. A few stop over in West Africa, mainly on passage through East Africa; only a vagrant to Canary and Seychelles islands. Most spend northern summer in central and south-eastern Africa where locally common.

SIMILAR SPECIES In size and plumage most resembles female **Amur Falcon** (p. 220) (grey above; thin 'moustache' streak; underparts all white with streaked chest and barred breast; cere, eye-ring and legs orange-red; usually in flocks with plain dark grey males; often hovers). In plumage, resembles larger and stockier juvenile **Peregrine Falcon** (p. 202) (pale cheek-patch not notched; no white on nape; underparts all white or buffy with narrow, black streaks). In habits, similar to slightly smaller, shorter-winged **African Hobby** (p. 216) (underparts, throat and cheeks all deep rufous; fine, black streaks on chest and breast only visible at close range; appears dark below, especially in flight) and **Sooty Falcon** (p. 228) (all grey (adult); grey with buffy underparts heavily spotted with grey (juvenile)).

Juvenile

ALTERNATIVE NAME: European Hobby

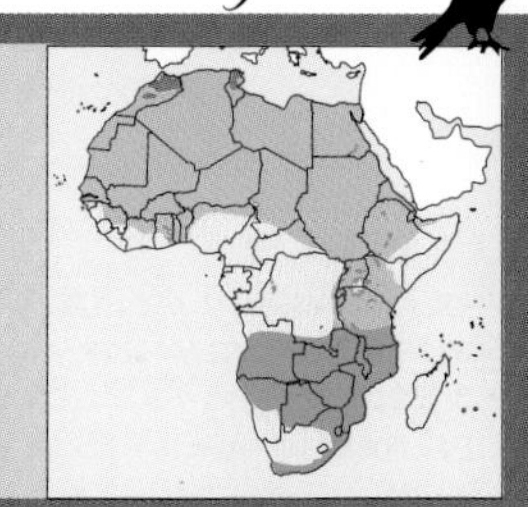

Slim, dark falcon. Black head with pale eyebrow and double-notched white patches. Below white with heavy black streaks. Deep rufous flanks and vent. Yellow cere, eye-ring and legs. In flight, note long, pointed wings (barred underneath), relatively short, narrow tail, speed and agility. Juvenile browner above, more heavily streaked and paler rufous below. Small (about 25 cm tall, 69–84 cm wingspan).

Amur Falcon (female), Peregrine Falcon

R. Cavignaux/BIOS

Amur Falcon

Falco amurensis

AT PERCH

Perches upright on short legs. Long, pointed wings reach tip of tail. Adult male appears all dark grey with contrasting orange face and feet. Rufous vent obvious in good light. Adult female appears pale grey above, with darker head; below white with dark streaks and bars. Dark eyes and 'moustache' obvious. Juvenile similar to female, often appears dull brown and untidy, with dark streaks on breast.

IN FLIGHT

Light and fast on long, pointed wings, often pausing to hover and fan the long, rounded tail. Adult male all dark grey with white underwing; rufous vent rarely visible. Adult female has barred body; wings and tail barred with broad, dark tips. Juvenile has paler, less obvious bars and dark tips, but body and underwing streaked.

DISTINCTIVE BEHAVIOUR

Usually found in flocks, perched, often on fences and utility wires, or flying about and hovering. Feeds mainly on insects picked up from the ground or caught and eaten on the wing. Roosts in large numbers in clumps of tall trees. Sometimes hunts and roosts in company of Red-footed Falcon and especially Lesser Kestrel.

ADULT Male dark grey all over, except for chestnut vent and white underwing. **Female** pale grey above, with broad, dark grey bars and cream forehead; below mainly white, but cheeks and throat plain, broad, dark grey streaks on chest, bars on breast, underwing spotted and barred and vent plain pale rufous. Female flight feathers and tail dark grey, barred above with pale grey and below with white, and with broad, dark tips. In both sexes, bill black with orange base; eye dark brown; cere, eye-ring and short, bare legs and weak feet orange-red. Sexes similar in size (male 97–155 g, female 111–188 g).

JUVENILE Both sexes similar to adult female, browner above with buff edges to feathers. Cheeks, throat and underparts pale buff, chest and breast streaked with grey-brown. Tail with narrower, more even bars. Dark eye patch and malar stripe obvious. Facial skin and feet pale orange.

DISTRIBUTION, HABITAT AND STATUS Long-distance non-breeding migrant to Africa from far-eastern Asia during the northern winter. Most enter Africa on the east coast from India and return via Arabia. Vagrant to Seychelles and Pemba. Flies in loose flocks to main wintering area in southeastern Africa. Common in open woodland and grass savanna.

Adult male resting

SIMILAR SPECIES Adult male very like adult male **Red-footed Falcon** (p. 222) (slightly paler cheeks and throat; dark grey underwing coverts in flight), **Grey Kestrel** (p. 224) (uniform dark grey all over; obvious yellow cere, eye-ring and feet), slender and long-winged **Sooty Falcon** (p. 228) (all dark grey; yellow facial skin and legs) and much larger, very rare dark form **Eleonora's Falcon** (p. 208) (all-dark sooty-brown; male yellow, female and juvenile with blue cere and facial skin). Adult female and juvenile most like longer-winged **Eurasian Hobby** (p. 218) (black back and head; double-notched cheek and side-of-nape patches; heavy 'moustache' stripe; heavily streaked underparts; deep rufous thighs and vent, facial skin yellow (adult) or blue (juvenile) and legs yellow; never hovers; usually solitary).

ALTERNATIVE NAMES: Eastern Red-footed Falcon, Manchurian Red-footed Falcon, Amur Red-footed Falcon

Small falcon, sexes differ. Adult male dark grey with chestnut vent. Adult female pale grey above, below white with black streaks and bars; dark eye and 'moustache'. Orange-red cere, eye-ring and feet. In flight, adult male underwing white, adult female and juvenile coverts streaked and flight and tail feathers barred. Usually in flocks, often hovers. Juvenile similar to adult female but browner. Small (about 20 cm tall, 65–75 cm wingspan).

Red-footed Falcon (male), Grey Kestrel, Eurasian Hobby

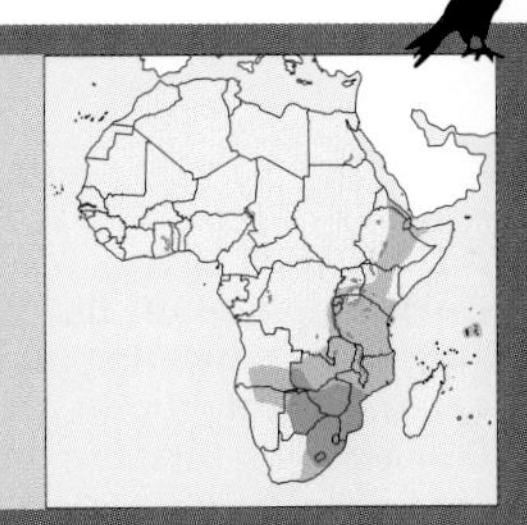

W. Tarboton

Red-footed Falcon

Falco vespertinus

AT PERCH

Perches upright on short legs. Long, pointed wings reach tip of tail. Adult male appears all dark grey with contrasting orange face and feet. Rufous vent obvious only in good light. Adult female appears mainly rufous, with pale grey back, dark eye-patch, white cheeks, and less obvious orange face and feet. Juvenile similar to female with dark streaks on head and breast.

IN FLIGHT

Light and fast on long, pointed wings, often pausing to hover and fan the long, rounded tail. Adult male all dark grey, including underwing coverts, but rufous vent rarely visible. Barred wings and tail with broad, dark tips most obvious on adult female and juvenile; pale rufous body and underwing (streaked in juvenile) and pale grey above with rufous crown.

DISTINCTIVE BEHAVIOUR

Usually found in flocks. Perches in open sites, often on fences and utility wires, or flies about and hovers nearby. Feeds mainly on insects picked up from the ground or caught and eaten on the wing. Gathers in large numbers to roost in clumps of tall trees. Sometimes occurs in the company of Amur Falcon and Lesser Kestrel.

Seitre/BIOS

Adult female hovering

ADULT Male dark grey all over, except for chestnut vent. Darkest on head, neck, flight feathers and tail. **Female** pale grey above; feathers have faint rufous tips and bars. Below (including underwing coverts) plain pale rufous. Cheeks white, but crown and hindneck dark rufous. Female flight feathers and tail grey, barred above with dark grey and below with white, with broad, dark tips. In both sexes bill black with orange base; eye dark brown; cere, eye-ring and short, bare legs and weak feet orange-red. Sexes similar in size (male 115–190 g, female 130–197 g).

JUVENILE Both sexes similar to adult female, with dark grey bars and broad, brown feather tips above. Head, neck and underparts pale rufous finely streaked with dark brown; cheeks and throat white. Tail has narrow, even bars. Dark eye-patch and malar stripe obvious. Facial skin and feet pale orange.

DISTRIBUTION, HABITAT AND STATUS Non-breeding migrant to Africa from Europe during the northern winter. Enters Africa across the eastern Mediterranean and Arabia. Vagrant to Cape Verde, São Tomé and Seychelles islands. Flies in loose flocks to main wintering area in semi-arid southwestern Africa. Common in open scrubland with scattered small trees, but migrates over moister habitats.

SIMILAR SPECIES Adult male very like adult male **Amur Falcon** (p. 220) (slightly darker cheeks and throat; white underwing coverts in flight), **Grey Kestrel** (p. 224) (uniformly dark grey all over; yellow cere, eye-ring and feet), slender, long-winged **Sooty Falcon** (p. 228) (all dark grey; yellow facial skin and legs) and much larger, very rare dark form **Eleonora's Falcon** (p. 208) (all-dark sooty-brown; male yellow, female and juvenile with blue cere and facial skin). Female and juvenile like short-winged **Red-necked Falcon** (p. 212) (broad, black subterminal band and white tail tip; back and underparts (adult) or flanks (juvenile) have narrow dark bars; cheeks pale rufous; facial skin and legs yellow; usually solitary or in pairs).

ALTERNATIVE NAME: Western Red-footed Falcon

Small falcon, sexes quite different. Adult male plain dark grey with chestnut undertail coverts. Adult female pale grey above, pale rufous below and on head and neck; black eye and 'moustache' patches accentuate orange-red cere, eye-ring and feet. In flight, adult male underwing plain grey, adult female and juvenile with barred dark-tipped flight and tail feathers. Usually in flocks; often hovers. Juvenile similar to adult female but browner and with dark streaks below and on head. Small (about 20 cm tall, 65–75 cm wingspan).

Amur Falcon (male), Grey Kestrel, Sooty Falcon, Red-necked Falcon

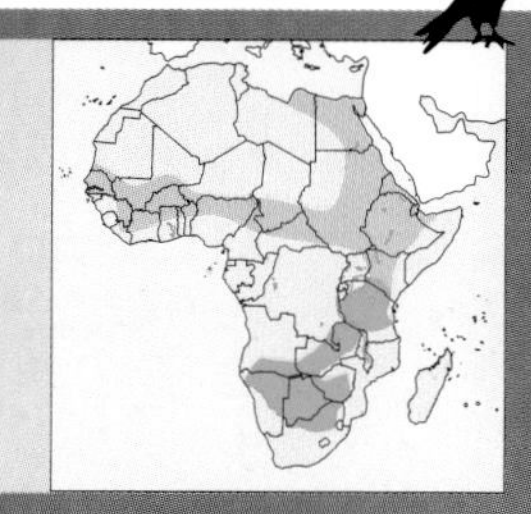

B. Van Damme/Wildlife Pictures

Grey Kestrel

Falco ardosiaceus

AT PERCH

Upright posture on long legs is emphasised by the long tail that projects well past the wing tips. Note stocky body, large head and heavy bill. Appears dark grey overall, with bright yellow facial skin around the dark eye. Juvenile appears browner.

IN FLIGHT

Note short, broad, pointed wings and long tail. Flies rather heavily with fast, shallow wingbeats but does not often glide, soar or hover. Appears dark grey overall, with black primaries and alula. Only faintly barred on undertail and underwing. Juvenile browner, especially on the pale, barred underwing coverts.

DISTINCTIVE BEHAVIOUR

Usually seen singly, perched on an open site in search of its main prey of small reptiles and insects. Dives down to take prey from the ground or low vegetation, less often in aerial pursuit. Changes perch regularly. Nests in holes in old Hamerkop nests or trees. Calls infrequently with repeated high notes 'kik kik kik'.

ADULT Dark grey overall, with fine black streaks on feather shafts. Primaries, eye and malar areas darker sooty-grey. Below slightly paler, with unstreaked, pale grey cheeks. Flight feathers and tail have faint, narrow, paler bars on inner edges. Bill black with yellow base; eye dark brown; cere, wide eye-ring and long, stout, bare legs and toes deep yellow. Sexes similar in plumage, female about 10% larger (male 205–255 g, female 240–300 g).

JUVENILE Similar to adult, but fine, brown feather tips produce a brown wash. Most evident on the head, neck and underparts, especially on the underwing coverts which also have narrow white bars. Facial skin blue-green at fledging but soon turns yellow.

DISTRIBUTION, HABITAT AND STATUS Breeding resident across western, eastern and north-central sub-Saharan Africa. Commonest in moist palm savanna and woodland; also found in openings in secondary and primary forest.

SIMILAR SPECIES Commonest all-grey falcon over its range. Adjoins in the southwest with very similar **Dickinson's Kestrel** (p. 226) (white head and rump; tail obviously barred with pale grey; wings much darker grey than body). Occasionally visited in southern summer along east of range by male **Red-footed** and **Amur falcons** (pp. 222 and 220) (dark grey overall; darker heads; chestnut undertail coverts; facial skin and legs orange-red; Amur has white underwing coverts) and vagrant **Sooty Falcon** (p. 228) (uniformly light grey (or in some birds dark grey); rump only slightly paler; dark primaries; black face obvious; slim build with long, slender wings; small head; juvenile with buff underparts spotted grey).

Juvenile

Dark, stocky, all-grey kestrel with a large head and heavy bill. Obvious yellow facial skin areas. In flight, grey overall with darker primaries. Only faint barring on underwing and undertail. Juvenile browner, with fine, white bars on underwing coverts. Small (about 25 cm tall, 70 cm wingspan).

Dickinson's Kestrel, Red-footed and Amur falcons (males), Sooty Falcon

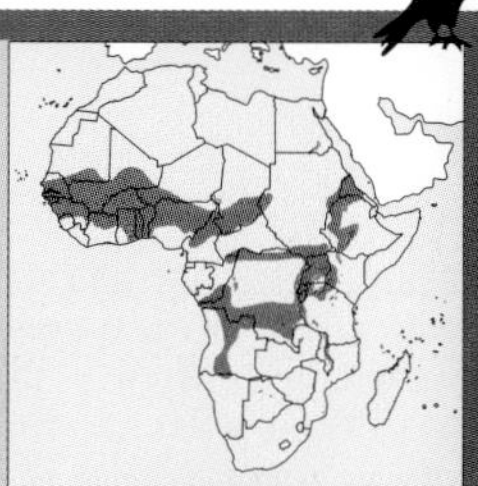

R. Henno/Wildlife Pictures

Dickinson's Kestrel

Falco dickinsoni

AT PERCH

Appears stocky and large-headed. Stands upright on relatively long legs. Dark wings and pale head obvious from afar, and dark eye within yellow skin at closer range. Juvenile has darker head than adult and blue-green facial skin when recently fledged, but later difficult to distinguish from adult.

IN FLIGHT

Short, broad, pointed wings and long tail. Flies with fast, almost parrot-like wingbeats. Pale rump and head obvious against dark grey body. Pale tail bands often seen when landing. Juvenile has uniformly grey body, head and rump, and buff bars in tail. Broad, dark subterminal tail bar and white tip often notable.

DISTINCTIVE BEHAVIOUR

Often hunts from open dead-tree perch, where it watches for its main prey of lizards, small birds and insects. These are caught after a fast strike to the ground, or sometimes after a short aerial pursuit. Attracted to hawk insects at bush fires but rarely hovers. Nests in holes in trees, palm stumps, old Hamerkop nests, even bridges. Calls infrequently with repeated shrill, high notes, 'kik kik kik'.

M. Goetz

Juvenile

ADULT Above dark grey, except for almost-white head, neck and rump. Possibly paler in male. Below grey with fine, dark feather shafts, especially on flanks. Flight feathers and tail sooty-grey, primaries have small, white spots. Tail has even, pale grey bars, broad black end-band and white tip. Heavy bill black; eye dark brown; cere, broad eye-ring and long, stout, bare legs and stubby toes deep yellow. Sexes similar in plumage and size (male 167–207 g, female 207–246 g).

JUVENILE Similar to adult, but slightly browner and with very fine brown tips to upperwing coverts. Head and rump only slightly paler than body. Pale tail bars buff. Cere and eye-ring blue-green for about 4 months after fledging, legs pale yellow.

DISTRIBUTION, HABITAT AND STATUS Resident with restricted range in south-central Africa, including Zanzibar and Pemba islands. Favours areas with open perches and sparse groundcover, such as palm-studded floodplains or dry, baobab-studded savanna.

SIMILAR SPECIES Northwest of range overlaps with very similar **Grey Kestrel** (p. 224) (uniformly dark grey all over; obvious bare, yellow cere and wide, yellow eye-ring). Visited during the southern summer by male **Red-footed** and **Amur falcons** (pp. 222 and 220) (dark grey overall with darker heads; chestnut undertail coverts; facial skin and legs orange-red; Amur has white underwing coverts) and vagrant **Sooty Falcon** (p. 228) (uniformly light grey; rump only slightly paler; dark primaries; some birds dark grey; slim build with long, slender wings; small head; juvenile with buff underparts spotted grey).

ALTERNATIVE NAME: White-rumped Kestrel

Stocky, grey kestrel with block-shaped head. Wings dark, head and rump pale grey, tail distinctly barred. Heavy, dark bill and dark eye. Extensive yellow facial skin. Fast, swerving parrot-like flight. Juvenile slightly browner with blue-green facial skin. Small (about 25 cm tall, 67 cm wingspan).

Grey Kestrel, Red-footed and Amur falcons (males), Sooty Falcon

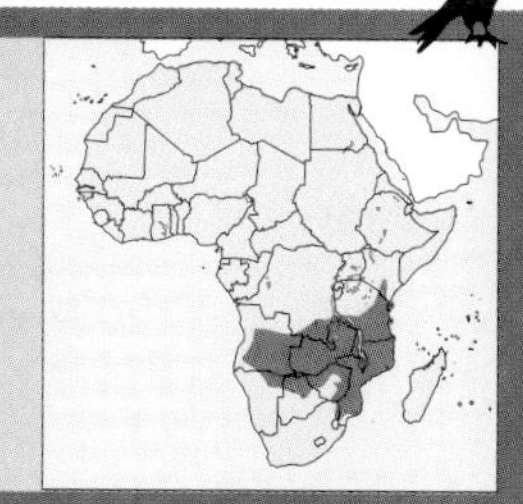

D. Balfour

Sooty Falcon

Falco concolor

AT PERCH

Stands rather horizontally on short legs, the long, pointed wings extending well past the tip of the tail. Appears light grey overall, with a dark face, often a buff throat and dark flight feathers. Note yellow facial skin and legs. Juvenile appears dull grey from behind; front buff with light spots and dark double-notched ear and 'moustache' stripes.

IN FLIGHT

Note long, narrow, pointed wings and relatively short tail. Fast and agile, with shallow wingbeats interspersed with glides. Seldom soars and never hovers. Dark primaries and pale rump most obvious from above. Note dark face and pale throat at close range. Juvenile has patterned head, streaked body, barred underwing, and barred tail with dark end-band and buff tip.

DISTINCTIVE BEHAVIOUR

Most active at dawn and dusk, hunting its main prey of small birds, bats and insects on the wing. Attacks from in flight or from a prominent perch. Hunts alone or in small groups, sometimes with Eleonora's Falcon. Nests solitarily or in colonies on rocky outcrops, cliffs, small islands, and sometimes in old stick-nests of other birds.

ADULT Light grey overall, with dark head and black eye- and malar patches. Rump pale grey, with fine, black shaft streaks. Most have buff throat, some dark grey. Primaries black above, rest of flight feather and tail surfaces grey. Tail with faint, narrow, paler bars on inner edges. Bill black with pale yellow base; eye dark brown; cere, eye-ring and slender, bare legs and long toes yellow. Sexes similar in plumage, female about 4% larger (about 210 g).

JUVENILE Above similar to adult, but with fine buff feather edges. Cream throat, cheek and side-of-nape patches, with dark grey eye-, ear- and 'moustache'-patches. Below buff, including underwing, with broad, grey stripes and tips, to appear more spotted than streaked. Flight feathers and tail paler grey and more clearly barred than those of adult. Tail with darker end-band and buff tip. Some birds darker overall with heavier streaking. Facial skin pale blue.

DISTRIBUTION, HABITAT AND STATUS Breeds in remote areas of the eastern Libyan and Egyptian deserts, and along the coast of and on islands in the Red Sea and Persian Gulf. Migrates during the southern winter to Madagascar, where common. Regular on coasts of Mozambique and northern South Africa. Rare vagrant elsewhere in southern Africa and to Mauritius and Seychelles. Recorded on passage through eastern and northeastern Africa.

SIMILAR SPECIES Coexists mainly with much larger and darker **Eleonora's Falcon** (p. 208) (even more slender and longer-winged; pale form has heavily streaked underparts and underwing; dark form much darker brownish-black; female and juvenile have blue cere and facial skin). On the African mainland, overlaps with similar-sized male **Red-footed** and **Amur falcons** (pp. 222 and 220) (dark grey overall; chestnut undertail coverts; facial skin and legs orange-red; Amur has white underwing coverts) and **Dickinson's Kestrel** (p. 226) (stocky build; large head; heavy bill; rather short wings and parrot-like flight; whitish head and rump; tail with prominent pale grey bars; wings much darker grey than body).

X. Eichaker/BIOS

Adult

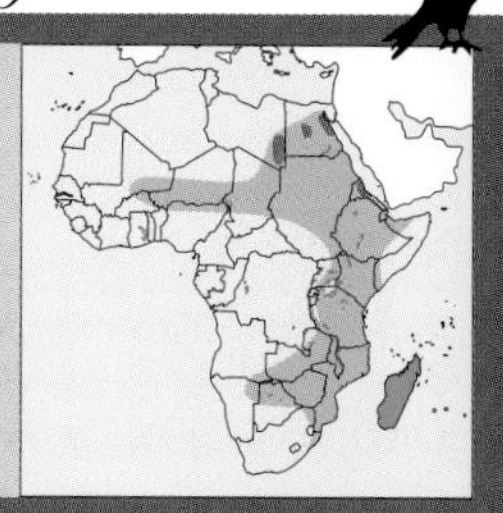

Small, slender, light grey falcon, a few dark grey. Darker primaries and face, but paler rump. Obvious yellow cere, eye-ring and legs. In flight, note long, slender wings. Juvenile buff below with grey streaks, flight feathers dark brown with buff bars. Small (about 25 cm tall, 78–90 cm wingspan).

Eleonora's Falcon, Red-footed and Amur falcons (males), Dickinson's Kestrel

P. Steyn/Photo Access

Banded Kestrel

Falco zoniventris

AT PERCH

Stocky, large-headed, heavy-billed kestrel. Stands rather upright. Relatively short, pointed wings do not extend to tip of long tail. Appears dark grey above and white below, with a darkly streaked chest and heavily barred breast. Note streaky throat and neck, pale yellow eye and yellow lower mandible at close range.

IN FLIGHT

Short, pointed wings and long tail are sparrowhawk-like, but wingbeats fast and shallow with little gliding or soaring. Not known to hover. Appears dark overall, with barred body and obvious narrow, white bars on flight feathers and tail.

DISTINCTIVE BEHAVIOUR

Usually seen perched on edge of forest or woodland, searching the ground, trunks, lower branches and foliage for main prey of lizards and insects. Snatches prey in fast, swerving flight. Occasionally chases small birds in aerial pursuit. Changes perches with fast, direct flight. Nests in tree cavities or the base of epiphytes. Call a harsh, rapid 'kek kek kek', often uttered in flight high over territory.

D. Halleux/BIOS

Adult

ADULT Above dark grey to grey-brown, with pale forehead and darker streaks on neck. Wing coverts and scapulars have broad, light grey bars; rump finely tipped with white. Below off-white, throat and cheeks finely streaked, slight dark 'moustache' stripe, chest heavily streaked, breast and vent with broad, dark grey-brown bars. Flight feathers and tail sooty-brown with narrow, white bars on inner edges and tips; underwings barred grey and white; wing tips grey; tail barred below and on sides, black on top. Individual variation in tone of grey-brown and contrast of markings. Heavy bill black with yellow base and lower mandible; eyes pale yellow; cere, eye-ring, stout, bare legs and stocky feet yellow. Sexes similar in plumage and size (180–240 g).

JUVENILE Similar to adult but browner, with fine, brown tips especially to upperwing coverts, back and rump. Flight feathers and tail barred pale rufous. Eyes dark brown. Facial skin dull blue-green at fledging, eyes brown.

DISTRIBUTION, HABITAT AND STATUS Throughout Madagascar except on the central plateau. Generally uncommon, favours the edges of rainforest, secondary forest and dry woodland up to about 2 000 m altitude.

SIMILAR SPECIES In markings, colours and size, most resembles the small-headed **Madagascar Sparrowhawk** (p. 162) (fine, dark bars from throat to vent; delicate black bill; long, slender legs and toes). The only other kestrel on the island is the smaller, slimmer, rufous **Madagascar Kestrel** (p. 240) (rufous above; white (pale form) or rufous (dark form) below with black streaks; dark brown eye; black bill). Much stockier than the migrant **Sooty Falcon** (p. 228) (uniformly light grey; slim with long, narrow wings; dark brown eye) and much larger and less common dark form of **Eleonora's Falcon** (p. 208) (uniformly sooty-brown; very long, slender wings; dark brown eye; blue facial skin in adult female and juvenile).

ALTERNATIVE NAMES: Barred Kestrel, Madagascar Banded Kestrel

Stocky grey-brown kestrel with a heavy bill. Grey streaks on chest and bars on breast. Barred flight and tail feathers. Note pale yellow eye of adult, also bright yellow lower mandible, cere, eye-ring and legs. Juvenile similar but browner and with dark eye. Small (about 25 cm tall, 65 cm wingspan). Restricted to Madagascar.

Madagascar Sparrowhawk

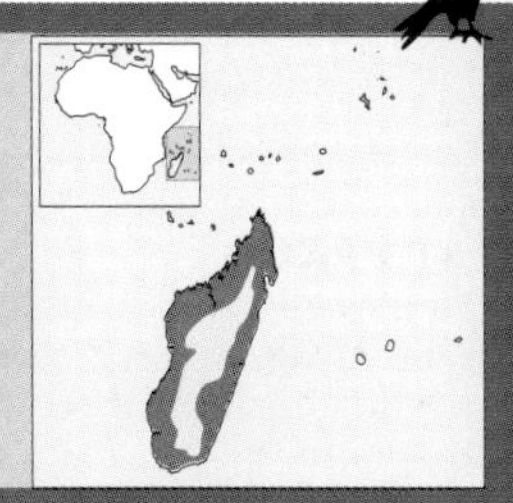

O. Langrand

Common Kestrel

Falco tinnunculus

AT PERCH

Short legs; pointed wings do not quite reach tail tip. Grey or brown head and neck and black-spotted rufous back most obvious, with paler, streaked underparts. Pale grey rump and single-banded (male) or multiple-banded (female) tail not always visible.

IN FLIGHT

Rather short, broad wings with rounded tips and long tail. Swerving flight with fast, choppy wingbeats, sometimes gliding; often soars, hangs or hovers. Rufous wings above contrast with dark primaries. Tail grey or rufous and generally barred (female) or plain (male), with broad, black end-band and white tip in both sexes. Underwing white with pale grey spots and bars.

DISTINCTIVE BEHAVIOUR

Often seen in flight, changing perches or soaring and hovering in light winds. Usually perches inconspicuously among branches, rocks or utility lines. Prey consists mostly of small vertebrates and insects, taken on the ground or less often on the wing. Nests on rock ledges or in old stick-nests of other birds in trees.

P. Ginn

Juvenile and adult *F.t. rupicolis*

ADULT Male head and nape grey with fine, black streaks. Wing coverts and back deep rufous with small, black spots; rump pale grey. Throat plain cream, with dark malar stripes. Below, including paler underwing coverts, buff to white spotted with black, heaviest on flanks but vent plain. Primaries sooty-brown above, white below, with dark grey bars on inner edges, secondaries rufous above with narrow, dark bars and broad tips. Tail pale grey with broad, black subterminal band and white tip, unbarred or (especially in southern Africa) with narrow, black bars on outer feathers. **Female** browner, more heavily spotted or even barred above. Head brown streaked with black, often grey in southern Africa. Below more heavily streaked, with narrowly barred rufous or grey tail. Resident sub-Saharan subspecies (*F.t. rupicolis and F.t. archeri*) generally darker and more rufous, female and male more similar, usually with grey heads and grey or rufous barred tails. Bill black with grey base; eye dark brown; cere, eye-ring and slender, bare legs and feet yellow. Female about 4% larger (male 136–252 g, female 154–314 g).

JUVENILE Both sexes resemble adult female, but more heavily and evenly barred above and streaked below, with buff tips to dark flight feathers. Rump and tail washed with rufous, with narrow, dark brown bars and broader, dark sub-terminal band. Cere greenish-yellow at fledging.

DISTRIBUTION, HABITAT AND STATUS Widespread across Africa except in dense forest and flat, treeless desert, extending to Cape Verde, Canary, Madeira and Socotra islands. Breeding resident throughout range except for some local movement. Also regular migration into western and eastern Africa of some European populations during the southern winter. Favours open habitats with short groundcover for hunting, and rockfaces or tree clumps for roosting and nesting sites. Residents only locally common in Africa despite wide range.

SIMILAR SPECIES Visited throughout range by slimmer, smaller, long-winged but otherwise very similar migrant **Lesser Kestrel** (p. 238) (broad, black subterminal tail band even more obvious; adult female and juvenile have heavily streaked buff underparts, rufous barred tail, dark 'moustache'; adult male has plain, bright rufous back; pale grey head, greater wing coverts and tail; lightly spotted buff breast; weak feet with white claws). Overlaps in western and northeastern Africa with larger **Fox Kestrel** (p. 236) (dark rufous, streaked with black all over; long, rounded, graduated, rufous tail with faint bars and narrow, black end-band; rarely hovers) and in eastern and southern Africa with **Greater Kestrel** (p. 234) (pale rufous all over; heavily barred above and on flanks with black; evenly barred grey (adult) or rufous (juvenile) tail with broad, white tip; cream eye in adult).

 ALTERNATIVE NAMES: European Kestrel, Eurasian Kestrel, Rock Kestrel, African Kestrel

Adult male has grey head; rufous back spotted and buff underparts streaked with black; tail plain grey or lightly barred, with broad, black end. Adult female and juvenile browner, more barred, usually with brown head; rufous or grey tail narrowly barred with black and with broad, black end and white tip. Small (about 25 cm tall, 65–82 cm wingspan).

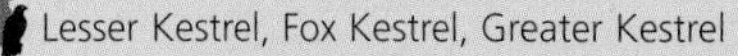
Lesser Kestrel, Fox Kestrel, Greater Kestrel

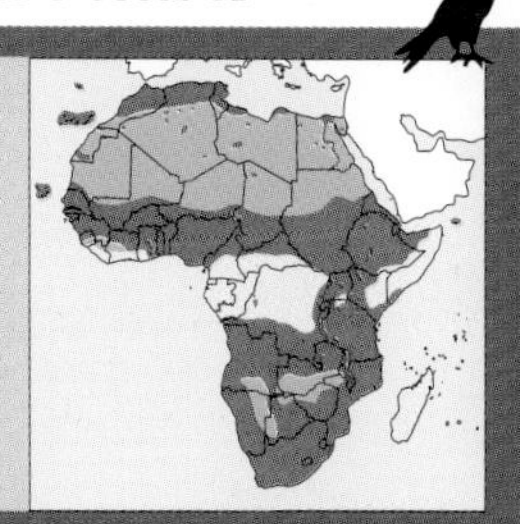

R. du Toit

Greater Kestrel

Falco rupicoloides

AT PERCH

Perches upright on long legs. Long, black flight feathers extend to tip of long, white-tipped tail. Large, neatly rounded head. Lacks any 'moustache' stripe. Pale eye of adult not always obvious in shade. Streaked flanks and rufous tail bars of juvenile only visible at close range.

IN FLIGHT

Long, pointed wings and long, rounded tail obvious. Buoyant in flight with steady wingbeats, also soars and frequently hovers. Above pale rufous and densely barred. Note grey-barred, white-tipped tail and black primary tips. Below, note silvery-white underwings, pale bars evident only when backlit. Body darker rufous, streaked and barred.

DISTINCTIVE BEHAVIOUR

Often seen perched on a prominent mound, rock, tree or utility pole. Flies frequently, mainly in light breezes, flapping and gliding between perches or soaring and hovering in search of prey. Performs fast flicker-diving territorial display with shrill, screaming calls, 'keeer keeer keeer'. Main diet is small birds, mammals and insects, taken mainly on the ground, less often after an aerial chase, rarely on the wing. Nests in old stick-nests of other birds, in low trees or on utility poles and pylons.

ADULT Overall pale rufous. Head, neck and underparts finely streaked with black. Wing coverts, back and flanks have broad, black bars. Rump pale grey with darker bars. Underwing coverts silvery white, undertail coverts plain pale rufous. Flight feathers sooty-brown with narrow paler bars, rufous above and buff below. Tail pale grey with darker bars and broad, white tip. Bill black with blue-grey base; eye pale cream; cere, eye-ring, long, bare legs and stocky toes yellow. Sexes identical in plumage, female about 2% larger (range 178–334 g).

JUVENILE Similar to adult, but usually more rufous; flanks and breast more heavily streaked with black; rump and tail rufous with dark grey bars, eyes dark brown. Facial skin blue-green at fledging, yellow by about 4 months, when barred flanks and grey bars on rump appear. Rufous tail only moults into adult colours after first year.

P. Ginn

Juvenile

DISTRIBUTION, HABITAT AND STATUS Main range in southern Africa, also in separate areas of eastern Africa and of northern Ethiopia and Somalia. Favours open grassland and semi-arid steppe with a few scattered trees. Common in southern Africa, more local and uncommon in northeastern Africa.

A. Kemp

Adult hovering

SIMILAR SPECIES Most often coexists with African races of smaller **Common Kestrel** (p. 232) (grey head; dark 'moustache'; black-spotted, deep rufous back; well-streaked buff underparts; tail with narrow bars and broad, dark subterminal band). Visited by smaller migrant female and juvenile **Lesser Kestrels** (p. 238) (slim build; broad, dark subterminal band to tail; short legs and weak feet with white claws; heavily streaked buff underparts; rufous, barred tail; dark 'moustache').

Large for a kestrel. Pale rufous. Barred black on wings, back and flanks. Finely streaked black on head and breast. Note pale cream eye of adult. In flight shows silvery-white underwings and grey barred tail and rump. Juvenile has streaked flanks, rufous barred rump and tail, and dark brown eyes. Small (about 25 cm tall, 84 cm wingspan).

Common Kestrel, Lesser Kestrel (female)

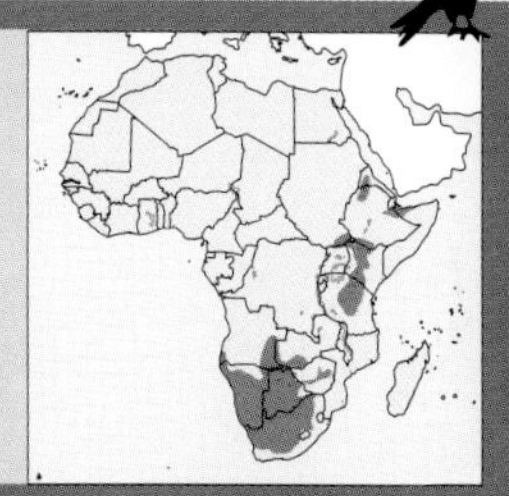

Fox Kestrel

Falco alopex

AT PERCH

Long, slender kestrel, rather upright on long legs. Long, graduated tail even longer than long, pointed wing tips. Appears rufous overall, finely streaked with black. Note dark 'moustache' patch, pale brown eye, indistinctly barred rufous tail and dark wing tips. Juvenile best distinguished by blue-green hue to facial skin and greenish-yellow feet.

IN FLIGHT

Note long, broad, pointed wings and especially very long, rounded, graduated tail. Flight buoyant, with slow flapping then gliding, or soaring easily. Rarely hovers. Appears rufous overall, including the tail; wings rufous above and white or pale rufous below, with black tips and black bars on secondaries. Juvenile more clearly barred with rufous on wings and black on tail.

DISTINCTIVE BEHAVIOUR

Spends long periods perched on low rocks or trees, in search of the small ground vertebrates and insects that are its main prey. Also attracted to bush fires, where it catches insects on the wing. Frequently soars and flies around rocky hills. Nests on rock ledges and in potholes, sometimes in a loose colony. Often noisy in nest area, with shrill high screams, 'kak kak kak'. Species little studied, so few detailed notes available on behaviour, plumage or softparts, or on age differences.

Seitre/BIOS

Juvenile

ADULT Overall rufous to deep chestnut, streaked with black. Streaks broad on nape, wings and back. Dark brown malar patch. Unstreaked throat, flanks, underwing and tail coverts. Greater underwing coverts and flight feathers white or pale rufous, primaries have dark brown ends, secondaries have black bars on inner edges. Tail rufous with faint, narrow, black bars and no broad, darker subterminal bar. Bill black with grey base; eye pale brown; cere, eye-ring and long, bare legs with stocky toes yellow. Sexes indistinguishable, female about 5% larger (250–300 g).

JUVENILE Very similar to adult, but tail more completely barred, and greater wing coverts and secondaries more clearly barred with rufous. Facial skin blue-grey and legs greenish-yellow.

DISTRIBUTION, HABITAT AND STATUS Narrow sub-Saharan band in the Sahel and northeastern Africa. Favours semi-desert and arid savannas, but also moves south into moister Guinea savannas and is a vagrant to the grasslands of eastern Africa. Found regularly only around isolated rocky hills. Generally sparsely distributed and uncommon, locally common only in Chad and northern Kenya.

SIMILAR SPECIES Overlaps with smaller **Common Kestrel** (p. 232) (grey head; dark eye; dark 'moustache'; black-spotted, deep rufous back; well-streaked buff underparts; grey tail with narrow bars (female) or plain (male) and in both sexes with broad, dark, subterminal band) and visited by smaller migrant female **Lesser Kestrel** (p. 238) (broad, dark subterminal band to tail; dark eye; short legs and weak feet with white claws; heavily streaked buff underparts; rufous, barred tail; dark 'moustache'). Most similar in size and build to **Grasshopper Buzzard** (p. 122) (rufous patch in wing; dark grey (adult) or pale rufous (juvenile) head; black line down throat; pale rufous below with fine, dark streaks; stocky legs and feet; pale yellow eye).

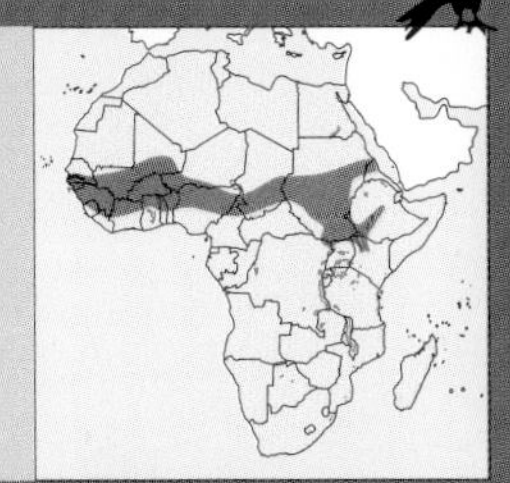

Large for a kestrel, slender, rufous, finely streaked with black. Note long, graduated tail. Eye pale brown. Juvenile more heavily streaked below, with narrow, black bars across tail. Small (about 30 cm tall, 90 cm wingspan).

Common Kestrel, Lesser Kestrel (female), Grasshopper Buzzard

Seitre/BIOS

Lesser Kestrel

Falco naumanni

AT PERCH

Slim build, with long, pointed wings almost reaching tip of long tail and extending well behind the feet. Stands close to perch on short legs. Male dapper and colourful, appears blue-grey and chestnut. Female and juvenile duller brown, with heavy barring above and streaking below. Black primaries and broad, black end to tail distinctive in both sexes.

IN FLIGHT

Buoyant on long, narrow wings and long, rounded tail. Frequently hovers with tail spread. Flight feathers above all-black in male, contrasting with pale grey and rufous back and coverts. Female with barred rufous body and wings, ending in black primaries. Male tail plain grey, with fine, dark bars when spread, female tail rufous with dark, even bars; both have grey rump and broad, black subterminal band to white-tipped tail. Juvenile like female but with rufous rump.

DISTINCTIVE BEHAVIOUR

Usually occurs in flocks and feeds, roosts and nests communally. Hunts from low perches or, often, from soaring and hovering. Main prey insects, with some small vertebrates when breeding. Drops lightly to the ground, or hawks flying prey, often rising to eat on the wing. During migration often feeds and roosts in stands of large trees with Red-footed and/or Amur falcons. Nests in cavities in rockfaces, buildings and ruins.

ADULT Male has pale grey head, rump and greater upperwing coverts. Back and lesser upperwing coverts plain bright rufous. Below cream or buff with small, black spots on breast and flanks. Underwing the same, or white and unmarked. **Female** pale rufous densely barred black above, with grey rump. Head and neck finely streaked with black; dark malar stripes. Below buff (plain or more usually streaked with black). Flight feathers all-black (male) or primaries black and secondaries rufous with black bars (female). Tail plain pale grey (male) or rufous with narrow, black bars (female); in both sexes tail has broad, black subterminal band and narrow, white tip. Bill black with pale grey base; eye dark brown; cere, eye-ring and short, thin, bare legs and feet deep yellow; claws white. Sexes similar in size (male 90–172 g, female 138–208 g).

I. Davidson

Adult male

JUVENILE Both sexes similar to adult female, but generally paler and duller, and with even less distinct malar stripe and rufous wash on rump. Sometimes more washed with grey (male) or rufous (female).

DISTRIBUTION, HABITAT AND STATUS Breeds, and part of population resides, in northwestern Africa. Most of European population migrates across a broad front to sub-Saharan Africa during the southern winter. Main population moves to grasslands of eastern Africa and highveld and Karoo steppe of South Africa. Locally very common at main roosts, where thousands congregate. Small flocks elsewhere and nomadic over a wide range of savanna and woodland habitats. Vagrant to Seychelles.

SIMILAR SPECIES Overlaps widely with dumpier **Common Kestrel** (p. 232) (grey head streaked with black; deep rufous back spotted with black; below buff streaked with black; tail plain grey (some males) or rufous with narrow bars (female), with broad, dark subterminal band). In western and northeastern Africa overlaps with larger **Fox Kestrel** (p. 236) (dark rufous, streaked with black all over; long, rounded, graduated rufous tail with faint bars and narrow, black end-band; rarely hovers) and in eastern and southern Africa with **Greater Kestrel** (p. 234) (pale rufous all over; heavily barred above and on flanks with black; evenly barred grey (adult) or rufous (juvenile) tail with broad, white tip; pale cream eye in adult).

Lesser Kestrel

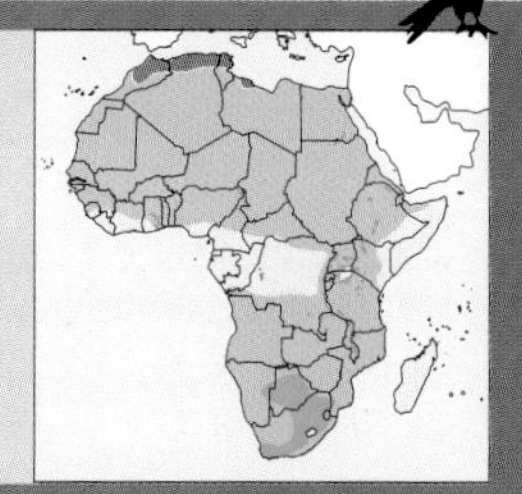

Small, slender kestrel, sexes different. Adult male pale grey above with plain rufous back and wing coverts; below buff with small, black spots. Adult female and juvenile pale rufous above with black bars and streaky head; below buff with broad, black streaks. In flight, primaries black, tail plain grey (male) or rufous with black bars (female, juvenile) and with a broad, black end. Small (about 25 cm tall, 58–72 cm wingspan).

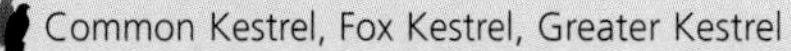

Common Kestrel, Fox Kestrel, Greater Kestrel

Madagascar Kestrel

Falco newtoni

AT PERCH

Small, dumpy kestrel, stands rather upright with pointed wing tips well short of tail tip. Deep rufous back, white or rufous underparts and grey, barred tail with broad, dark end most obvious features. Note grey (male), tan (female) or dark (dark form) head; pale form has slight 'moustache'.

IN FLIGHT

Barred wings rather short, broad and pointed; tail rather long and broad. Flies parrot-like with fast, shallow wingbeats, also glides, soars and hovers. Dark flight feathers contrast with rufous of body and wing coverts above, and with white or rufous below. Note pale base to primaries and grey, barred tail with broad, dark end.

DISTINCTIVE BEHAVIOUR

Hunts from a low perch, hovering or hawking on the wing. Pumps tail up and down when excited. Main prey insects, with more small vertebrates when breeding, taken on the ground or in flight. Shrill, fast 'kiky kiky kiky' call. Often hunts at dawn and dusk, with Eleonora's and Sooty falcons. Nests in holes in trees, rockfaces or buildings, on epiphytes or in old stick-nests of other birds.

ADULT Above rufous, head streaked, back and upperwing coverts spotted, and greater coverts and scapulars barred with black. Head grey (male), pale rufous (female) or dark grey (rufous male). Cheeks and throat plain buff. Below off-white (pale form, 80% of population) or rufous (dark form, 20%), with fine, black streaks becoming bars on flanks. Thighs and vent plain or spotted. Underwing coverts white or rufous with black spots in both colour forms. Aldabran birds smaller, less spotted, some almost plain below. Flight feathers sooty-brown above, primaries have white bases and faint bars below. Tail sooty-brown barred with grey (light form) or rufous (dark form) above, white and grey below, with broad, dark subterminal band and buff tip. Bill black with grey base; eye dark brown; cere, eye-ring and short, bare legs and toes yellow. **Female** often browner and more heavily marked below; about 5% larger (male 90–117 g, female 131–153 g).
JUVENILE Very similar to adult, but with fine, buff edges to flight feathers and usually more heavily streaked below.

I. Sinclair

Adult pale form

DISTRIBUTION, HABITAT AND STATUS Widespread and common on Madagascar, uncommon and local on Aldabra, vagrant to Anjouan (Ndzuani), Comoros. Thrives in open secondary forest and cultivation, where palm trees and buildings have increased nest and roost sites. Uncommon in forested areas of Madagascar, where replaced somewhat by Banded Kestrel.
SIMILAR SPECIES Only kestrel on Aldabra and the Comoros; only rufous kestrel on Madagascar. Dark form might, in poor light, be mistaken for larger, heavier **Banded Kestrel** (p. 230) (dark grey above; broadly barred grey and white below; yellow eye (adult); no tail-pumping). Rufous form resembles **Broad-billed Roller** (*Eurystomus glaucurus*) in colour (but yellow bill).

R. Bergerot/BIOS

Adult pale form feeding

ALTERNATIVE NAMES: Newton's Kestrel, Aldabra Kestrel, Malagasy Kestrel

Small, delicate, rufous kestrel with 2 forms. Pale form rufous above, white below with black spots; head blue-grey (male) or pale rufous (female) with dark streaks and buff cheeks. Dark form all-rufous with black spots and streaks. Head grey. Flight feathers and tail black (male) or brown (female) with grey or rufous bars. Small (about 20 cm tall, 55 cm wingspan).

Only rufous kestrel on Madagascar and Aldabra

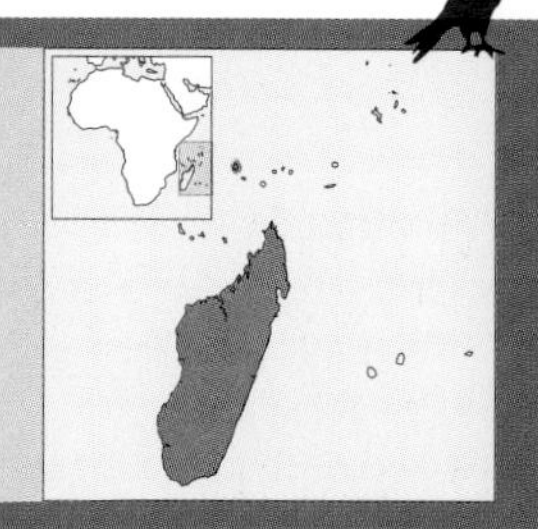

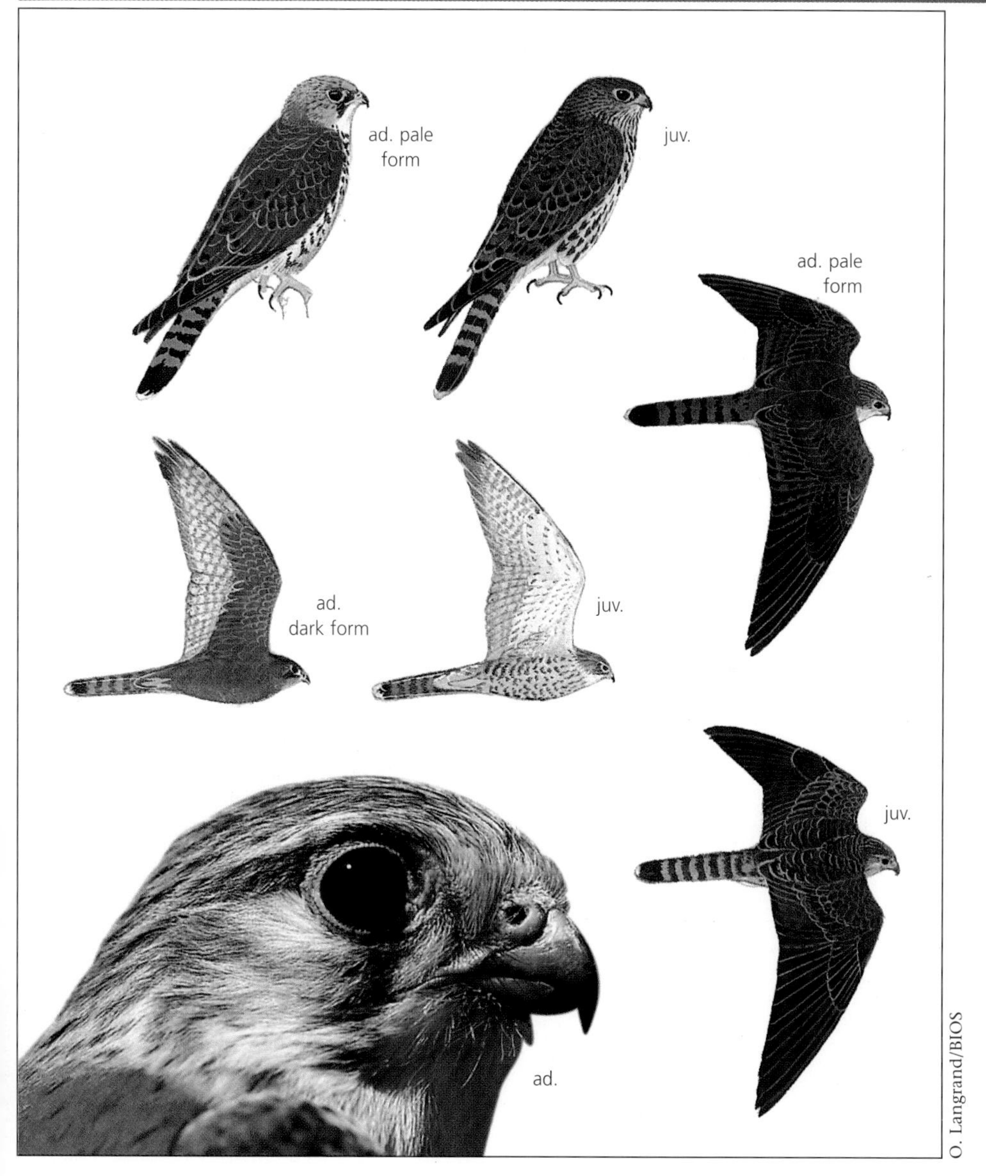

O. Langrand/BIOS

Mauritius Kestrel

Falco punctatus

AT PERCH

Usually stands rather upright on long legs, with short wing tips well above end of long tail. Note rufous upperparts barred black, and white or buff underparts spotted with black.

IN FLIGHT

Wings relatively short, broad and round-tipped for a kestrel, tail long and rounded. Light, shallow wingbeats, glides and soars easily, sometimes hovers. Dark wings and tail contrast with rufous above and white or buff below with black spots. Note rufous bars on secondaries and especially on tail, the latter with a broad, dark end-band.

DISTINCTIVE BEHAVIOUR

Often perches on small branches within canopy of forest trees. Hunts by short, fast dashes onto branches, into foliage or along the ground. Less often glides and soars low over canopy, hovers, or makes short aerial chases. Main prey arboreal lizards, small birds and insects. Nests in tree or rock cavities, more recently in nest boxes.

ADULT Above rufous, head and neck streaked, back and wing coverts barred with black. Tail narrowly barred with black except for broad, black subterminal band and white tip. Often buff forehead, eyebrow and cheeks; black malar stripe below eye. Below white, throat and thighs plain, flanks barred and rest spotted with black. Flight feathers sooty-brown above and white below, barred with rufous above and with grey below. Bill black with yellow base to lower mandible; eye dark brown; cere, eye-ring and long, slender, bare legs and toes yellow. Sexes very similar in plumage and size (male 178 g, female 231 g).

JUVENILE Very similar to adult, but flight feathers finely edged with buff and underparts more streaked than spotted and base colour on underparts often more buffy. Facial skin blue-grey at fledging and for several months thereafter.

DISTRIBUTION, HABITAT AND STATUS Found only on Mauritius. Originally a kestrel of low- and highland forest, latterly restricted to a few highland patches, more recently introduced to secondary forest and lightly wooded lower slopes. Now locally common through excellent conservation management, but was among the rarest birds in the world in 1974.

SIMILAR SPECIES None in range.

Small, plump, short-winged kestrel. Rufous above, barred with black. White or buff below, spotted with black. Long, rufous tail, narrowly barred with black and with broad, dark end-band. Very small (about 15 cm tall, 50 cm wingspan).

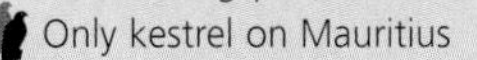

Only kestrel on Mauritius

D. Steele/Photo Access

Seychelles Kestrel

Falco araea

AT PERCH

Upright with rather large head. Stocky body exaggerated by short, pointed wings, the tips well above the end of the tail. Note plain, pale underparts against dark rufous back and especially dark grey head. Tail barred grey, with broad, dark subterminal bar and white tip.

IN FLIGHT

Fast wingbeats, rather short, pointed wings and short, rounded tail. Not known to hover. Dark primaries above with pale bases and barred secondaries below to contrast with pale, spotted underwing and body. Note broad, dark subterminal bar and white tip to tail.

DISTINCTIVE BEHAVIOUR

Small size makes detection difficult as it perches on rockfaces and buildings or among palm fronds and tree branches. Diet mainly lizards, geckos and insects, and sometimes small birds, taken from branches, foliage or (less often) the ground or on the wing. Nests mainly in rock cavities, also at base of palm fronds, on buildings or in tree holes in developed areas.

ADULT Above deep rufous with dark brown spots forming bars on scapulars and secondaries. Head and cheeks dark grey with fine, black shaft streaks. Back rufous, rump grey, secondaries with buff edges. Below pale rufous, palest on throat between dark brown malar stripes, darkest on flanks and thighs, often with small, black spots; vent plain. Underwing cream, either plain or with dark brown spots. Primaries sooty-brown with paler, slightly barred bases above, below white with pale grey brown bars. Tail grey above and paler below, barred dark brown or black with a broad, dark subterminal band and white tip. Bill black with yellow base; eye dark brown; cere, eye-ring and small, thin, bare legs and toes yellow. **Female** often paler; sexes similar in size (male 73 g, female 87 g).
JUVENILE Similar to adult, but often darker and with rufous head. Above more barred, below streaked with brown, especially on lower breast. Rump rufous with grey feather bases. Tail dark brown with paler bars.
DISTRIBUTION, HABITAT AND STATUS Seychelles, on main island of Mahé. Recently (1981) re-introduced to Praslin. Favours forest among granite outcrops, but also widespread in inhabited areas of the lowlands with scattered palms and buildings.
SIMILAR SPECIES None in range.

Very small, stocky kestrel. Rufous above, spotted with black. Head grey. Cheeks and breast plain pale rufous. Tail and rump grey. Tail narrowly barred black with broad, black subterminal bar and white tip. Juvenile has rufous head and is streaked below. Very small (about 15 cm tall, 45 cm wingspan).

Only kestrel on Seychelles

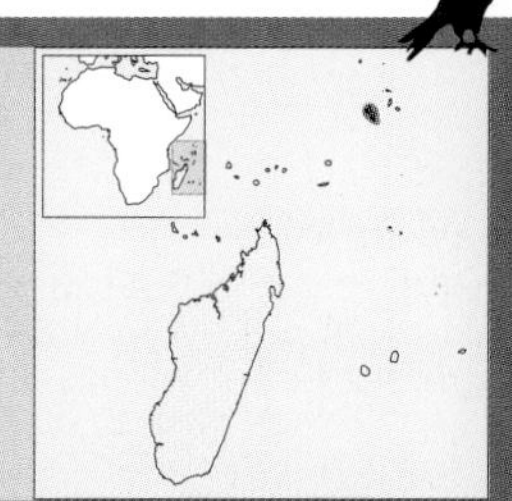

African Pygmy Falcon

Polihierax semitorquatus

AT PERCH

Small size and dumpy body accentuated by upright stance. Wings short and pointed, with tips crossing just below rump. Tail short and spotted. Appears pale and shrike-like at a distance. Note orange facial skin and legs. Black wings and tail and dark back of female obvious from behind. Juvenile not easy to distinguish except in good light.

IN FLIGHT

Bursts of fast, shallow wingbeats, with dip and glide between, produce undulating flight path like that of a shrike or woodpecker. When changing perches, drops close to the ground and then swings up to the new site. Note black wings and tail with white spots, against pale body and wing coverts, and white rump. Dark back of female conspicuous.

DISTINCTIVE BEHAVIOUR

In southern Africa, lives in and around large, haystack-like nest structures of Sociable Weaver (*Philetairus socius*); rarely nearby White-browed Sparrow-weaver (*Plocepasser mahali*). In eastern Africa uses spiky nest clumps of White-headed Buffalo Weaver (*Dinemellia dinemelli*). Most often seen perched on open twigs and branches, from where it takes its main prey of insects and small lizards from the ground below. Usually alone, in pairs or in small family groups. Often noisy, with high-pitched calls and song 'tee tee twip', accompanied by head-bobbing and tail-pumping. Nests in a chamber of its weaver host, the rim of the entrance becoming coated in pinkish-white droppings.

ADULT Male pale grey above, except for white collar and rump. Eyebrow, throat and underparts white. Flight feathers and tail black, with white spots and tips. **Female** similar, but mantle is deep chestnut. Eastern African birds slightly darker and more richly coloured. Bill black with pale grey base; eye dark brown; cere, broad eye-ring and stout, bare legs and toes deep orange. Sexes similar in size (male 59–64 g, female 54–67 g).

JUVENILE Similar to adult of same sex, but washed with rufous on breast and grey areas of upperparts, and with buff feather tips. Facial skin and legs slightly paler orange.

DISTRIBUTION, HABITAT AND STATUS Occurs in 2 discrete populations. One is associated with the Sociable Weaver in southwestern Africa, the other with the White-headed Buffalo Weaver in eastern Africa. Favours semi-arid areas with sparse groundcover and scattered large trees (or utility poles in southern Africa).

SIMILAR SPECIES No raptors as small, or similar in colour, within its range. Most similar is much larger, long-winged **Black-shouldered Kite** (p. 194) (pale grey above and white below; black shoulder patches; red eye; yellow cere and short, stocky legs; plain, dark grey wings; short, all-white tail; pumps tail).

A. Weaving

Adult male

Adult

ALTERNATIVE NAME: Pygmy Falcon

Small, shrike-like falcon, sexes different. Male pale grey with white underparts. Female has chestnut back. In both sexes, facial skin and legs deep orange. White rump, collar and spots on black flight feathers and tail. Juvenile similar to adult of same sex, but with rufous wash. Very small (about 15 cm tall, 37 cm wingspan). Lives in bulky nests of sociable and buffalo weavers.

Much smaller than all other species.

N. Dennis

Milky Eagle-owl

Bubo lacteus

AT PERCH

Stands upright, often with the head deeply withdrawn and just the stubby, dark ear-tufts evident. Appears pale grey overall with dark flight feathers. Dark eyes and facial rim obvious in contrast. Pink eyelids distinctive at close range.

IN FLIGHT

Mainly nocturnal and silent, but does emerge by day and especially at dusk. Flies with deep, strong beats of the long, broad wings, the dark flight feathers contrasting with the paler body and coverts.

DISTINCTIVE BEHAVIOUR

Roosts amongst dense foliage in large trees, often a pair in close proximity. Emerges at dusk, less often by day, to hunt main prey of medium-sized birds and mammals. Often perches in the open and attracted to any animal disturbance. Calls with deep, gruff grunts 'hok hok-hok hok', in irregular sequence and often a pair in uneven duet. White throat obvious when calling. Chick begs with prolonged, querulous whine. Nests mainly on old stick-nests of other birds, when it is often obvious while incubating; less frequently in tree hollows.

P. Funston

Adult sleeping

Adult

ADULT Overall pale grey-brown with fine, white bars. Scapulars have white outer webs. Broad, dusky-brown rim to facial disc and front of 'ears'. Black bristles at base of bill. Flight feathers and tail a dark, dusky-brown with broad, paler grey bars. Bill cream; cere blue-grey; eyes dark brown; bare eyelids deep pink; heavy, powerful feet pale brown. Sexes similar in plumage, female about 13% larger (male 1 615–1 960 g, female 2 475–3 115 g).

JUVENILE Juvenile body and head down grey-brown with broad, dark bars, ear-tufts barely evident at fledging. Moults into plumage similar to that of adult.

DISTRIBUTION, HABITAT AND STATUS Widespread in sub-Saharan Africa, from arid semi-desert to moist, tall woodland. Only absent from treeless desert and continuous forest. Less common and more local in western than in eastern and southern Africa. Common in open woodland.

SIMILAR SPECIES Overlaps most widely with 'eared' but smaller, more lightly built **Spotted Eagle-owl** (p. 256) (darker grey; more mottled with dark grey and white; dark blotches on chest; black bill; white, feathered eyelids; eye yellow in central and southern Africa; dark brown in eastern and western Africa). In mountainous areas also overlaps with 'eared' and more similar-sized **Cape (Mackinder's) Eagle-owl** (p. 250) (orange eyes; dark horn-coloured bill; rufous-brown with dark brown blotches, especially on chest).

ALTERNATIVE NAMES: Giant Eagle-owl, Verreaux's Eagle-owl

Very large, heavily built 'eared' owl. Pale grey with fine, white bars. Dark flight feathers, rim to facial disc and stubby ear-tufts. Large, dark brown eyes. Bare, pink eyelids. Pale creamy bill. Call a series of deep grunts 'hok hok-hok hok' or a long, shrill whine. Large (about 50 cm tall, 143 cm wingspan).

Spotted Eagle-owl, Cape Eagle-owl

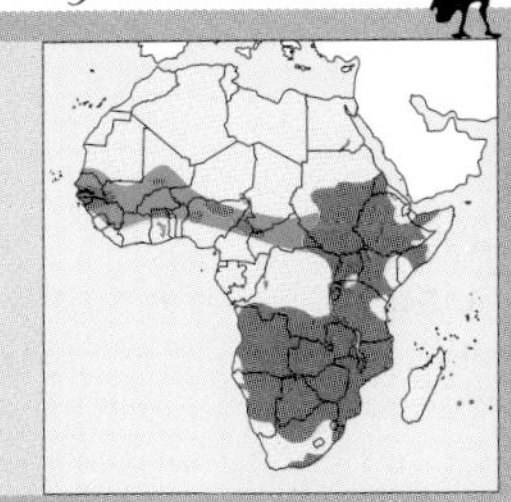

Cape Eagle-owl

Bubo capensis

AT PERCH

Often seen crouched on a rock ledge, less often standing upright on a perch. Note heavy build, long 'ears' and dark chest-patches. Overall appears dark and blotched with rufous and white. Orange eyes and heavy feet obvious at close range.

IN FLIGHT

Nocturnal and silent, only seen in flight by day if disturbed. White underwing coverts contrast with dark, barred flight feathers. Flies with long, broad rounded wings and strong, deep wingbeats.

DISTINCTIVE BEHAVIOUR

Emerges just before nightfall to open perches, calling after dark with a series of deep, hooting notes repeated at intervals, 'hoo hoo ho' or 'hoo ho'. White throat pulses when calling, often heard again around midnight and before dawn. Feeds mainly on medium-sized mammals, such as hyrax, hares and molerats, leaving ossuaries of their bones at feeding sites. Roosts in wooded valley or rocky gorge, moving far over surrounding grass- and woodlands to hunt. Nests on ledges or in potholes, often in wooded valley near water. Utters harsh quack of alarm 'wak wak'.

Adult Head and mantle dark brown with pale rufous spots. Back and rump dark brown with pale rufous and buff bars. Upperwing coverts dusky brown, greater coverts with white tips and scapulars with broad, white ends. Facial disc pale buff with dark edges. Dark edges to long ear tufts and crown. Throat white. Below cream with dark, sooty blotches on sides of chest and broad, buff and sooty bars on breast. Flight feathers and tail dark brown with irregular buff bars, paler below with white underwing coverts. Bill dark horn colour; eyes orange; heavy feet dark brown. Sexes similar in plumage, female about 5% larger. South African subspecies *B.c. capensis* smallest (905–1 360 g); darker, more blotched central African *B.c. mackinderi* about 10% larger (1 220–1 800 g) and similar in size to even darker *B.c. dilloni* of northeastern Africa.

Juvenile Similar to adult, but 'ears' shorter; grey juvenile down finely barred with rufous still obvious on body for several months after fledging.

Distribution, habitat and status Occurs in 3 populations, in mountains of South Africa and southern Namibia (*B.c. capensis*), from Zimbabwe northwards to southern Kenya (*B.c. mackinderi*) and on the Ethiopian highlands (*B.c. dilloni*). Local and generally uncommon to rare, but also easily overlooked. Favours low cliffs or wooded valleys for roosting and nesting, adjacent to open grassland, rocky slopes or granite domes over which to hunt.

Similar species Overlaps over most of range with similar-sized (South Africa) or smaller (central and northeastern Africa) **Spotted Eagle-owl** (p. 256) (greyer; less blotched and more barred; weak feet; yellow (southern) or brown (northern) eyes, except for uncommon rufous form with more orange eyes and more rufous coloration). Overlaps in northern Ethiopian highlands with similar-sized **Pharaoh Eagle-owl** (p. 252) (light sandy-brown; dark streaks and pale blotches above; finely barred below; small feet; prefers more arid lowlands).

B.c. capensis adult

Alternative name: Mackinder's Eagle-owl (*B.c. mackinderi*)

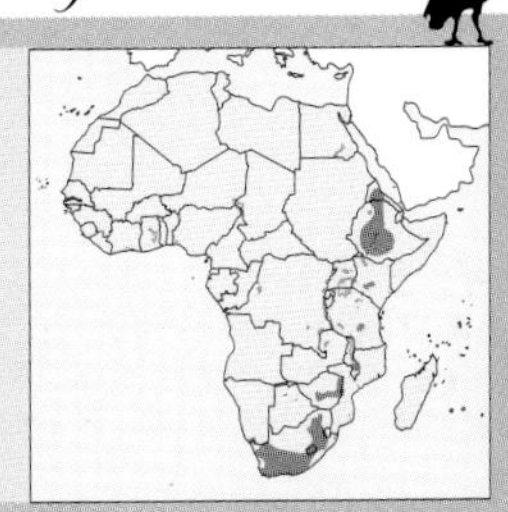

Large, dark brown and rufous 'eared' owl. Dark blotches on sides of chest. Orange eyes. Heavy feet. Calls with mellow, hooting notes 'hoo hoo ho' or 'hoo ho', or a loud 'wak-wak' in alarm. Large (about 45 cm tall, 125 cm wingspan).

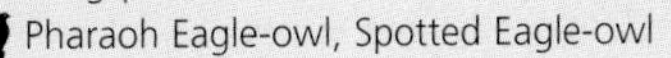

Pharaoh Eagle-owl, Spotted Eagle-owl

B.c. capensis ad.

B.c. capensis ad.

B.c. capensis juv.

B.c. capensis ad.

B.c. mackinderi ad.

B.c. capensis ad.

N. Dennis

Pharaoh Eagle-owl

Bubo ascalaphus

AT PERCH

Crouches horizontally on the ground or stands upright on a tree or rock perch. Appears light brown and mottled above, with long, dark 'ears' and a pale face. Blotched chest, lightly barred breast, orange eyes and dark bill evident at close range.

IN FLIGHT

Nocturnal and silent, seen in flight by day only if disturbed, when it flies far and strongly with deep, slow beats of broad, bowed wings.

DISTINCTIVE BEHAVIOUR

Emerges at dusk to perch in the open and after nightfall to call with deep, hooting phrases – 'ho hoo' – repeated at regular intervals. Puffs out white throat as it calls, often also vocal in the middle of the night and before dawn. Roosts among rocks or trees, often on the ground. Utters a loud quack of alarm on sighting intruders. Main diet is small animals such as gerbils and hares. Nests in a scrape on the ground or on a rockface, sometimes down wells or in tree hollows. Large chicks beg with a loud, swishing hiss.

ADULT Sandy-rufous above with pale rufous spots and dark rufous streaks (*B.a. desertorum*) or even paler, with white areas (*B.a. ascalaphus*). Face plain pale sandy colour with dark rim. Long ear-tufts edged with dark brown. Throat white. Below pale tawny to white, the chest blotched and breast finely barred with dark brown and white. Flight feathers and tail tawny-rufous with irregular sooty-brown bars. Bill black; eyes orange; skin of relatively small feet dark brown. Sexes similar in plumage, female about 6% larger (male 1 900 g, female 2 300 g).

JUVENILE Similar to adult, but 'ears' shorter, and barred and blotched juvenile down remains for several months after leaving the nest.

DISTRIBUTION, HABITAT AND STATUS Throughout the Sahara desert and adjacent semi-arid mountains of northwestern Africa, wherever there are rocky outcrops and sufficient prey. Extends southwards just into the dry Sahel savanna across western Africa.

SIMILAR SPECIES Overlaps in northwestern Africa and along the Nile valley with rare and larger **Eurasian Eagle-owl** (p. 254) (darker; browner; more heavily streaked; heavy, powerful feet). Overlaps in northern Ethiopian highlands with similar-sized **Cape Eagle-owl ssp. *dilloni*** (p. 250) (dark brown and rufous; heavy feet; dark blotching on sides of chest; usually in grassland habitat), but prefers more arid lowlands.

Adult taking off

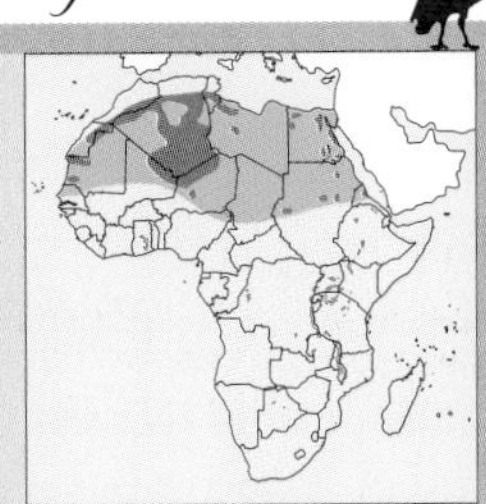

Large, sandy-brown 'eared' owl, darkly streaked above, and dark blotches on the chest. Orange eyes; dark bill. Call a deep, hooting 'hoo hoo' at regular intervals or utters a loud quack if alarmed. Large (about 50 cm tall, 135 cm wingspan).

Eurasian Eagle-owl, Cape Eagle-owl

B.a. ascalaphus ad.

B.a. desertorum ad.

B.a. desertorum ad.

B.a. ascalaphus ad.

M. Gunther/BIOS

Eurasian Eagle-owl

Bubo bubo

AT PERCH

Often stands very upright. Long, dark ears obvious even in silhouette. Appears dark brown and rufous with pale blotches above and fine bars and streaks below. Note heavy build and, at close range, orange eyes and blotched chest.

IN FLIGHT

Usually nocturnal and silent unless disturbed. Flies on long, broad, bowed wings, the pale rufous underwing coverts, speckled with brown, obvious against the darkly barred flight feathers. Often attracts smaller birds if flies by day.

DISTINCTIVE BEHAVIOUR

Roosts on the ground among rocks or on large limbs among forest trees. Often calls on emerging at nightfall, usually a series of 2 or 3 deep, hooting notes 'u huu huu' every 8–10 seconds. Sometimes calls again around midnight and pre-dawn, the white throat becoming inflated and obvious when calling. Main diet includes hares, rodents and gamebirds. Barks out a high, sharp 'quack' call when sighting danger or attacking an intruder. Lays its eggs in a shallow scrape on a rockface or in a tree hollow.

M. Danegger/NHPA

Adult

ADULT Above deep rufous with broad, dark brown streaks and paler rufous spots on sides of feathers creating blotched and mottled effect. Facial disc plain rufous with slightly darker brown rim. Throat white. Below tawny, with narrow, dark brown bars. Chest blotched with sooty-brown; base of feathers white. Flight feathers and tail deep buff or deep rufous with irregular dark brown bars. Bill black; eyes orange; skin of powerful feet dark brown. Sexes similar in plumage, female about 6% larger (male 1 835–2 810g, female 2 280–4 200 g).

JUVENILE Similar to adult, but 'ears' shorter and has stubby, barred or blotched juvenile down on upperparts for the first few months after leaving nest.

DISTRIBUTION, HABITAT AND STATUS Breeding resident from wooded northern slopes of Rif Mountains in Algeria, an extension of the Spanish population. Always was rare and may even be extinct. Also a rare vagrant from the Middle East to the Nile valley in Egypt.

SIMILAR SPECIES Very similar in form and markings to smaller **Pharaoh Eagle-owl** (p. 252) (pale tawny-rufous; lightly streaked above, barred and much paler below), whose more arid desert range it adjoins in Algeria and near the Nile valley in Egypt. Often considered to be 2 forms of the same species.

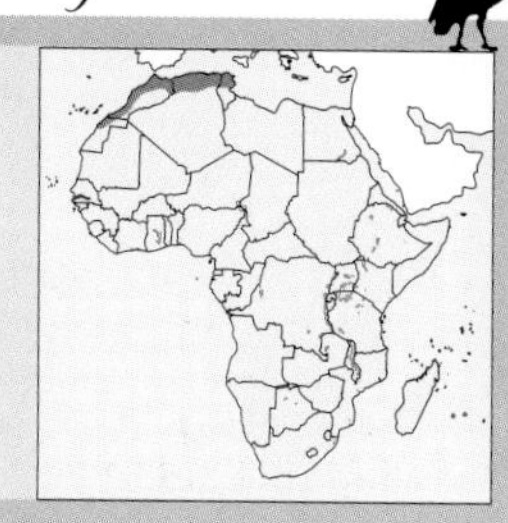

Large, brown 'eared' owl. Darkly blotched chest. Orange eyes. Dark bill. Utters deep, hooting call 'u huu' every 8–10 seconds or, if alarmed, a loud quack. Large (about 55 cm tall, 150 cm wingspan). Rare in Nile Valley and mountains of northwestern Africa.

Pharaoh Eagle-owl

M. Leach/NHPA

Spotted Eagle-owl

Bubo africanus

AT PERCH

Crouches down on ground perches, showing mottled back, or stands upright on tree branches, with long ear-tufts always erect. Becomes remarkably slender-bodied and large-headed with slitted eyes if disturbed. White tips to greater coverts form pale line across wing. Note dark chest-patches, yellow or orange (southern) or brown (eastern, western) eyes, and at close range black edges to facial disc and 'ears'. Brown form with orange eyes.

IN FLIGHT

Nocturnal; silent when flushed by day or seen at dusk, flaps with deep, strong wingbeats and shows broad, fanned tail. Darkly barred wings and tail, finely barred pale grey and white underwing coverts.

DISTINCTIVE BEHAVIOUR

Emerges at dusk and often calls with a mellow, hooting 'whoo hoo', descending in pitch, or 'who are you', rising and falling in pitch, sometimes as a duet. Utters loud 'keeow' note in alarm, also bill-snaps. Often visits nest area again around midnight and before dawn. Hunts a wide variety of small animals, usually taken on the ground, sometimes running about in pursuit. Regularly encountered on roadside perches or standing in tracks. Nests in a scrape on a rockface, a building, the ground or in a tree hollow.

N. Myburgh

B.a. africanus adult landing

B.a. africanus hunting on the ground

ADULT Above grey-brown with white spots, particularly on mantle. Back and rump grey with buff bars. Face white with fine, grey bars and black rim. Long, black-sided 'ears'. Below white with fine, grey bars. Throat and legs white; underwing coverts white finely barred with grey. Sides of chest have dark, sooty blotches. Flight feathers and tail dark grey with broad, paler bars above, below white with dark grey bars. Bill black; eyes yellow; weak, bare feet dark brown. Brown form with orange eyes most common in arid areas. *B.a. cinerascens* of East and West Africa has dark brown eyes. Sexes similar in plumage, female about 2% larger (male 487–620 g, female 640–850 g).

JUVENILE Similar to adult, but generally browner below and less spotted above; juvenile down dark grey with fine, darker bars.

DISTRIBUTION, HABITAT AND STATUS Widespread across sub-Saharan Africa except for the densest forest and driest desert areas. The commonest owl in a wide variety of habitats, especially where a mixture of low, rocky hills adjoins productive scrub and grasslands.

SIMILAR SPECIES Most like **Cape Eagle-owl** (p. 250) (similar size in southern Africa, larger in central and northeastern Africa; underparts heavily barred and blotched; orange eyes; browner plumage, except in rufous form; large, heavy legs and feet). Much larger than grey 'eared' scops owls

ALTERNATIVE NAME: African Eagle-owl

Large, grey 'eared' owl, blotched with white above, and dark patches on the chest. Eyes yellow (most southern birds) or brown (eastern and western birds). Uncommon southern brown form has orange eyes. Call a mellow hooting 'whoo you' or 'who are you', sometimes a high quack of alarm. Medium-sized (about 40 cm tall, 113 cm wingspan).

Cape Eagle-owl

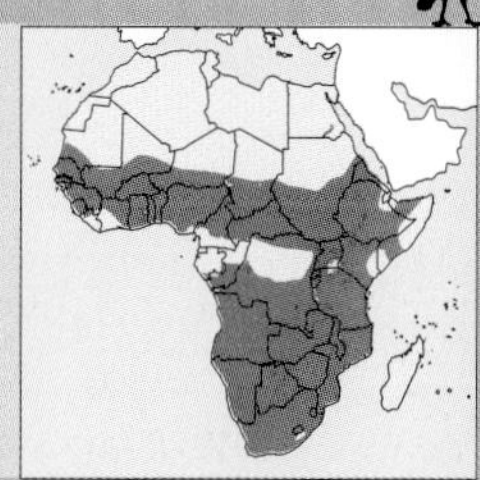

P. Chadwick

Akun Eagle-owl

Bubo leucostictus

AT PERCH

Sits upright and appears dark brown with broad bars on white below. White lines on forehead and ears, white spots on nape, and pale yellow eyes and bill obvious at close range.

IN FLIGHT

Nocturnal and silent, only seen flying by day if flushed from roost.

DISTINCTIVE BEHAVIOUR

Hunts from dead snags or looped lianas in mid-stratum of forest, often near clearings, roads or trails along which it also flies low over the ground. Feeds mainly on insects, hawked on the wing, taken from foliage or from the ground. Emerges at dusk at the same time as many insects. Calls less often than many other owls, mainly an accelerating series of deep, short clucks 'tok tok kok-ok-ok-ok-ok'; young begs with querulous wail, adult utters quack of alarm. Roosts deep within foliage of tall forest trees. Nests on the ground.

ADULT Above brown with broad, dusky-brown bars. Forehead, eyebrows and base of 'ears' white. Nape spotted white. Wing coverts finely barred with buff. Facial disc, front of 'ears', throat and neck light rufous with fine, paler grey-brown bars. Below white with broad, dark brown and rufous bars at feather ends. Legs heavily barred. Flight feathers and tail dusky-brown with pale bars and paler below; tail has white tip. Much individual variation in density of colour and barring. Bill and cere pale greenish-yellow; eyes pale yellow; bare eyelids dark brown; skin of small feet pale yellow. Sexes similar in plumage, female about 1% larger (male 486–536 g, female 524–607 g).

JUVENILE Body and head down white with fine, dense, reddish-brown bars; down retained for at least 8 months. Contrasts with dark flight feathers, wing coverts and tail. First full feather plumage similar to adult.

DISTRIBUTION, HABITAT AND STATUS From Sierra Leone to Cameroon and across the Congo basin. Widespread in lowland rainforest and tall secondary growth, including forest edge and along rivers. Uncommon, but also easily overlooked.

SIMILAR SPECIES Most similar in lowland rainforests to more heavily built **Fraser's Eagle-owl** (p. 260) (dark rufous-brown above with broad, buff bars; finely barred below; face with dark rim and long, dark 'ears'; lightly barred legs; dark brown eyes; blue bill and eyelids) and along edges of range to **Spotted Eagle-owl** (p. 256) (grey-brown above with fine, grey bars below; dark brown eyes; black bill). Also overlaps in lowland forest with larger **Shelley's Eagle-owl** (p. 326) (brown; heavily barred above and below; dark brown eyes; heavy yellow bill). Call also distinctive.

ALTERNATIVE NAME: Sooty Eagle-owl

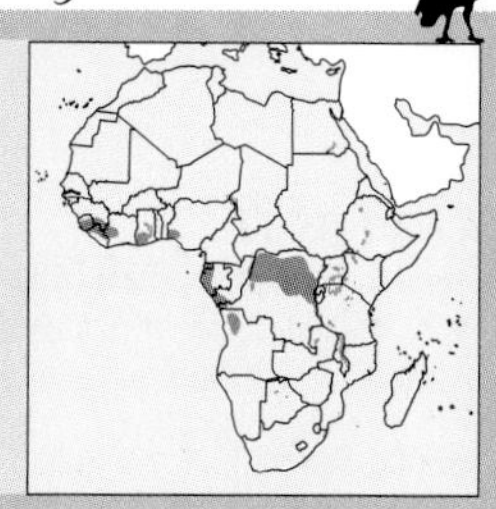

Large, dark brown 'eared' owl. Blotched with heavy, brown bars below. White forehead, eyebrow and base of 'ears'. Yellow eyes and bill distinctive. Calls with accelerating deep clucks 'tok tok kok-ok-ok-ok-ok'. Medium-sized (about 40 cm tall, 105 cm wingspan).

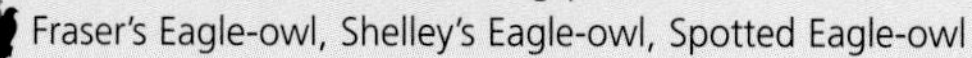

Fraser's Eagle-owl, Shelley's Eagle-owl, Spotted Eagle-owl

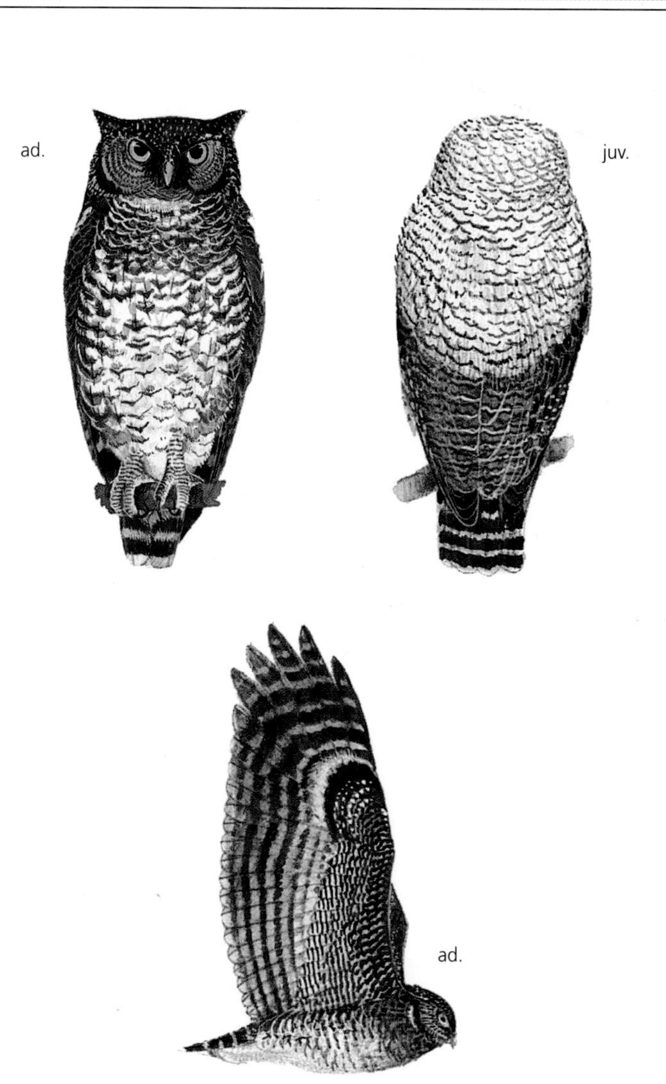

Fraser's Eagle-owl

Bubo poensis

AT PERCH

Stands upright with obvious long 'ears'. Appears dark overall with pale line across shoulder and broad, pale bars below. Dark eye obvious in paler rufous face, and in contrast to blue bill and eyelids.

IN FLIGHT

Nocturnal and silent, only likely to be seen if flushed from roost.

DISTINCTIVE BEHAVIOUR

Roosts deep within foliage of tall forest trees, emerging at dusk to hunt small animals and arthropods. Calls with prolonged, mellow, double hoot 'twoo wooot', the second part higher and wheezier, the whole call repeated every 3–4 seconds. Also makes single hoots and steady loud, purring notes 'put put put ...' like a small engine. Calls of Usambara Eagle-owl said to differ but not conclusively studied. Nests on the ground or in a tree hollow.

ADULT Rufous above with dusky-brown bars. Upperwing coverts dusky-brown. Scapulars with white outer webs. Facial disc rufous with dark brown rim and edges to long ear-tufts. Below white with fine, dark brown and rufous bars. Upper breast washed with rufous and mottled with dusky-brown ends to feathers. Underwing pale rufous with fine, dark brown bars. Flight feathers and tail rufous with dusky-brown bars, paler below. Individual variation in density of rufous and of barring. Separate eastern Tanzanian population (Usambara Eagle-owl (*B.p. vosseleri*)) larger, browner above, with less dark barring below but heavier chest blotching. Bill, cere, bare eyelids and strong feet pale blue-grey; eyes dark brown. Sexes similar in plumage, female about 6% larger (male 575 g, female 685–1 052 g).

JUVENILE Juvenile down present for almost a year; very pale rufous with fine, dark brown bars. Contrasts with dark, adult-like flight feathers, tail, facial disc and 'ears'. Scapulars much whiter than adult in first plumage.

DISTRIBUTION, HABITAT AND STATUS Lowland rainforest from Liberia eastwards, including Bioko island, to western Uganda and across the Congo basin to northern Angola (*B.p. poensis*). A second population on eastern arc montane forests of East Usambaras and Ulugurus in Tanzania (*B.p. vosseleri*), sometimes considered a separate species. Often found on forest edge in clearings, secondary forest or plantations.

SIMILAR SPECIES Most similar in lowland rainforests to more lightly built **Akun Eagle-owl** (p. 258) (paler brown above; wide, brown bars below; white forehead and front of 'ears'; darkly barred legs; yellow eyes and bill) and in north-eastern Tanzania to **Spotted Eagle-owl** (p. 256) (grey-brown above with fine, grey bars below; black bill; yellow eyes). Also overlaps in lowland forest with rare, larger, dark-eyed **Shelley's Eagle-owl** (p. 326) (brown; heavily barred below; heavy creamy-yellow bill) in lowland forest, and in eastern Tanzania with **Milky Eagle-owl** (p. 248) (pale grey; fine white bars all over; black facial rim; stubby ear-tufts; pink eyelids; cream bill). Call higher and softer than congeners.

ALTERNATIVE NAMES: Nduk Eagle-owl, Usambara Eagle-owl

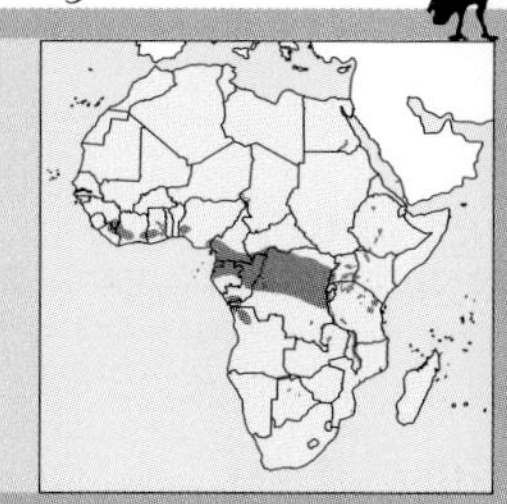

Large, rufous 'eared' owl. Broad, white and buff bars below. Rufous face. Long 'ears'. Pale blue eyelids and bill. Dark brown eyes. Calls with a mellow double hoot 'twoo wooot'. Larger form in eastern Tanzania (Usambara Eagle-owl) sometimes considered a separate species. Medium-sized (about 40 cm tall, 100 cm wingspan).

Akun Eagle-owl, Shelley's Eagle-owl, Spotted Eagle-owl

B.p. poensis ad.

B.p. vosseleri ad.

B.p. poensis ad.

B.p. poensis ad.

D. Roberson/BirdLife International

Pel's Fishing-owl

Scotopelia peli

AT PERCH

Stands upright, usually with head retracted. Appears rufous with large, dark eyes and bill. Bare legs and markings evident at close range. Rounded, 'earless' head notable, and appears very large and fluffy when feathers raised.

IN FLIGHT

Flushes from riverine trees with strong, deep, noisy beats of the long, broad wings. Note the large, rounded head, long, broad wings and short tail.

DISTINCTIVE BEHAVIOUR

Roosts in deep shade of large trees, usually within 50 m of the water's edge, but can roost up to 1 km away. Emerges in late dusk to hunting perches over or alongside open water. Catches fish after short plunges or long strikes into the water. Catfish heads, pellets of fish scales and moulted feathers often reveal roost or nest area. Calls with very deep humming hoot 'hoom hoo' or a longer 'hooommm hoet', often a pair in duet. Juvenile begs with prolonged, high whine, 'miaaow', descending in pitch. Nests in tree holes and hollows.

D. Balfour

Adult

ADULT Rufous overall, with fine, dark brown bars above and streaks below ending in chevron-shaped spots, especially on flanks. Density of markings varies considerably between individuals. Face and underparts paler rufous than above. Facial disc not well-developed. Flight feathers rufous with broad, dusky-brown bars. Bill black; cere dark grey; eyes large and very dark brown; bare, powerful legs and feet pale yellow-brown with long, curved horn-coloured claws. Sexes similar in plumage and size, but female often paler (female 2 055–2 325 g).

JUVENILE Body and head down white with pale rufous wash, lightly barred on back and streaked on flanks. Flight feathers similar to those of adult. Moults into adult-like plumage by about 10 months of age.

DISTRIBUTION, HABITAT AND STATUS Widespread along rivers, small lakes and marshes of sub-Saharan Africa. Most common in Congo basin and Zambezi catchment area, including swamps of central Africa. Uncommon elsewhere, rare and local in western Africa, linear along larger rivers in eastern and southern Africa.

SIMILAR SPECIES Overlaps in lowland rainforest of central African Congo basin with smaller **Vermiculated Fishing-owl** (p. 264) (below white or buff with clear, dark streaks; crown streaked with black; pale yellow bill and cere; single, higher resonant 'ooo' call-note) and in western African forests with smaller and rare **Rufous Fishing-owl** (p. 266) (unmarked deeper rufous mantle; streaked rather than spotted below; black bill; yellow cere). Same size and often shares riverine trees with **Milky Eagle-owl** (p. 248) (pale grey; finely barred with white; 'eared'; dark rim to obvious facial disc; pink eyelids).

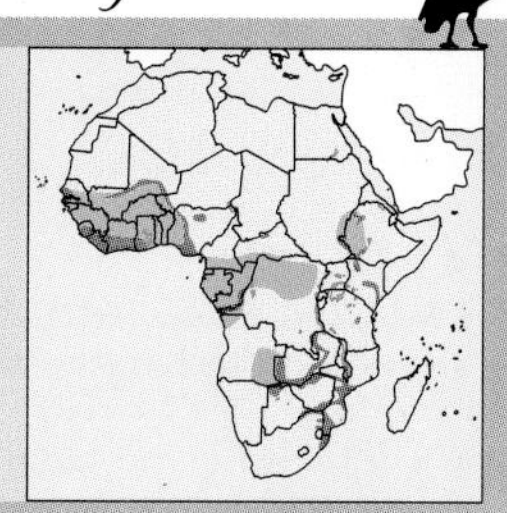

Very large, rufous, 'earless' owl. Dark brown bars above. Chevron-shaped spots below. Large, dark eyes. Heavy, dark bill. Powerful legs and feet unfeathered. Call a series of deep, resonant hoots 'hooommm hoet', chick begging a high, drawn-out whine. Large (about 55 cm tall, 153 cm wingspan).

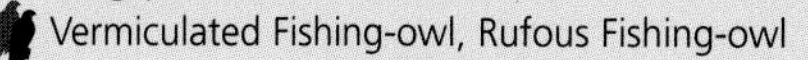

Vermiculated Fishing-owl, Rufous Fishing-owl

A. Bannister/ABPL

Vermiculated Fishing-owl

Scotopelia bouvieri

AT PERCH

Large, rounded, 'earless' head obvious, with pale face and pale, streaked underparts. Appears dark above but markings not obvious. Dark eyes and pale yellow bill and cere notable at close range.

IN FLIGHT

Nocturnal and rarely seen in flight unless flushed.

DISTINCTIVE BEHAVIOUR

Little known. Hunts for fish mainly along edges of large rivers at least 10 m wide, perched 1–2 m above the water. Flies back into forest to perch at 5–20 m up if disturbed. Also lives along small rivers (2–4 m wide) and pools in swamp forest. Roosts near the water. Call an accelerating series of 6–8 deep hoots 'krook krook-ook-ook-ook-ook', sometimes given by a pair in duet. Also a high whine from juveniles soliciting food. Nests in old stick structures of other birds, such as ibis, 2–3 m above the water.

ADULT Above cinnamon to rufous with fine, dark brown vermiculations, becoming more streaked on head and neck. Small facial disc pale rufous with dark rim. Below pale rufous with broad, dark brown streaks and white feather edges. Variation in intensity of rufous and extent of markings. Thighs and belly plain white. Flight feathers and tail rufous with broad, dark brown bars. Bill and cere pale yellow; large eyes dark brown; bare legs and feet pale yellow-brown. Sexes similar in plumage and size.

JUVENILE Body and head down white with cinnamon wash over head and neck, more streaked above and more finely streaked below than adult. Retains down for at least 3 months after leaving the nest.

DISTRIBUTION, HABITAT AND STATUS Restricted to gallery forest of large and small rivers within the primary forest of the Congo basin. Locally common.

SIMILAR SPECIES Overlaps only with much larger **Pel's Fishing-owl** (p. 262) (barred rufous head and mantle; chevron-spotted rufous underparts; black bill; dark grey cere; very deep, double-noted hoots or high whine). Other large forest owls are all 'eared'.

Large, cinnamon-coloured 'earless' owl. Fine, wavy, dark bars above. Below off-white with broad, dark brown streaks. Large, dark eyes. Yellow bill and cere. Bare, yellow-brown legs and feet. Call an accelerating series of 6–8 deep hoots 'krook krook-ook-ook-ook-ook'. Large (about 45 cm tall, 100 cm wingspan).

Pel's Fishing Owl

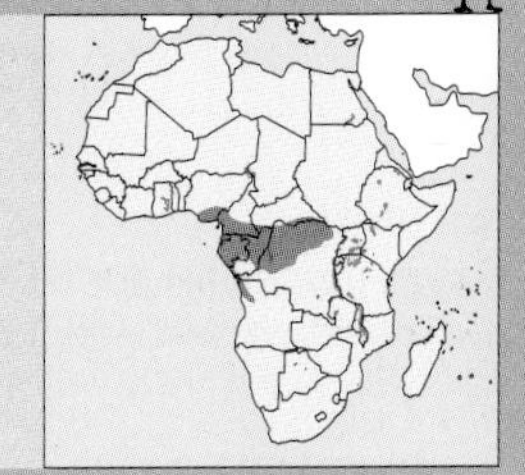

Rufous Fishing-owl

Scotopelia ussheri

AT PERCH

Stands rather upright. Dark rufous above contrasts with pale, streaked underparts. Broad bars on flight feathers and scapulars notable. Broad, flat-sided head without 'ears'. Very large, dark brown eyes. Black bill with yellow cere.

IN FLIGHT

Emerges by day only if flushed, flying with deep, strong beats of the long wings. Dark flight feathers contrast with pale, streaked underparts and underwing, so bird appears almost rufous and white.

DISTINCTIVE BEHAVIOUR

Little studied. Flushed from tangles near rivers at about 2 m above ground, where it probably roosts. Deep, dove-like hoot 'ooo' uttered at intervals of about 1 minute, similar to call of White-crested Tiger Heron (*Tigriornis leucolophus*). Hunts for fish from perches on or next to open water and also seen resting on sandbars at night. Nesting undescribed.

ADULT Above dark rufous, with dark brown streaks on crown and nape. Below much paler fawn with long, rufous streaks. Facial disc unmarked pale rufous. Flight feathers and tail dark rufous, with dusky bars on primaries and faint, dark bars on tail. Bill black; cere yellow; eyes dark brown (reports of yellow eyes unsubstantiated); strong, bare legs and feet pale yellow-brown. Sexes similar in plumage and size.

JUVENILE Body and head down white washed with buff, moulting after several months into adult-like plumage with paler head.

DISTRIBUTION, HABITAT AND STATUS Known from gallery forest on the banks of both large and small rivers, and around lagoons. Occurs in primary and even in secondary forest and plantations of western Africa. It is a rare species, and not often sighted.

SIMILAR SPECIES Overlaps only with the much larger **Pel's Fishing Owl** (p. 262) (finely barred head and mantle; chevron-spotted underparts; dark grey cere; very deep, double-noted hoots or high whine). Other large forest owls in its range are all 'eared'.

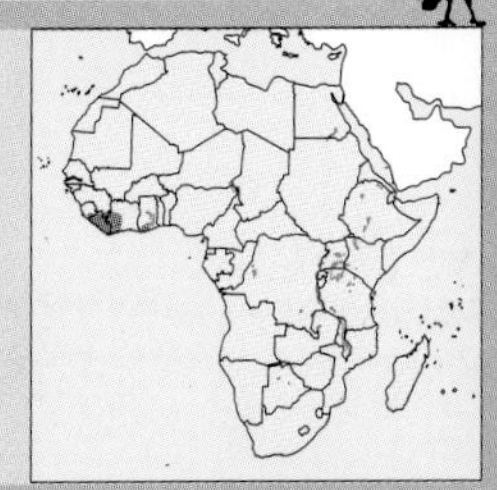

Large, rufous 'earless' owl. Unmarked on mantle. Streaked with dark brown on crown and nape. Below buff with dark brown streaks. Large, dark brown eyes. Black bill with yellow cere. Bare, pale yellow legs and feet. Call a single, deep, resonant dove-like hoot, 'ooo'. Large (about 45 cm tall, 100 cm wingspan).

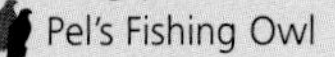
Pel's Fishing Owl

BirdLife International

Marsh Owl

Asio capensis

AT PERCH

Often stands rather horizontally. Long wings reach tip of short tail. Note pale legs. Appears plain dark brown overall, with dark eyes and bill obvious in the pale face. Short 'ears' difficult to see.

IN FLIGHT

Flies buoyantly on long wings, with deep, slow wingbeats and short periods of gliding. Pale bars on flight feathers obvious against dark plumage, especially as golden-buff patch at base of outer primaries. Wings held up in slight V-shape when gliding, showing buff undersides with dark bars.

DISTINCTIVE BEHAVIOUR

The only African owl regularly seen flying in the late afternoon, less often at dawn. Travels low over open habitats, coursing back and forth with occasional twists to the ground after prey. Feeds mainly on rodents, with some insects and small birds. Stops to perch in the open or on the ground. Roosts on the ground, frequently next to a grass tuft and often communally. Pellets long and twisted like small carnivore scats. Main call a grating croak 'squeerk'. Displays with high, zig-zag flights, including calling and wing-claps, mainly before breeding. Nests on the ground in a small hollow among grass and weed stems.

J. Carlyon

Adult

L. Hes

Adult

ADULT Above dark brown, with fine, buff vermiculation on head, neck and rump. Facial disc pale buff or grey, with dark rim and eye-rings. Below brown; plain chest grades into pale buff bars on breast. Vent and feathered legs and feet plain pale buff. Flight and tail feathers dark brown above with broad, buff bars, especially at base of outer primaries; below buff with dark bars near tips only. Bill black; eyes dark brown; feet dull brown. Sexes similar in plumage, female about 2% larger (male 243–340 g, female 305–376 g).

JUVENILE Head and body down buff with dark brown bars. Moults to adult-like plumage at 10 weeks, but scapular and back feathers have buff edges and facial disc is dark brown with broad, black rim.

DISTRIBUTION, HABITAT AND STATUS Widespread across sub-Saharan Africa and Madagascar. Also found in northwestern Africa, where rare, and rare vagrant to Canary Islands (and Iberian Peninsula). Locally common in areas of tall, moist, open grassland and weedy marshes, including cereal fields and open, grassy savanna. Nomadic; patchily distributed; colonises remote areas under suitable conditions.

SIMILAR SPECIES North of the equator, similar in flight and habitat to larger **Short-eared Owl** (p. 270) (buff with dark brown streaks; yellow eyes; pale buff patch at base of primaries; black carpal patch and wing tips; similar small 'ears'; usually silent). Overlaps widely by habitat in eastern and southern Africa with **African Grass Owl** (p. 284) ('earless'; heart-shaped, white or pale rufous facial disc; small, black eyes; buff underparts with small, dark spots; glides on bowed, broad wings; appears large-headed; call a single long screech or frog-like clicks).

Plain brown owl. Above dark brown, below paler. Facial disc pale brown or grey. Large, dark brown eyes. Black bill. Small 'ears' relatively close together. Often seen flying by day. Buff patches at ends of the long wings. Call a harsh croak. Medium-sized (about 30 cm tall, 90 cm wingspan).

Short-eared Owl, African Grass Owl

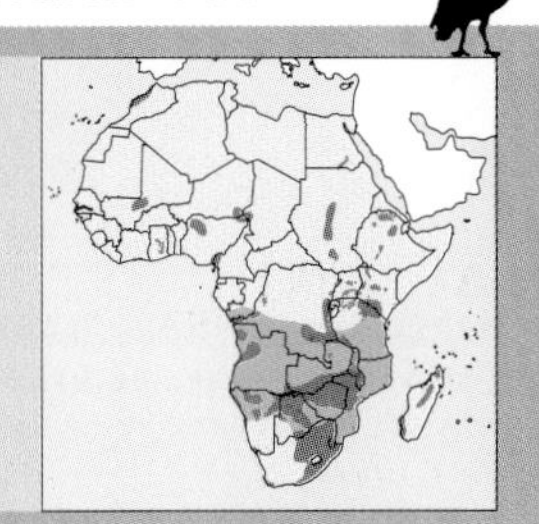

J. Laurie/Photo Access

Short-eared Owl

Asio flammeus

AT PERCH

Often stands rather horizontally. Long wings extend past tip of short tail. Appears pale brown with dark streaks and pale, plain facial disc. Note dark bill and black-rimmed yellow eyes at close range. Short 'ears' not usually obvious.

IN FLIGHT

Flies buoyantly on long, narrow wings with deep, slow wingbeats, interspersed with short glides with the wings held up in a slight V-shape. Note striped body, pale underwings with dark tips and carpal patch, and buff patch in outer primaries.

DISTINCTIVE BEHAVIOUR

Often seen hunting at dusk and dawn, flying low over open vegetation and twisting down after small-rodent and insect prey. Usually perches and always roosts on the ground, sometimes in scattered groups. Sometimes hunts from low, open perches.

A. Williams/NHPA

Adult hunting on the wing

ADULT Overall dark buff, broadly streaked with dusky-brown. Above more barred on mantle and wing coverts. Facial disc pale buff with dusky-brown rim and eye-rings. Small, dark, central 'ears'. Vent and feathered legs and feet plain buff. Flight and tail feathers dusky-brown with broad, buff bars, especially at base of black-tipped outer primaries. Underwing pale buff with dusky-brown primary coverts. Bill black; eyes yellow; skin of feet grey. Female often more buff and about 3% larger than male (male 320–385 g, female 400–430 g).

JUVENILE Head and body down dark brown with buff tips. Moults within a few months to adult-like plumage. Juvenile plumage not likely to be seen in Africa as it has usually been lost by the time of migration.

DISTRIBUTION, HABITAT AND STATUS Uncommon non-breeding migrant to the northern African coast and sub-Saharan Africa during the northern winter. Vagrant to Canary and Cape Verde islands. Favours open marsh and moist grassland habitats, but crosses desert and open savanna on migration. Locally common along the Mediterranean coast, uncommon in western Africa and rare in eastern Africa.

SIMILAR SPECIES Overlaps most in habitat and flight behaviour with smaller, resident **Marsh Owl** (p. 268) (appears a uniform dark brown; pale buff primary patch; dark brown eyes). Most similar in plumage to woodland **Eurasian Long-eared Owl** (p. 276) (even more streaky; orange eyes; long, dark 'ears'; shorter, more rounded wings; glides with wings level or slightly bowed).

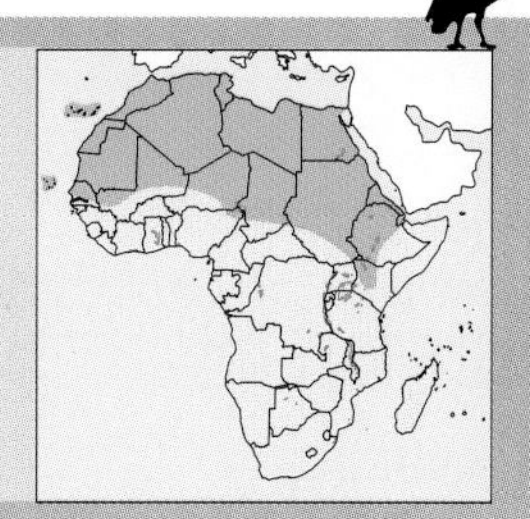

Buff owl with dark brown streaks. Pale plain facial disc. Obvious yellow eyes. Small centrally placed 'ears'. Often seen flying by day. Long wings with black tips and buff outer patches. Medium-sized (about 30 cm tall, 95 cm wingspan).

Marsh Owl, Eurasian Long-eared Owl

Abyssinian Long-eared Owl

Asio abyssinicus

AT PERCH

Stands upright and slender, the long, dark 'ears' arising more from the centre of the forehead. Appears reddish-brown or dark brown with rufous or grey-brown facial disc and dark mottling and streaking. Orange eyes obvious at close range.

IN FLIGHT

Long-winged and short-tailed, flying with deep, strong, slow beats and appearing buoyant and agile on the wing. Paler underwing coverts contrast with dark, barred flight and tail feathers.

DISTINCTIVE BEHAVIOUR

Little known. Roosts amongst dense foliage or on old crow nests, sometimes in groups of up to 30 among giant heaths. Emerges after dark, only rarely on the wing in daylight. Main call suspected to be a deep, hooting 'oooo-oooom', rising slightly in pitch. Probably hunts mostly over open country on the wing, feeding mainly on small mammals. Breeding undescribed, probably breeds on old stick-nests of other birds.

ADULT Overall golden-brown or dark brown with red-brown and grey vermiculations. Above streaked with dark brown, especially on crown. Facial disc rufous or grey with dark brown rim and white eyebrows. Long, fairly centrally placed 'ears' dusky-rufous or brown. Below heavily streaked with brown and barred with white. Chequered pattern in dark brown and grey on breast. Vent and feathered legs and feet plain buff. Flight and tail feathers dark brown with broad, rufous bars. Individual variation in hues and markings, East African birds greyer and more heavily streaked with black. Bill black; eyes orange; skin on feet dark brown. Sexes similar in plumage, female about 5% larger.

JUVENILE Undescribed.

DISTRIBUTION, HABITAT AND STATUS A rare resident on the Ethiopian Highlands, Mount Kenya and the Ruwenzori and Kabobo Mountains of Congo. Found in patches of forest adjacent to open moor- and grassland. In Ethiopia, also uses stands of exotic plantations in unforested grassland areas.

SIMILAR SPECIES Very similar to **Eurasian Long-eared Owl** (p. 276) and often considered to be the same species. Sub-Saharan populations occur most commonly with slightly smaller 'earless' **African Wood Owl** (p. 278) (large, dark brown eyes; red-rimmed eyelids; rich, reddish-brown with white mottling; pale grey facial disc), and may also overlap with larger, heavier 'eared' **Cape Eagle-owl** (p. 250) (dark brown mottled with buff; dark blotches on sides of chest; orange eyes; dark rim to pale facial disc; 'ear-tufts' at side of head) and **Spotted Eagle-owl** (p. 256) (dark grey mottled with white; dark brown eyes; black rim to pale facial disc; ear-tufts at side of head; white feathered legs).

P. Smitterberg

Adult at roost

ALTERNATIVE NAME: Abyssinian Owl

Large, slender, 'long-eared' golden-brown owl. Blotched above with dark grey and brown. Below blotched on chest and spotted on breast with dark brown and grey. Orange eyes. Call apparently a deep, hooting 'oooo-oooom', rising slightly in pitch. Large (about 35 cm tall, 100 cm wingspan).

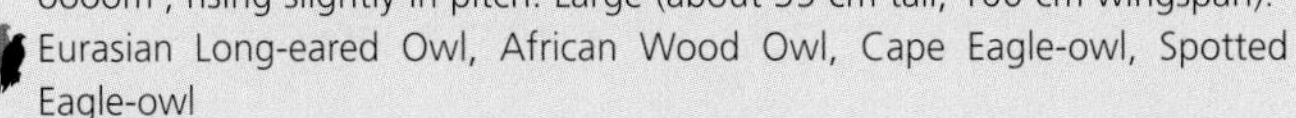

Eurasian Long-eared Owl, African Wood Owl, Cape Eagle-owl, Spotted Eagle-owl

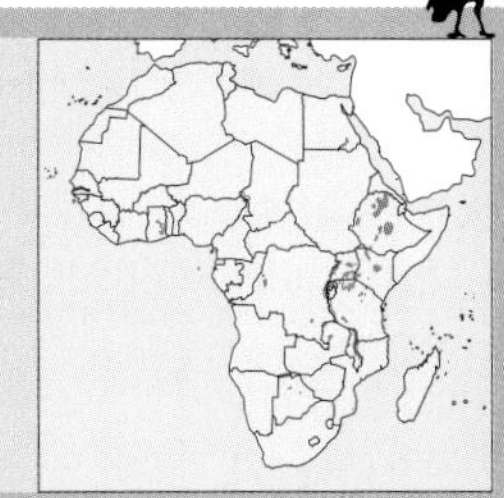

Madagascar Long-eared Owl

Asio madagascariensis

AT PERCH

Sits very upright, the long 'ears' seeming to sprout from the centre of the forehead. Appears dark brown mottled with light tan. Orange eyes and dark rim to facial disc obvious at close range.

IN FLIGHT

Nocturnal and silent unless flushed. Wings long and relatively narrow, and tail rather short. Flies with deep, strong beats and buoyant flight. Large feet prominent below tail, as if the bird might be carrying prey.

DISTINCTIVE BEHAVIOUR

Roosts within dense cover and usually only emerges after dark. Call a series of rhythmic, lilting but harsh notes 'han-kan han-kan hankan ...'. Often calls when taking to the wing or while in flight. Hunts for part of the time in flight. Nests in old stick-nests of other birds.

ADULT Above dark brown mottled with buff, especially on mantle. Face plain buff, with dark brown around eyes, along rim and on long, centrally placed 'ears'. Below buff, flecked and streaked with dark brown. Vent and feathered legs and feet plain buff. Flight feathers and tail dark brown with buff and grey-brown bars. Bill black, eyes orange, large, heavy feet brown. Sexes similar in plumage, female about 5% larger.

JUVENILE Head and body down white, in marked contrast to black facial disc and dark brown flight feathers. 'Ears' short. Soon moults into adult-like plumage.

DISTRIBUTION, HABITAT AND STATUS Madagascar. Common and widespread in deciduous, gallery and rainforests at low and medium altitudes, but uncommon in the drier southern parts of the island.

SIMILAR SPECIES In size and flight behaviour most resembles **Marsh Owl** (p. 268) (uniform dark brown; buff wing-patches in flight; dark brown eyes; short ear-tufts close together; call a harsh croak) and **Common Barn Owl** (p. 282) (pale golden-brown above; white or cream below; prominent, white, heart-shaped facial disc; small, black eyes; call a single long screech). Shares some habitats with smaller **Madagascar Hawk-owl** (p. 288) (prominent pale eyebrows; dark brown eyes; heavily barred underparts; calls with hooting then yelping notes).

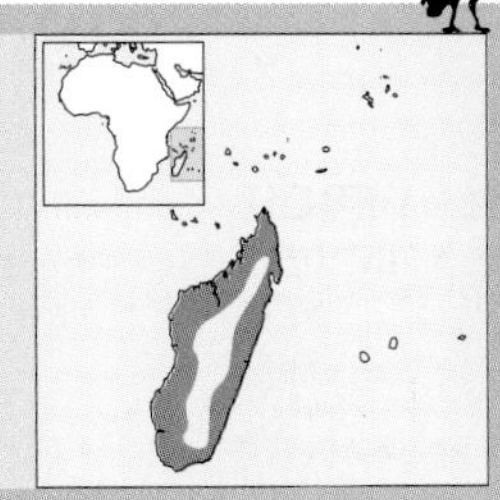

Long, slender, 'eared' owl. Buff with dark brown streaks. Long, centrally placed 'ears'. Orange eyes. Call a series of harsh, rhythmic notes 'han-kan han-kan han-kan ...', given when flushed or in flight. Medium-sized (about 35 cm tall, 90 cm wingspan). Restricted to Madagascar.

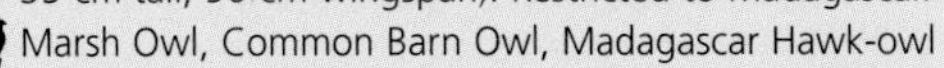

Marsh Owl, Common Barn Owl, Madagascar Hawk-owl

O. Langrand/BIOS

Eurasian Long-eared Owl

Asio otus

AT PERCH

Usually stands upright and slender, the long 'ears' arising from the centre of the forehead. Appears pale brown with dark mottling and streaking. Dark rim to face and orange eyes obvious at close range.

IN FLIGHT

Long-winged and short-tailed, flying with deep, strong, slow beats and appearing buoyant and agile on the wing. Pale underwing coverts contrast with dark carpal patches and dark, barred flight feathers. Pale buff patches at base of primaries.

DISTINCTIVE BEHAVIOUR

Roosts amongst dense foliage, usually against a trunk or branch and only rarely on the ground. Emerges after dark, only occasionally seen on the wing in daylight. Main call a single repeated, dove-like hoot 'whoo', often calls in duet at different pitches prior to breeding, or a loud bark or squeak of alarm. Hunts mainly over open country on the wing, feeding mainly on small rodents and less often on birds and insects. Makes high, zig-zag flights and loud wing-claps before breeding. Nests on old stick-nests of other birds.

G. Lacz/NHPA

Adult

Adult at roost

ADULT Overall buff and brown, with brown and grey vermiculations. Above streaked with dark brown, crown freckled with golden-brown, often has white eyebrows. Facial disc pale grey finely barred with brown and with dark brown rim. Long, dark brown, centrally placed 'ears' with buff edges. Below pale buff or rufous, streaked and barred with brown and grey on chest, lightly streaked with brown on breast, and plain on legs and vent. Underwing pale buff with dusky-brown carpal patch on white primary coverts. Flight and tail feathers dark brown with broad, buff bars. Considerable individual variation in hues and markings from overall dark grey to golden-brown. Bill black; eyes orange; feet dark brown. Sexes similar in plumage, female about 5% larger (male 220–280 g, female 250–370 g).

JUVENILE Head and body down brown with fine, cream bars, soon moulted to adult-like plumage.

DISTRIBUTION, HABITAT AND STATUS Breeding resident on the northern slopes of the Atlas Mountains in northwestern Africa, and on Canary Islands. A rare vagrant to the Nile Delta in Egypt. Favours clumps of woodland and patches of forest in which to roost and nest. Uncommon.

SIMILAR SPECIES Most similar in size and flight behaviour to **Short-eared Owl** (p. 270) (tiny, centrally placed 'ears'; even longer-winged; even lighter on the wing; usually flies with wings held out in a slight V-shape; often active in daylight) and, in Morocco, to **Marsh Owl** (p. 268) (plain dark brown; dark brown eyes; tiny, closely spaced 'ears'; often hunts in daylight; very light and agile in flight; long, narrow wings; buff primary patches).

ALTERNATIVE NAME: Long-eared Owl

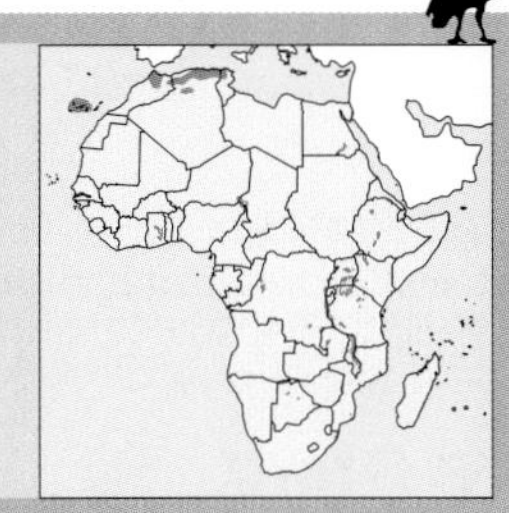

Large, slender, 'long-eared' buff and brown owl. Streaked with dark brown and grey. Orange eyes. Call a single repeated, dove-like hoot 'whoo', or a bark or squeak of alarm. Large (about 30 cm tall, 90 cm wingspan). Uncommon in northern Africa.

Short-eared Owl, Marsh Owl

African Wood Owl

Strix woodfordii

AT PERCH

Roosts standing upright with head withdrawn, but rarely narrows eyes. Appears dark brown with paler underparts. Large, dark eyes very prominent in pale facial disc. White spots across shoulder and side of wing often obvious.

IN FLIGHT

Nocturnal and silent but if flushed flies with deep, soft beats of the long wings and appears large-headed and long-tailed. Note white bars on flight feathers and tail.

DISTINCTIVE BEHAVIOUR

Usually roosts deep within a tangle of creepers, 1–30 m above ground. Emerges after dark to hunt for arthropods and small vertebrates, from the ground or foliage, or hawked on the wing. Lives as territorial pairs. Repeats either a single howl 'weow' or a rhythmic series of hoots 'whu-hu whu-hu-hu hu-hu'. Mates usually reply, each sex at a different pitch, probably male high, female low. Nests in tree holes, rarely on the ground among roots.

N. Dennis

Adult pair at roost

ADULT Much individual and regional variation in hue and density of colours and markings within and between various dubiously defined subspecies. Above dark reddish-brown barred with white to produce flecked effect. Upperwing coverts and scapulars have broad, white tips. Facial disc white to pale rufous with fine, dark bars, often a dark outer rim and dark around eyes. Below white broadly barred with rufous. Flight feathers and tail reddish-brown with broad, white or buff bars. Bill and cere deep yellow; eyes large and dark brown; eyelid rims red; feet dull yellow. Sexes similar in plumage and size (male 242–269 g, female 285–350 g).

JUVENILE Head and body down pale rufous with white bars. Moults into adult-like plumage after about 5 months.

DISTRIBUTION, HABITAT AND STATUS Widespread across sub-Saharan Africa in woodland and forest, including Bioko Island. Occupies any forest, from lowland tropical rainforest to cold montane forest, including riverine and coastal forests. The commonest forest and woodland owl throughout Africa.

SIMILAR SPECIES Most similar in size to 'eared' **Maned Owl** (p. 286) (deep rufous; long, floppy crown feathers; long, white-tipped 'ears'; streaked below; yellow eyes). Larger than any 'eared' ***Otus*** **scops owls** (pp. 304–324) (grey or rufous forms; yellow eyes; most with dark bills) or 'earless' ***Glaucidium*** **owlets** (pp. 294–300) (small, yellow eyes; white bar across shoulder; whistling calls), and smaller than 'eared' ***Bubo*** **eagle-owls** (pp. 248–260) (black or pale yellow bills; brown or yellow eyes; deep, hooting calls).

Subadult hunting

Medium-sized, dumpy, 'earless' owl. Rich brown colour, flecked and barred with white and rufous. Pale facial disc. Large, dark brown eyes with red-rimmed eyelids. Deep yellow bill and cere. Call either a single howl 'weow' or a rhythmic series of hoots 'whu-hu whu-hu-hu hu-hu'. Small (about 25 cm tall, 79 cm wingspan).

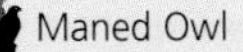
Maned Owl

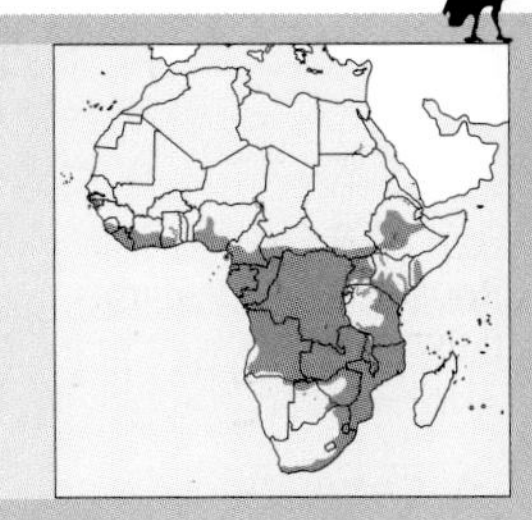

P. Pickford/SIL

Madagascar Red Owl

Tyto soumagnei

AT PERCH

Usually seen hunched up during the day, the overall orange colour and paler, browner facial disc most obvious. Small, dark eyes often closed to slits. Short, broad wings barely extend below perch and tail, can completely enclose the short tail or tail can extend up to 2 cm below wings.

IN FLIGHT

Flies during the day only if flushed. Short, broad, rounded wings and short tail probably appear unbarred. Orange colour should be obvious.

DISTINCTIVE BEHAVIOUR

Strictly nocturnal; perches amongst dense foliage during the day. Calls regularly at night, a single, high, scratchy screech, sometimes repeated. Hunts deep within forest and normally in mid-stratum to canopy. Frogs included in diet.

ADULT Uniform orange-brown overall, slightly paler below. Fine, black spots on feather tips, especially on upperparts and chest. Wide, heart-shaped facial disc, pale cinnamon-brown with darker orange to brown rim and cream chin line. Flight feathers and tail orange-brown, with traces of dark bars and small, black terminal spots. Bill cream; small eyes dark brown; sparsely feathered legs and bare toes pale grey-brown. Sexes apparently similar in plumage and size.
JUVENILE Undescribed.
DISTRIBUTION, HABITAT AND STATUS Eastern and northeastern Madagascar, rare and confined to primary evergreen forest and adjacent secondary forest. Recently found in various localities and may be more widespread and common than previously thought, especially if tolerant of some secondary forests.
SIMILAR SPECIES Most similar to larger **Common Barn Owl** (p. 282) (white facial disc; mottled grey and golden-brown above; white or pale buff below with fine, dark spots; screech call deeper but very similar; rarely if ever hunts or calls within rainforest). Similar in colour to much smaller rufous form of **Madagascar Scops Owl** (p. 316) (yellow eyes; obvious 'ear' tufts).

ALTERNATIVE NAMES: Madagascar Owl, Red Owl, Soumagne's Owl

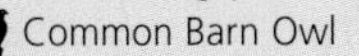

'Earless' and large-headed. Overall orange-brown with fine, black spots. Wide, plain, cinnamon-brown facial disc with darker orange to brown rim; small, dark eyes; cream bill. Call a high screech. Small (about 25 cm tall, 70 cm wingspan). Rare and confined to Madagascar.

Common Barn Owl

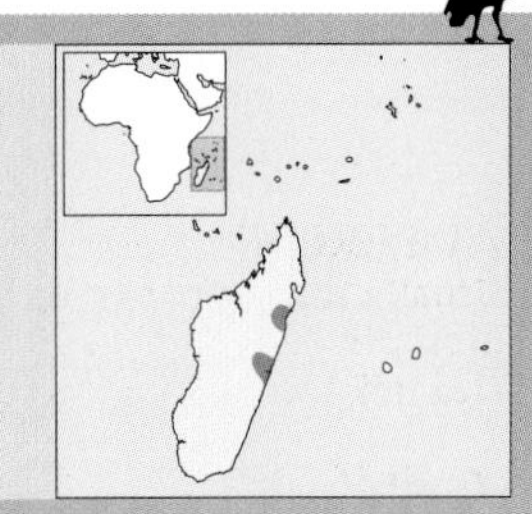

Common Barn Owl

Tyto alba

AT PERCH

Stands rather upright on long, slender legs, but the dumpy body and long primaries usually obscure the short tail. Note the large, rounded head, lack of ear-tufts, white or pale buff heart-shaped facial disc and dark eyes. Appears pale grey-brown above and white or pale buff below.

IN FLIGHT

Flies silently and buoyantly on long, broad wings. Appears large-headed, with the short tail obscured by the long legs and feet. White below, like a ghost at night. Facial 'mask' obvious on the mobile head. Appears mainly grey above with obvious pale brown barring in the primaries.

DISTINCTIVE BEHAVIOUR

Mostly nocturnal. Main call is a single long, tremulous scream 'schreez'. Also utters various snoring, hissing, wheezing and squeaking calls around the roost and nest. Sometimes hunts from a perch, but more often from low, coursing flight. Main prey small rodents, and in some areas birds or reptiles. Produces squat, compact pellets, often coated in dark, shiny mucous. Breeds in a variety of natural and man-made cavities.

J.J. Brooks/Photo Access

Adult landing at nest

ADULT Above golden-brown or pale rufous with grey vermiculations and small, black-rimmed, white spots. Heart-shaped facial disc with prominent centre ridge white or pale buff, rimmed with black and with a dark brown patch in front of the eye. Below white or pale buff with fine, dark brown spots. Legs and vent plain. Flight and centre tail feathers golden-brown above and white below with grey bars, outer tail feathers plain white. Bill pale pinky-cream; eyes very dark brown; long, slender, sparsely feathered legs and bare toes grey. Sexes similar in plumage and size (226–470 g). Plumage colour varies according to location. North African populations have all-white underparts and face; other minor island variations.

JUVENILE Similar to adult, but slightly darker, more rufous and greyer above, and washed with buff below. Chicks clad in plain white down.

DISTRIBUTION, HABITAT AND STATUS Pale European *T. a. alba* extends to northern Africa. Darker spotted *T. a. affinis* occurs throughout sub-Saharan Africa and on offshore islands of Zanzibar, Pemba, Bioko and Príncipe. Extinct on Aldabra, introduced to Seychelles. Slightly larger *T. a. hypermetra* found on Madagascar and the Comoros, *T. a. schmitzi* (paler with larger spots below) on Madeira, *T. a. gracilirostris* on Canary and *T. a. detorta* on Cape Verde Islands. Quite different population (*T. a. thomensis*) on São Tomé, grey above, brown below and spotted black and white all over. Variation between subspecies slight relative to distinctiveness of species as a whole. Widespread in a variety of natural and man-made habitats. Often common locally. Resident, or nomadic in response to rodent outbreaks. A vagrant to true desert and evergreen forest, except around oases and in forest clearings, where it is more common.

SIMILAR SPECIES Very similar in body form to slightly larger **African Grass Owl** (p. 284) (plain dark brown above; facial disc longer, rounder, often buff-coloured; below pale ochre with only a few spots; clicking call in flight). Overlaps in Madagascar with **Madagascar Red Owl** (p. 280) (smaller; uniformly red-brown; fine, black spots all over) and in central Africa with rare **African Bay Owl** (p. 326) (smaller; rufous with brown spots; grey forearm and shoulder). Overlaps widely in habitat with **Marsh Owl** (p. 268) (uniformly brown all over; pale ochre 'window' in primaries; large, dark eyes; dark bill; small forehead ear-tufts).

ALTERNATIVE NAME: Barn Owl

Pale and 'earless'. Above pale golden-brown or pale rufous mottled with grey and flecked with white. Below white or buffy, either plain or finely spotted with brown. Note the white or pale buff heart-shaped facial disc, small, dark eyes, cream bill and long, slender legs. Main call a single long, tremulous screech. Medium-sized (about 30 cm tall, 76–91 cm wingspan).

African Grass Owl, Madagascar Red Owl, African Bay Owl

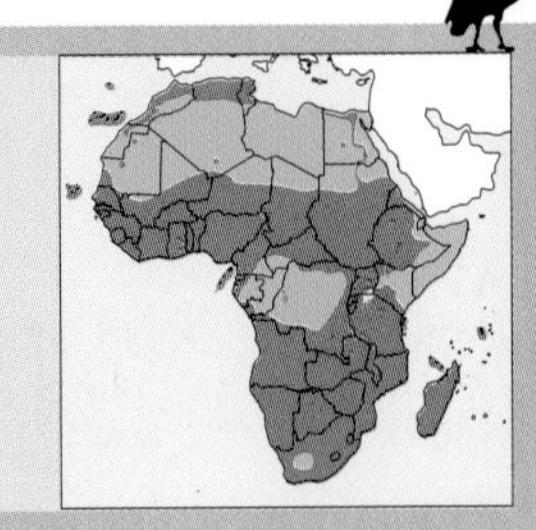

L. Hes

African Grass Owl

Tyto capensis

AT PERCH

Usually stands on the ground in dense vegetation, very rarely on an open perch. Note long legs, small, dark eyes and pale bill. Appears dark brown above and white below, with large, almost circular facial disc.

IN FLIGHT

Flies rather slowly on long, broad, rounded wings, clearly bowed, with large head very obvious. Short tail usually obscured by long legs and feet, unless fanned, when it appears white with dark centre. Appears dark brown above, with ochre primary bases, and buff below with slight dark bars.

DISTINCTIVE BEHAVIOUR

Mainly nocturnal but sometimes hunts in the early mornings. Silent but for bursts of soft, frog-like, clicking notes in flight, or an occasional wavering screech. Courses back and forth, usually over marshy vegetation or its surroundings, hesitating or hovering briefly before dropping on prey. Main diet rodents (especially *Otomys* vlei rats) and small birds. Roosts among tall, dense grass and weeds. Forms landing platforms and tunnels, one of which may lead to a nest chamber. Produces squat, compact pellets, often coated in dark, shiny mucous. Flies only a short distance if flushed.

P. Ginn

Adult female at nest

ADULT Above dark sooty olive-brown with tiny, white spots. Large facial disc white or pale buff, with dark brown eyebrows, eye-ring and chin line. Below and side of neck pale ochre with fine, dark brown spots. Flight feathers, greater coverts and alula, above sooty-brown with buff bars and long, dark ends, below white with faint dark bars. Tail sooty-brown in centre grading to white on outside. Bill pale pinkish-cream (appears almost white); small eyes very dark brown; very long, sparsely feathered legs and toes pinkish cream. Sexes similar in plumage and size (both sexes 266–470 g).

JUVENILE Similar to adult, but feather bases more ochre and spotted with white above; buff facial disc and nestling down.

DISTRIBUTION, HABITAT AND STATUS Eastern to southern Africa, from Ethiopian highlands and Kenya to South Africa; Cameroon highlands. Found in areas of tall grass and weeds, mainly in marshes but also lowland fynbos. Nowhere common; easily overlooked. Patchy, localised distribution.

SIMILAR SPECIES Very similar in body form to slightly smaller **Common Barn Owl** (p. 282) (pale golden-brown above spotted with white and mottled grey; facial disc white or pale buff and heart-shaped; underparts white or pale buff with dark spots; no special flight call but similar screech). Overlaps in habitat with **Marsh Owl** (p. 268) (uniformly brown; ochre bars in primaries form pale 'window'; long, narrow wings; large, dark eyes; dark bill; small forehead ear-tufts).

ALTERNATIVE NAME: Grass Owl

Dark, 'earless' and large-headed; pale buff or white facial disc; small, dark eyes; cream bill. Above plain dark sooty-brown with white spots; below pale ochre with dark spots. Tail white with dark centre. White underwings unbarred. Call a continuous frog-like clicking in flight, rarely a single screech. Medium-sized (about 35 cm tall, 95 cm wingspan).

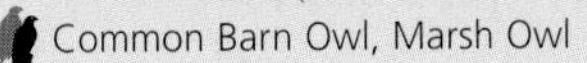
Common Barn Owl, Marsh Owl

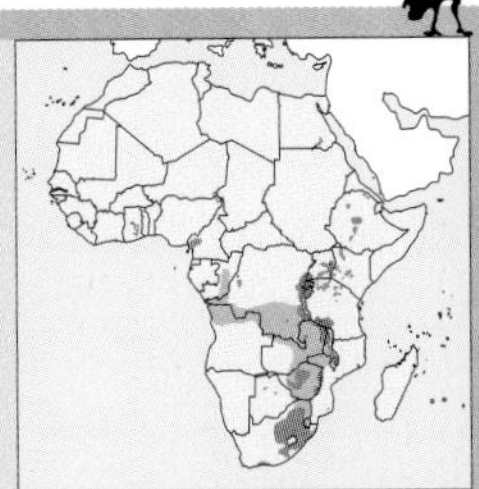

Maned Owl

Jubula lettii

AT PERCH

Perches upright. Long crown and 'ear' feathers most obvious. Note pale lines across wing and shoulder, dark rim to face and white eyebrows; at close range note yellow bill and eyes.

IN FLIGHT

Nocturnal and silent, not normally flushed by day.

DISTINCTIVE BEHAVIOUR

Little known. Favours areas in forest with dense creepers, among which it often roosts, as low as 2 m above ground. Emerges to open perch at dusk, sometimes along rivers. Insects appear to be main prey. Call unknown. Nesting undescribed.

ADULT Above deep chestnut with fine, wavy, sooty bars. Darkest on lesser wing coverts and head, in contrast to white forehead and white tips to 'ears'. Scapulars and greater coverts have buff bases and white tips. Very long crown, nape and 'ear' feathers. Facial disc rufous with broad, dusky and black rim, throat white. Chest rufous with white vermiculation, breast more buff with broad, dusky-brown streaks. Legs and vent plain rufous or buff. Flight feathers and tail rufous with broad, dusky-brown bars. Much individual variation in hues and markings. Bill yellow grading into greenish-yellow cere; eyes yellow; feet yellow. Sexes similar in size, female usually darker and more heavily marked (single female specimen 183 g).

JUVENILE Head down white, body buff with fine rufous bars. Later moults into adult-like plumage.

DISTRIBUTION, HABITAT AND STATUS Known only from tropical lowland rainforests of western and central Africa. Secretive and may not be as uncommon as it appears.

SIMILAR SPECIES Most similar in size to 'earless' **African Wood Owl** (p. 278) (dark brown and rufous barred with white; rounded head with pale facial disc; large, dark brown eyes with red-rimmed lids; deep yellow bill). Larger than any 'eared' ***Otus*** **scops owls** (pp. 304–324) (grey or rufous forms; yellow eyes; most with dark bills) or 'earless' ***Glaucidium*** **owlets** (pp. 294–300) (large, rounded head; small, yellow eyes; white bar across shoulder; whistling calls), and smaller than 'eared' ***Bubo*** **eagle-owls** (pp. 248–260) (black or pale yellow bills; brown or yellow eyes; deep, hooting calls).

Medium-sized, rufous owl. Long, floppy, white-tipped crown and 'ear' feathers. Appears especially large-headed. Heavy, dark streaks below. Yellow eyes, bill and feet. Call unknown. Small (about 25 cm tall, 85 cm wingspan).
African Wood Owl

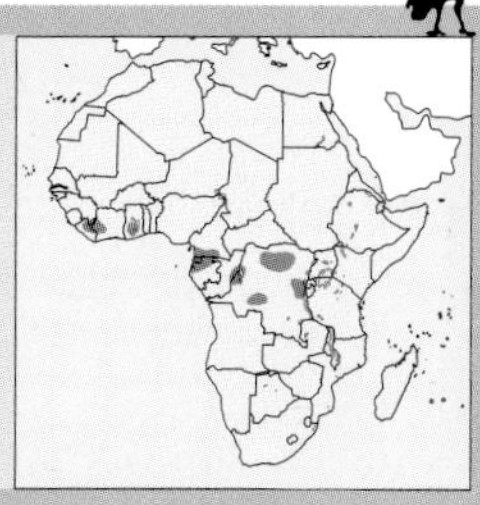

Madagascar Hawk-owl

Ninox superciliaris

AT PERCH
Dumpy owl with large, rounded head. White eyebrow and barred breast most obvious. Dark eyes prominent in pale face. Stands upright and appears large-headed and short-tailed.

IN FLIGHT
Nocturnal. Strong, fast flier on rather long wings and short tail. Wings notably dark and with prominent white spots.

DISTINCTIVE BEHAVIOUR
Roosts in dense cover. Emerges at nightfall to perch over open sites. Flies out to catch insects, and a few birds and bats, which it generally takes on the wing. Main call a muffled hooting 'hoo-oo hoo-oo' followed by a series of up to 15 strident 'kiang kiang kiang ....' notes, rising in pitch and volume. Mates often call together and neighbours respond to one another. Nests on the ground.

ADULT Above brown with white spots on crown, mantle and wing coverts. Face grey-brown, with prominent white eyebrows and buff chin and throat. Below buff, broadly barred with dark brown. Underwing, legs and vent pale buff. Flight feathers brown with incomplete white bars forming large spots. Tail plain brown. Bill and cere off-white, eyes dark brown, feet pale yellow. Sexes similar in plumage and size.
JUVENILE Similar to adult.
DISTRIBUTION, HABITAT AND STATUS Madagascar. Widespread at lower altitudes. Commonest in drier woodland and gallery forest on the west and south of the island, including in secondary forest. Uncommon and more local in the moister forests of the northeast. Also in cultivated areas, up to about 800 m a.s.l. in sparsely populated areas.
SIMILAR SPECIES Larger than the 'eared' **Madagascar Scops Owl** (p. 316) (slender form; plainer above; reddish-brown plumage; yellow eyes; call a series of soft, croaking notes). Similar in size and flight form to **Common Barn Owl** (p. 282) (paler golden-brown above; white or cream underparts; prominent, heart-shaped facial disc; small, black eyes; call a single, drawn-out screech). Shares white eyebrow with yellow-eyed and -legged **Henst's Goshawk** (p. 182).

Adult perched in roost hole

ALTERNATIVE NAME: White-browed Owl

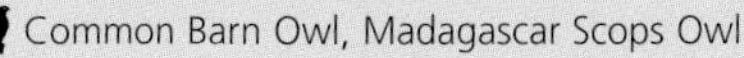

Medium-sized stocky 'earless' brown owl. Prominent pale eyebrows. Above spotted with white. Below buff with broad, brown bars. Dark brown eyes. Pale horn-coloured bill. Call a muffled 'hoo-oo hoo-oo' followed by a strident 'kiang kiang kiang kiang'. Small (about 25 cm tall, about 70 cm wingspan). Restricted to Madagascar.

Common Barn Owl, Madagascar Scops Owl

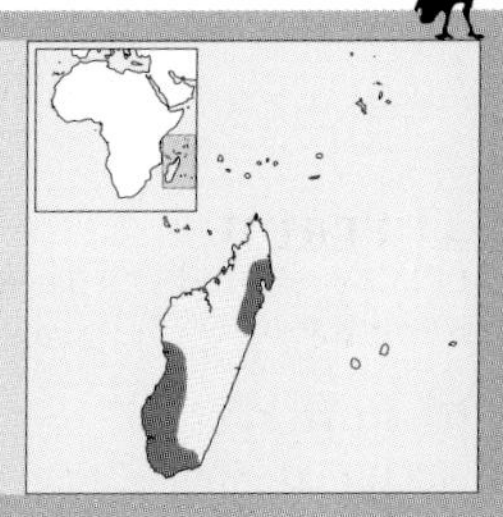

Hume's Owl

Strix butleri

AT PERCH

Appears sandy-brown above and white below without obvious markings except for pale bar across shoulders. Orange eyes and plain facial disc with rufous rim evident at close range.

IN FLIGHT

Nocturnal and silent; seen on the wing by day only if flushed. Flies directly on broad, rounded wings and often lands nearby if disturbed.

DISTINCTIVE BEHAVIOUR

Roosts in caves, on rock ledges or among boulders, less often in palm trees. Calls just before dark, any time at night and sometimes on into dawn. Main call a series of slow deep, double hoots over about 2 seconds 'whooo woohoo-woohoo', the second part lower in pitch, repeated at about 15–60-second intervals. Hunts mainly small mammals and insects on the ground. Rarely hawks on the wing. Nests in caves and rock crevices.

ADULT Above dusky-fawn or mid-brown, nape and mantle mottled with rufous. Scapulars dusky brown with pale rufous outer edges. Facial disc dusky-fawn to white with rufous rim. Below pale cinnamon to white, with narrow dusky-brown bars throughout underparts. Flight feathers and tail dark brown with broad, pale rufous-fawn bars. Much individual variation in hues and markings, some fawn above and almost white below. Bill dark brown merging into slate-grey cere; eyes pale orange; feet slate-grey. Sexes similar in plumage and size (female 214–225 g).

JUVENILE Paler and less clearly marked.

DISTRIBUTION, HABITAT AND STATUS In Africa known only from Red Sea Mountains of eastern Egypt and Sinai Peninsula. Desert equivalent of Tawny Owl, widespread in Middle East. Occupies rocky gorges, often near water. Uncommon and very local.

SIMILAR SPECIES Occurs with similar-sized but longer-winged **Common Barn Owl** (p. 282) (pale golden-brown above; plain white or cream below; white, heart-shaped facial disc; small, dark brown eyes; main call a single, high screech) and much larger 'eared' **Pharaoh Eagle-owl** (p. 252) (also sandy-brown; pale brown and streaked below; orange-eyed; call a deep 'oo hoo'). Much larger than 'earless' **Little Owl** (p. 302) (yellow eyes; spotted with white above; long legs; call a single, high yelp 'krooh') or 'eared' **Pallid Scops Owl** (p. 312) (yellow eyes; finely streaked; pale shoulder-bar; call a single 'hoop' note repeated).

ALTERNATIVE NAMES: Hume's Tawny Owl, Hume's Desert Owl

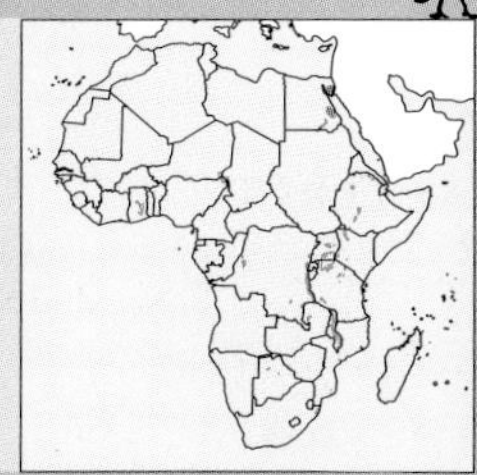

Medium-sized, dumpy, fawn-coloured 'earless' owl. Orange eyes; dark brown bill. Call a series of slow, deep hoots 'whooo woohoo-woohoo' repeated at intervals. Small (about 25 cm tall, 80 cm wingspan). Restricted to north-eastern Egypt.

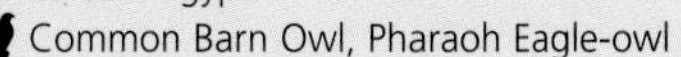

Common Barn Owl, Pharaoh Eagle-owl

M. Gunther/BIOS

Tawny Owl

Strix aluco

AT PERCH

Usually stands with head drawn in. Appears grey or brown above with dark bars and white below with grey or brown streaks. Pale blobs forming bars across shoulder and side of wing often notable. Large, rounded head with dark eyes and pale bill obvious at close range.

IN FLIGHT

Nocturnal and silent; only rarely emerges by day or at dusk.

DISTINCTIVE BEHAVIOUR

Roosts well-concealed within foliage, against branches or in tree holes. Emerges after dark to hunt main prey of small mammals, birds and insects. Catches most prey on the ground or in foliage, hunting on the wing over more open areas. Main calls repeated at intervals, a complex bubbling 3-part series of mellow hoots 'hu-hu-hu-hoo...ho...hoo-hoo-hu-hu-hu hoooooo', or a loud 'kee-wick' contact call. Nests in tree hollows.

P. Steyn

Adult perched within roost

Adult at nest hole

ADULT Above grey or brown with dark grey or brown bars and vermiculations, especially on crown and mantle. Scapulars and greater wing coverts have white ends. Facial disc white with fine, grey bars and broad, dark grey rim. Throat and underparts white and brown, with dark grey streaks on chest and breast and with bars on flanks. Flight feathers and tail grey-brown with broad, buff and brown bars. Bill and cere pale brown; eyes dark brown; eyelids rimmed pink; feet dull brown. Sexes similar in plumage, female about 4% larger (male 342–540 g, female 418–650 g).

JUVENILE Head and body down of juvenile white; moults soon after fledging into plumage similar to adult's, but browner and less clearly marked.

DISTRIBUTION, HABITAT AND STATUS Restricted to area north of the Atlas Mountains in Morocco, Algeria and Tunisia. Pairs occupy patches of woodland, deciduous forest and plantation. Common.

SIMILAR SPECIES Occurs in the same areas as similar-sized **Eurasian Long-eared Owl** (p. 276) (brown and buff; without obvious markings; slender build; long 'ears'; yellow eyes) and longer-winged **Common Barn Owl** (p. 282) (pale golden-brown above; plain white or pale buff below; white, heart-shaped facial disc; small, dark brown eyes, main call a single high screech).

ALTERNATIVE NAME: Eurasian Tawny Owl

Medium-sized, dumpy, grey or brown 'earless' owl with buff to cream underparts. Barred above and streaked below with dark grey or brown. Dark brown eyes. Pale brown bill. Call a complex 3-part series of hoots 'hu-hu-hu-hoo...ho...hoo-hoo-hu-hu-hu hoooooo', or a loud 'kee-wick' contact call. Small (about 30 cm tall, 90 cm wingspan). Restricted to extreme northwest of Africa.

Common Barn Owl, Eurasian Long-eared Owl

A. Williams/NHPA

Red-chested Owlet

Glaucidium tephronotum

AT PERCH

Roosts in tree holes. When it emerges by day, grey head, plain brown back without pale shoulder-bar, darkly spotted underparts and white spotted tail are most obvious. Note yellow softparts at close range.

IN FLIGHT

Short, dark, rounded wings and relatively long, spotted tail are distinctive. Very direct, rapid flight with whirring wingbeats.

DISTINCTIVE BEHAVIOUR

Reported to roost in tree holes, but also sometimes active by day. Call a series of up to 20 high, whistling notes, usually at 1–2-second intervals with short breaks between series. Pair may call together, giving notes at faster and slower rates. Difficult to detect when it calls high up in the forest, 10–15 m above ground level, but does call from among tangled undergrowth 1–2 m above the ground. Diet mainly arthropods and a few small birds, caught off foliage or on the wing. Nest undescribed.

ADULT Head and neck dark grey with broad, white tips to nape feathers; otherwise dusky-brown with rufous wash on wing coverts. Facial disc and throat pale grey flecked with white. Below white or pale rufous, washed with deeper rufous on chest and flanks. Breast has large, dark brown spots. Flight feathers dusky-brown with faint, paler bars, tail dusky-brown with 3 broad, incomplete, white bars forming large spots. Bill greenish-yellow; cere, eyes, feet and claws deep yellow. Sexes similar in plumage and size (male 80–95 g, female 80–103 g).
JUVENILE Undescribed.
DISTRIBUTION, HABITAT AND STATUS Lowland tropical rainforests, western and central Africa. Uncommon and local with a patchy distribution.

SIMILAR SPECIES Overlaps with much larger **African Barred Owlet** (p. 296) (pale shoulder-bar; face and chest barred; calls with repeated purring notes) – most similar forms in western Africa (plain rufous-brown above; almost unbarred flight and tail feathers) and in eastern Africa (faintly barred back; barred head; narrowly barred flight and tail feathers). Also overlaps with **Albertine Owlet** (p. 328) (spotted head; plain back; pale shoulder-bar; call undescribed). Occurs in Congo Basin with much larger **Chestnut-backed Owlet** (p. 300) (barred head, neck and underparts; plain chestnut back; rufous underparts; call consists of a series of a few slow, purring notes).

Very small, brown 'earless' owl. Plain dark grey head. White tail spots. Plain brown back. No pale shoulder-bar. Below rufous and white with dark brown spots. Note yellow bill, eyes and feet at close range. Call a series of up to 20 high, whistling notes. Very small (about 14 cm tall, 35 cm wingspan).
African Barred Owlet, Albertine Owlet, Chestnut-backed Owlet

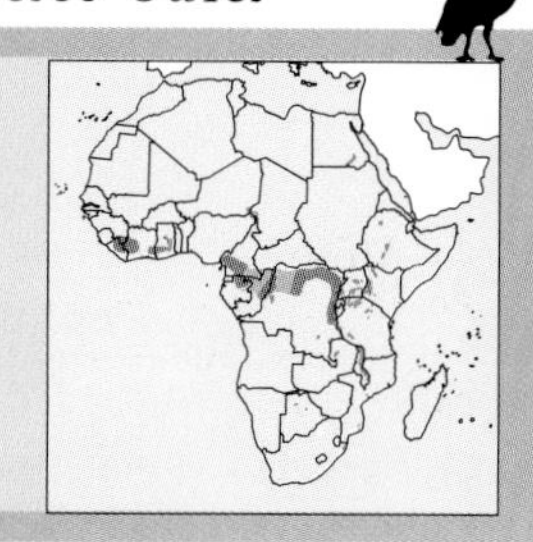

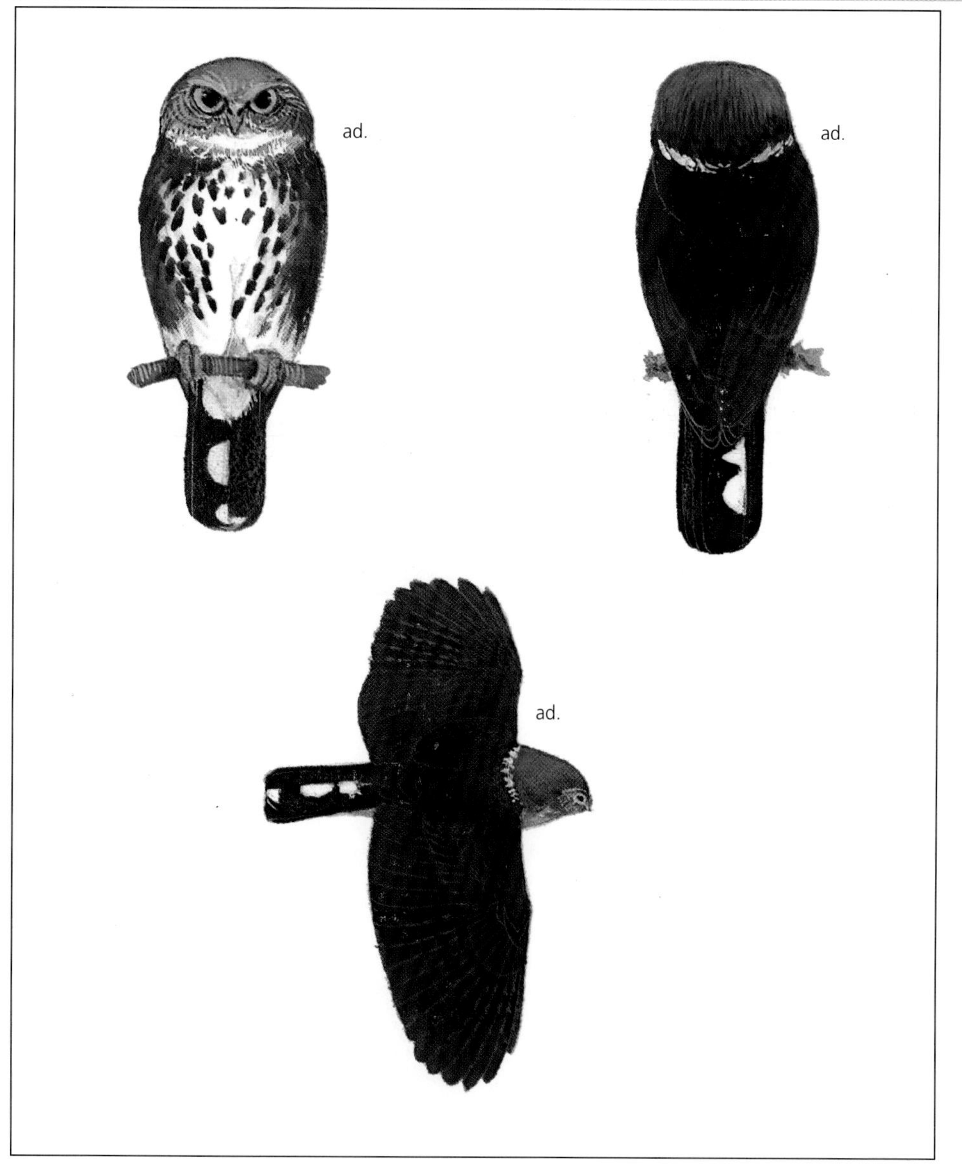

African Barred Owlet

Glaucidium capense

AT PERCH

Large, rounded 'earless' head enhances dumpy build. Appears brown with white shoulder-bar at a distance. Note fine barring and yellow eyes. Wing tips much shorter than long tail.

IN FLIGHT

Sometimes active by day. Fast wingbeats audible, interspersed with short, dipping glides. Wings appear short and rounded, with white shoulder-stripes; tail rather long; overall finely barred.

DISTINCTIVE BEHAVIOUR

Often perches within dense cover, but at times emerges by day and sits in the open. Call a series of purring notes at one pitch 'wirow wirow wirow wirow', also double-noted 'prr-purr prr-purr prr-purr', and other softer, more mellow notes. Same call across whole range. Feeds mainly on insects with more small mice and birds when breeding. Nests in tree holes.

ADULT In southern and central Africa (*G.c. capense* and *G.c. ngamiense*), grey-brown above with fine, buff bars and thin, white eyebrow. In East Africa (*G.c. scheffleri* and *G.c. castaneum*), head and neck have fine, buff bars; rest often plain rufous. In western Africa (*G.c. etchecopari*), all deep rufous-brown and unbarred. Face always finely barred. Scapulars and greater wing coverts have white outer webs and dark brown tips. Below, chest brown finely barred with buff. Breast and flanks white spotted with dark brown feather tips. Underwing coverts, legs and vent white or coverts buff. Rufous wash below in eastern and western Africa. Flight and tail feathers dark brown with narrow, rufous bars, markings much reduced in western Africa. Individual and regional variation in density of hues and markings. Bill and cere dull greenish-yellow; eyes pale yellow; feet dull yellow to pale brown. Sexes similar in plumage and size (male 83–132 g, female 93–139 g). Smallest in eastern and western Africa.
JUVENILE Similar to adult but less barred above and spotted below, with very short tail at fledging. Mouth and tongue black.
DISTRIBUTION, HABITAT AND STATUS Widespread in woodland and densely treed savanna of southern Africa, with forest population in western Africa and Congo–Uganda border. Common in patches of woodland. Local and uncommon in more open savanna, along rivers or in wooded gullies. Rare in forest.
SIMILAR SPECIES Overlaps widely in savanna and woodland with slightly smaller **Pearl-spotted Owlet** (p. 298) (smaller head; dark, false nuchal 'eye' spots; spotted back and head; streaked breast; off-white facial disc; heavy bill and feet; call a series of whistles rising in pitch and volume), in eastern forests with **Albertine Owlet** (p. 328) (spotted head and nape; plain maroon-brown back; call undescribed), and in western forests with **Red-chested Owlet** (p. 294) (plain grey head; dark maroon-brown back; no pale shoulder-bar; below deep rufous and white, boldly spotted with brown; white-spotted tail; call a series of high whistling notes with breaks between series). Dumpier than any of the small 'eared' ***Otus* scops owls** (pp. 304–324) (slender build; small bill and feet; mainly grey or rufous; calls widely spaced single notes).

I. Hes

Adult *G.c. ngamiensis* perched at night

ALTERNATIVE NAMES: African Barred Owl, Barred Owlet, Chestnut Owlet, Ngami Owlet, Scheffler's Owlet

Small, dumpy, large-headed 'earless' owl. Brown above, finely barred with buff. Prominent white shoulder-bar. Below white with brown spots. Note small, barred face and yellow eyes. Call a series of purring notes at one pitch 'wirow wirow wirow wirow'. Very small (about 17 cm tall, 40 cm wingspan).
Pearl-spotted Owlet, Red-chested Owlet, Albertine Owlet

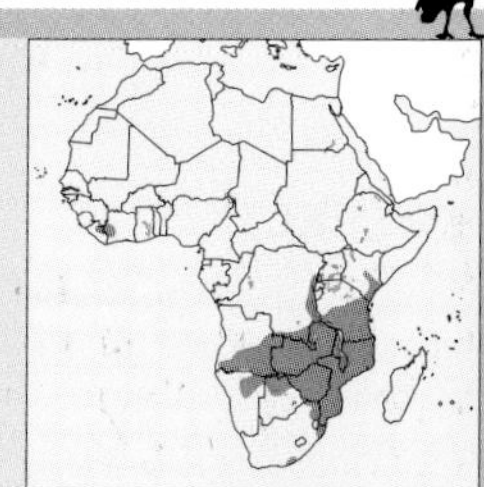

Pearl-spotted Owlet

Glaucidium perlatum

AT PERCH

Usually looks squat and dumpy, with a large, rounded head. Wing tips much shorter than long tail. If disturbed at roost, may tighten plumage to slim form with smaller head and slight, rounded 'ear' bumps. Appears brown above with pale shoulder-bar or white below with dark streaks. Pearl-like spots and 'cross'-looking face obvious at close range.

IN FLIGHT

Emerges regularly by day. Flaps with fast, audible bursts of the short, rounded wings. Intersperses glides, often close to the ground, before swinging up to a perch. Note white-spotted flight, tail and rump feathers. Appears tailless during moult.

DISTINCTIVE BEHAVIOUR

Each pair usually territorial and responsive to imitation of its main call, a long series of penetrating whistles, rising in pitch and volume to a crescendo of longer notes '... peu-peu-peu-peu peeuu peeuu peeuu'. Often jerks tail to one side. Hunts a wide range of small animals, mainly from a perch but less often in flight. Often emerges by day to hunt or call. Nests in a tree hole, female keeping up a monotonous peeping before laying.

N. Dennis

Adult with 'ear' tufts evident

M. Goetz

Adult

ADULT Above cinnamon-brown with small, white spots, except for broad, white tips to scapulars and greater wing coverts. Back of head with sooty-brown nuchal patches, rimmed with white and rufous. Facial disc and eyebrows white. Underparts white with broad, brown streaks, especially on chest. Flight feathers and tail dark brown with incomplete white bars forming white spots. Individual and regional variation in density of colour, spots and rufous-brown wash. Heavy bill dull yellow; cere brown; eyes yellow; strong feet brown. Adults usually moult tail after breeding so may also appear tailless. Sexes similar in plumage and size (male 68–86 g, female 77–147 g).

JUVENILE Similar to adult, even at fledging, but tail initially very short. False 'eyes' on nape or nuchal 'face' may be more prominent. Mouth and tongue pink.

DISTRIBUTION, HABITAT AND STATUS Widespread in savanna and woodland of sub-Saharan Africa. In many areas the commonest and most conspicuously diurnal small owl. Only absent from dense woodland and forest, and from treeless desert and grassland.

SIMILAR SPECIES Overlaps widely in more wooded areas with slightly larger **African Barred Owlet** (p. 296) (finely barred back, head and chest; even larger rounded head; face appears small; lacks false nuchal 'eyes'; spotted breast; flight, tail and rump feathers barred with pale rufous; call a series of even-pitched purring notes). Dumpier than any of the small 'eared' ***Otus* scops owls** (pp. 304–324) (slender build; small bill and feet; overlapping species grey or rufous; calls of single, widely spaced purring notes).

Very small, brown 'earless' owl. Spotted with white above. Below white streaked with brown. Large, rounded head. Yellow eyes and bill. Back of head has black, white-rimmed, false 'eyes'. Call a series of penetrating whistles which rise in pitch and volume 'peu-peu-peu-peu peeuu peeuu peeuu'. Often active by day. Very small (about 15 cm tall, 37 cm wingspan).
African Barred Owlet

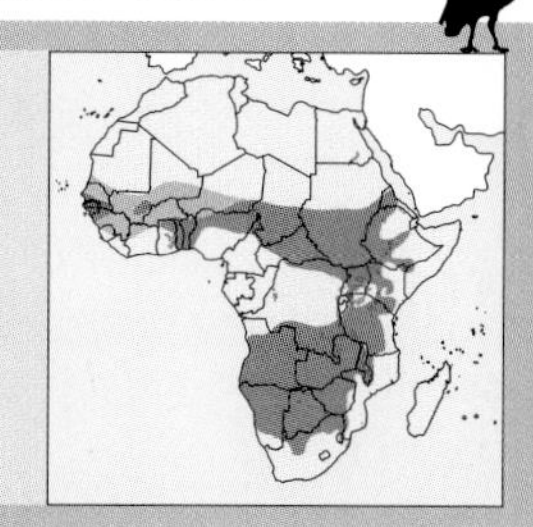

C. Paterson-Jones

Chestnut-backed Owlet

Glaucidium sjostedti

AT PERCH
Large, rounded, finely barred head most obvious feature. Note yellow softparts at close range. Appears dark overall with fine, lighter barring.

IN FLIGHT
Nocturnal but audible, with short, rounded wings and a rather long tail. Flushes readily by day if approached too close.

DISTINCTIVE BEHAVIOUR
Roosts deep and low at 2–6 m inside cover. Pairs are territorial and mates often call together. Main call a series of 2–4 deep, purring notes at half-second intervals, repeated after a short pause of about 1 second. Calls mainly at dusk and dawn, often very high in and above forest canopy. Usually found hunting close to the forest floor, at 1–2 m, but does not usually perch in open sites. Feeds mainly on insects and other small animals. Nests in tree holes.

ADULT Head and neck dusky-brown with fine, white bars. Back and wing coverts deep chestnut. Outer edge of scapulars has fine, buff bars, ends of greater wing coverts white. Throat white. Below cinnamon, breast has fine, dark brown bars, legs and vent plain and paler. Flight and tail feathers dusky-brown with narrow, white bars. Bill, cere, eyes, feet and claws pale yellow. Sexes similar in plumage and size.
JUVENILE Similar to adult, but paler with fine, buff bars on scapulars and upperwing coverts and only faint, dark bars on breast.
DISTRIBUTION, HABITAT AND STATUS Restricted to tropical lowland primary rainforest from Cameroon eastwards to the Congo Basin. Uncommon and local.
SIMILAR SPECIES Range overlaps widely with much smaller **Red-chested Owlet** (p. 294) (plain grey head; dark brown back; below rufous and white, boldly spotted with brown; white spots on tail; call a series of high, whistling notes with breaks between series). Dumpier and larger-headed than the much smaller 'eared' **Sandy Scops Owl** (p. 314) (rufous with white spots; pale shoulder-bar; slender build; small bill and feet; call a repeated long, falling whistle) or the slightly larger 'eared' **Maned Owl** (p. 286) (deep rufous; long, floppy crown feathers; long, white-tipped 'ears'; streaked below; yellow eyes; call undescribed). Smaller than 'earless' **African Wood Owl** (p. 278) (mottled and barred with red-brown and white; large, dark brown eyes; red-rimmed eyelids; call either a single howl or a rhythmic series of 7 hoots).

Swanepoel/Transvaal Museum

captive adult

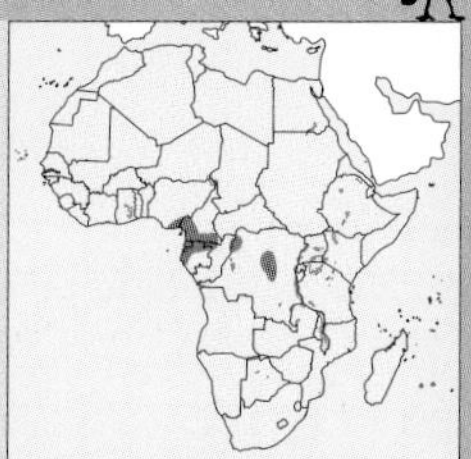

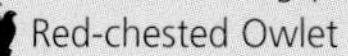

Medium-sized, boldly coloured, 'earless' owl. Head and neck dark brown with white bars, eyebrows and throat. Back deep chestnut. Below cinnamon, breast barred with dark brown. Bill, eyes and feet pale yellow. Call a repeated series of 2–4 deep, purring notes at half-second intervals. Small (about 20 cm tall, 60 cm wingspan).

Red-chested Owlet

Swanepoel/Transvaal Museum

Little Owl

Athene noctua

AT PERCH

Stands very upright on long legs. Note flat-topped head, often withdrawn, and short tail. Appears light sandy-brown or darker brown with pale streaks. Yellow eyes obvious at close range.

IN FLIGHT

Often active by day. Flaps with short bursts of fast wingbeats, dipping in between with an undulating flight path. Wings and tail lightly barred, short and rounded.

DISTINCTIVE BEHAVIOUR

Roosts in bushes or among crevices in rocks and buildings. Often emerges in daylight, especially in open desert habitats. Main call a loud, high-pitched yelp – 'kroooh' – given singly or repeated at different frequencies. Often first noticed by a sharp 'chik' or soft whistle of alarm. Hunts from open perches, catching prey on the ground, often after a chase on foot. Eats mainly insects and some small vertebrates. Nests in rock, tree or man-made holes.

ADULT Above cinnamon-brown to dark brown, head and neck streaked and the rest spotted with white. Face plain cinnamon-brown, throat white. Below white, cream or pale rufous, the breast with broad cinnamon-brown streaks and the belly finely streaked. Flight and tail feathers dusky-cinnamon with buff bars. Bill and cere pale greenish-yellow; eyes yellow; feet dull yellow. Sexes similar in size and plumage (male 108–210 g, female 120–207 g).

JUVENILE Similar to adult, but browner with less distinct markings.

DISTRIBUTION, HABITAT AND STATUS Widespread across coastal northern and northeastern Africa, extending inland to rocky mountains in the Sahara. Commonest in open, wooded savanna with sparse groundcover, or in desert areas with rocky outcrops or buildings. Very rare vagrant to sub-Saharan western Africa.

SIMILAR SPECIES Only other small owls within its range are the slender 'eared' **Eurasian Scops Owl** (p. 310) (grey or grey-brown with fine, darker streaks; call a single high, croaking note every 2 seconds) and, in northeastern Egypt, **Pallid Scops Owl** (p. 312) (sandy-brown, grey-brown or grey with fine, darker streaks; call a 'whoop' note about every 0.6 seconds).

K. Taylor/Bruce Coleman Ltd

Adult landing

F. Coleman/Planet Earth Pictures

Adult with prey

Small, 'earless' long-legged owl. Pale or dark brown, spotted and streaked with white. Large, yellow eyes. Call a single loud, high-pitched yelp 'kroooh' or a sharp 'chik' of alarm. Very small (about 20 cm tall, 45 cm wingspan). Only in northern Africa.

Eurasian Scops Owl, Pallid Scops Owl

S. Dalton/NHPA

White-faced Scops Owl

Otus leucotis

AT PERCH

Stands upright within dense cover or against a branch, the long 'ears' very prominent and the large eyes often slitted if approached too closely. Overall appearance grey with fine, dark streaks. Black-rimmed white face and orange eyes (if open) are distinctive.

IN FLIGHT

Nocturnal, silent and seen in flight by day only if flushed.

DISTINCTIVE BEHAVIOUR

Inconspicuous, despite being twice as large as other scops owls. Best located by call or by searching at night. Population north of the equator calls with a disyllabic mellow hoot 'po proo', those to the south with a series of bubbling notes starting fast and rising in pitch at the end 'popo-popopopopreoo'. Breeds in old stick-nests of other birds, the long 'ears' often obvious above the rim, or in open tree holes. Main prey of small mammals, birds and insects taken from the ground.

ADULT Above grey with dark grey vermiculations and streaks. Facial disc pale grey to white with broad, black rim and black outer edges to long ear-tufts. Outer end of scapulars white. Below grey with white feather bases, dark grey shaft streaks and fine, grey bars. Flight feathers and tail grey with dark grey bars. Bill pale grey; eyes dark yellow to deep orange; partly feathered toes grey-brown. Sexes similar in plumage and size (150–250 g).

JUVENILE Similar to adult, but paler, browner and less heavily marked, especially on the throat and breast. Eye yellow. Nestling down grey with darker bars.

M. Goetz

Adult in daytime: alert pose

DISTRIBUTION, HABITAT AND STATUS Widespread across sub-Saharan Africa. Occurs in wide range of savanna and woodland, usually with sparse ground cover, from semi-desert thorn scrub to moist tall miombo or forest clearings. Only absent from treeless areas and continuous forest. Generally uncommon, but locally common at rodent outbreaks.

SIMILAR SPECIES Most resembles much smaller 'eared' **African Scops Owl** (p. 306) and **Eurasian Scops Owl** (p. 310) (yellow eyes; grey, unrimmed facial disc; mottled and streaked dark grey plumage; single purring call-notes repeated at regular intervals). Also overlaps widely with much larger 'eared' **Spotted Eagle-owl** (p. 256) (small spotted 'ears'; brown or yellow eyes; mottled above and barred below with dark grey).

I. Hes

Adult in relaxed pose

ALTERNATIVE NAME: White-faced Owl

Large for a scops owl. Grey and white. Long 'ears'. Large, orange eyes. White facial disc with a broad, black rim. Calls either a repeated, liquid hooting 'po proo' (north of equator) or a bubbling 'popopopopopreuu' that starts as a stutter and rises in pitch. Small (about 20 cm tall, 68 cm wingspan).

African Scops Owl, Eurasian Scops Owl, Spotted Eagle-owl

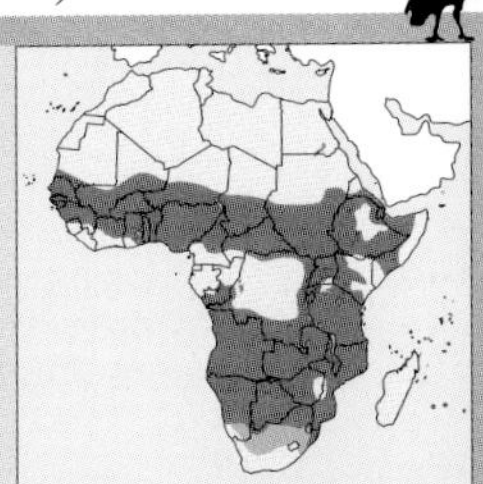

J. Carlyon

African Scops Owl

Otus senegalensis

AT PERCH

During the day, keeps itself upright, long and slender with erect ear-tufts and, when approached, closes the eyelids to slits. Turns the head slowly to follow movement. At night, tucks the ear-tufts away and becomes dumpy, round-headed and wide-eyed.

IN FLIGHT

Nocturnal, silent and rarely seen in flight unless flushed. Flies a short distance with fast wingbeats and undulating flight.

DISTINCTIVE BEHAVIOUR

Often roosts in same site each day, among tangled branches or in the open against a vertical branch or trunk. Very cryptic, especially when eyes closed. At night, best located by its single monotonously repeated, purring call-note, 'kruup', given every 5–8 seconds. Feeds mainly on insects caught on the ground. Nests in natural holes in trees, chick down pale grey.

ADULT Grey overall, with fine, black centre streak and dark grey vermiculations on each feather. Heaviest streaks on head and mantle, facial disc unstreaked with fine, black rim. Below slightly lighter grey, underwing coverts buff with slight grey streaks. Scapulars with white tips. Extent of rufous wash, especially on upperwing coverts, and colour tone, varies individually between grey and rufous forms. Flight feathers dark grey barred with white to pale buff, tail grey with indistinct dark bars. Bill brownish-black; eyes yellow; slender, sparsely feathered legs and small, bare feet dull pale brown. Sexes similar in size and plumage (45–97 g). Island population of Annobon (*O.s. feae*) darker and more broadly streaked with black.

JUVENILE Very similar to adult, but sometimes slightly paler with brown wash, ear-tufts shorter and grey nestling down still evident at fledging.

DISTRIBUTION, HABITAT AND STATUS Widespread in savanna and woodland of sub-Saharan Africa. Probably the scops owl seen on Socotra. Call of darker form on Annobon (Pagulu) unrecorded, possibly a separate species. Absent only from treeless desert and grassland and from dense forest. Most common in well-wooded savanna, with up to 12 individuals audible from a single spot, but difficult to find by day.

SIMILAR SPECIES In western and northeastern Africa visited by very similar, slightly larger, migrant **Eurasian Scops Owl** (p. 310) (less boldly marked; higher, pure 'to' call-note uttered every 2 seconds, but usually silent on migration) – probably indistinguishable by sight in the field. Similar in form to much larger **White-faced Scops Owl** (p. 304) (underparts pale grey with fine, black streaks; white facial disc with broad, black rim; orange-red eyes; long, black-edged ear-tufts; bubbling double- or multiple-noted call). Small ***Glaucidium*** **owlets** (pp. 294–300) are dumpier and lack 'ears'.

T. Dressler/Planet Earth Pictures

Adult in cryptic roosting pose

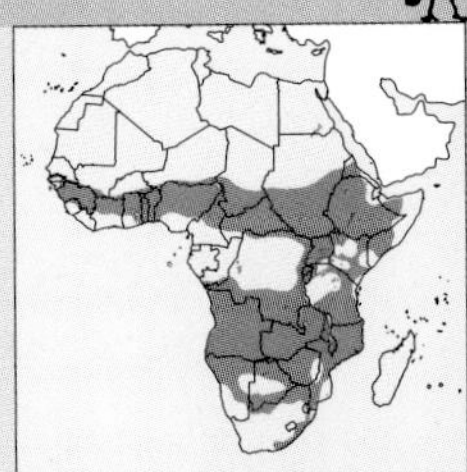

Delicate, slender little 'eared' owl, with 2 forms, either grey or rufous-grey all over. Fine, black streaks and mottling. Yellow eyes. Call a single monotonous, tremulous 'kruup' note, repeated at 5–8-second intervals. Very small (about 15 cm tall, 45 cm wingspan).

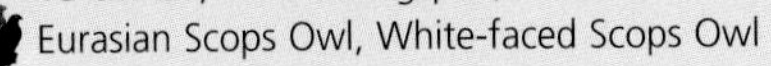

Eurasian Scops Owl, White-faced Scops Owl

HPH Photography/Photo Access

Sokoke Scops Owl

Otus ireneae

AT PERCH

Roosts standing rather upright among branches or against trunks, the 'ears' erect and plumage sleeked down. Pale yellow eyes and cere contrast with rather even plumage colour with only fine spots and streaks.

IN FLIGHT

Nocturnal, silent and rarely seen in flight unless flushed.

DISTINCTIVE BEHAVIOUR

Calls with a series of 5–10 high, whistling 'who' notes repeated at 1-second intervals, similar to a tinker barbet *Pogoniulus* sp. Normally found roosting in the lower canopy of trees. Feeds mainly on small insects.

A. Weaving

Adult pair (rufous form) at roost

ADULT Grey-brown, dark brown or rufous overall, depending on form, with fine, darker vermiculations. Above, crown streaked with dark brown, rest with pale bases and dusky tips for spotted effect. Facial disc buff with dusky rim. Below lighter with rufous wash on chest, and white and dusky spots. Outer primaries barred with white, inner primaries, secondaries and tail barred with darker brown. Bill dark horn colour to black; cere pale yellow; eyes pale yellow; thin, feathered legs and bare feet dull yellow. Sexes similar in plumage and size (male 46–50 g).

JUVENILE Very similar to adult, but feet greenish-grey, facial bristles more obvious and begging call a long, drawn-out, soft trill.

DISTRIBUTION, HABITAT AND STATUS Restricted to southeastern Kenya in Arabuko–Sokoke forest and to the lowlands north of the East Usambara range in northeastern Tanzania. Mainly in *Cyanometra-Manilkara* forest, with vagrants in adjacent miombo *Brachystegia* woodlands. Locally common within its restricted ranges.

SIMILAR SPECIES Most similar to larger resident **African Scops Owl** (p. 306) (calls with long purring notes at 5–8-second intervals; greyer; more streaked and mottled with black; dark bill and cere) and rare migrant **Eurasian Scops Owl** (p. 310) (whistles lower and slower than 1 second apart; greyer; more streaked with black). Small ***Glaucidium*** **owlets** (pp. 294–300) are dumpier and lack 'ears'.

ALTERNATIVE NAME: Morden's Scops Owl

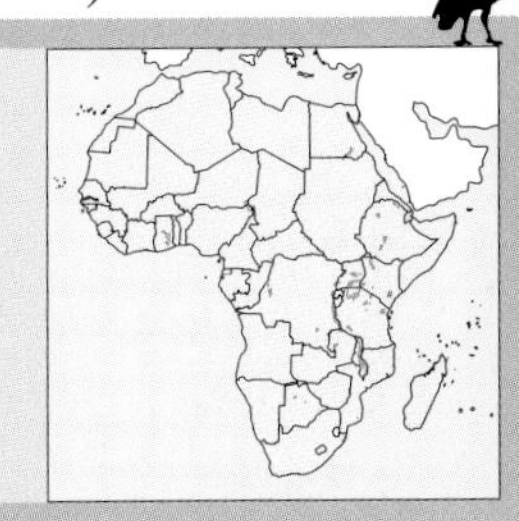

A very small 'eared' owl, with 3 forms. Grey-brown, dark brown or rufous overall. Streaked with white and sooty-brown. Pale yellow eyes and cere. Call a high whistle 'who' repeated 5–10 times at 1-second intervals. Very small (about 12 cm tall, 35 cm wingspan). Restricted to eastern Kenya/Tanzania.

African Scops Owl

Eurasian Scops Owl

Otus scops

AT PERCH
During the day, keeps itself upright, long and slender with erect ear-tufts and, when approached, closes the eyelids to slits. Turns the head slowly to follow movement. At night, tucks the ear-tufts away and becomes dumpy, round-headed and wide-eyed.

IN FLIGHT
Nocturnal, silent and flies by day only if flushed.

DISTINCTIVE BEHAVIOUR
Roosts cryptically in a tangle of branches or against a vertical trunk. Roosts in numbers in the Sahara on migration, at oases or, if necessary, on the ground. Calls at night with a single pure, high croaking 'to' note, given every 2 seconds. Feeds mainly on insects caught on the ground. Nests in natural holes in trees, chick down pale grey.

ADULT Grey or rufous-grey overall, depending on form, with fine, black streaks and dark grey vermiculations on each feather. Streaks heaviest on head and mantle, facial disc unstreaked with fine, black rim. Below slightly lighter, underwing coverts buff with fine, grey streaks. Scapulars tipped white. Flight feathers dark grey barred with white, tail grey with indistinct dark bars. Bill dark horn colour or black; eyes yellow; slender sparsely feathered legs and small, bare feet dull brown. Sexes similar in size and plumage (range 54–91 g). Resident northwestern population (*O.s. mallorcae*) plainer and shorter-winged than migrant western (*O.s. scops*) or eastern (*O.s. pulchellus, O.s. turanicus*) European populations.
JUVENILE Very similar to adult, but sometimes slightly paler with brown wash, ear-tufts shorter and grey nestling down still evident at fledging.
DISTRIBUTION, HABITAT AND STATUS Breeds in open woodland in northwest Africa. Migrates across the Sahara in a broad front, mainly to western Africa but some as far south as northern Tanzania. Uncommon in western Africa, rare elsewhere, but easily overlooked since silent and so similar to local species. Vagrant to Madeira and the Canary Islands. A ship-board vagrant near Seychelles was probably this species.
SIMILAR SPECIES Overlaps widely in sub-Saharan Africa with resident, slightly smaller **African Scops Owl** (p. 306) (more boldly marked; softer, more purring 'kruup' call-note uttered every 5–8 seconds; calls regularly) – probably indistinguishable by sight in the field. Similar in form to much larger **White-faced Scops Owl** (p. 304) (underparts pale grey with fine, black streaks; white facial disc with broad, black rim; orange-red eyes; long, black-edged ear-tufts; bubbling double- or multiple-noted call). Rare in northeastern Egyptian range of **Pallid Scops Owl** (p. 312) (sandy-grey or grey-brown; lightly marked; obvious pale shoulder-bar). Small ***Glaucidium* owlets** (pp. 294–300) are dumpier and lack 'ears'.

ALTERNATIVE NAMES: European Scops Owl, Common Scops Owl

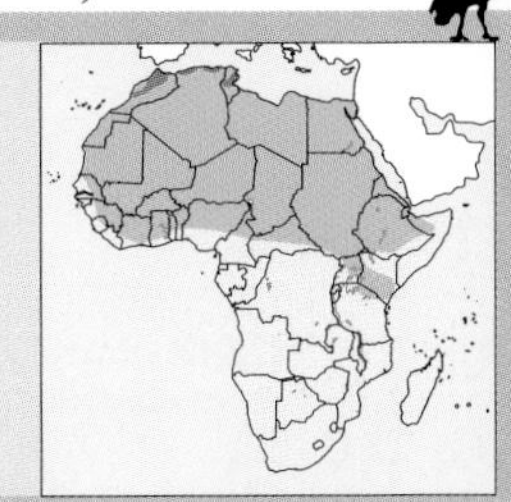

Small, slim 'eared' owl, with 2 forms. Plumage grey or rufous-grey, finely streaked with black and mottled with grey. Yellow eyes. Call a single high, croaking note given about every 2 seconds. Very small (about 15 cm tall, 50 cm wingspan). Breeds in northwestern Africa, an uncommon migrant to western and northeastern Africa.

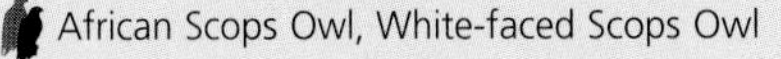

African Scops Owl, White-faced Scops Owl

A. Williams/NHPA

Pallid Scops Owl

Otus brucei

AT PERCH
Usually stands upright and sleeks itself down if approached, raises the ear-tufts and closes the eyes to slits. Pale buff or grey colour and pale cream bar across the wing are distinctive.

IN FLIGHT
Sometimes hunts by day. Sandy upperparts and white underwings contrast with darkly barred buff flight feathers.

DISTINCTIVE BEHAVIOUR
Utters a single repeated, liquid 'whoop' note at 0.6-second intervals, sometimes at irregular 3–5-second intervals. Roosts in tree-holes amongst foliage or against tree trunks. Usually nocturnal, hunting mainly insects on the ground, from foliage or, rarely, on the wing. Occasionally hunts by day.

ADULT Above buff, light sandy-brown or grey, depending on form, with fine grey vermiculations and fine, dusky-brown streaks, especially on head. Scapulars have broad, cream ends and fine, dusky-brown tips. Facial disc with dusky rim; 'ears' and throat white or grey. Below grey- or sandy-buff with dark brown streaks. Flight feathers and tail buff and grey with dusky-brown bars, underwing coverts white. Bill black with yellow base; eyes yellow; slender, feathered legs and bare toes with dusky-brown skin. Sexes similar in plumage and size (male 100 g).

JUVENILE Very similar to adult, but slightly paler buff, with buff nestling down evident at fledging. Bill horn-coloured, feet deep blue-grey.

DISTRIBUTION, HABITAT AND STATUS Rare in northeastern Egypt, extending from adjacent Middle East, where quite common. Occurs among small patches of vegetation or in rocky gorges with scattered trees.

SIMILAR SPECIES Likely to be confused only with scarce migrant **Eurasian Scops Owl** (p. 310) (dark and light grey or rufous-grey; streaked and mottled with black; more obvious pale shoulder-bar).

ALTERNATIVE NAMES: Striated Scops Owl, Bruce's Scops Owl

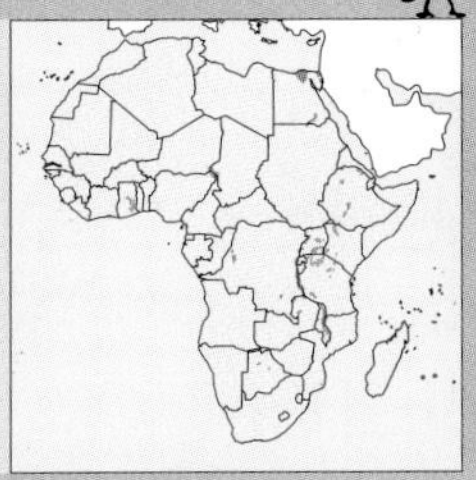

Small, slight 'eared' owl, with 2 forms, either grey- or sandy-brown overall. Finely streaked with dusky-brown. Cream shoulder-bar. Yellow eyes. Call a rapidly repeated 'whoop' note about every 0.6 seconds. Very small (about 15 cm tall, 45 cm wingspan). Restricted to northeastern Egypt.

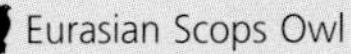
Eurasian Scops Owl

Sandy Scops Owl

Otus icterorhynchus

AT PERCH

Pale rufous colour overall with plain facial disc; pale yellow bill and eyes are distinctive. From the side or behind, white shoulder-bar is obvious. 'Ears' usually upright during the day.

IN FLIGHT

Usually nocturnal, silent and rarely seen on the wing, but will emerge in the middle of the day. Flies fast and directly with rapid, shallow, accipiter-like wingbeats.

DISTINCTIVE BEHAVIOUR

Roosts in tree holes. At night, calls with a long 'twooo' whistle that drops in tone and volume, lasts about a second and is repeated every 3–4 seconds. Eats small insects. Otherwise virtually unknown.

ADULT Above pale cinnamon to rufous-brown, finely spotted with white and barred with buff. Facial disc plain pale cinnamon with cream eyebrows. Below paler and more rufous, with white spots and buff bars on abdomen, and darker buff streaks on breast. End of scapulars white with brown tips. Flight feathers and greater upperwing coverts rufous with dark brown bars; primaries with white bars. Tail barred dark and pale rufous. Much individual variation in tone and markings. Bill, cere and eyes pale yellow; slender, feathered legs and bare toes cream. Sexes similar in plumage and size (male 69–80 g, female 61–80 g).

JUVENILE Similar to adult, but more finely barred above and below, juvenile down pale tawny. Facial disc more rufous. Softparts even paler, almost white.

DISTRIBUTION, HABITAT AND STATUS Known only from scattered localities across lowland evergreen rainforests of western and central Africa (Liberia, Ivory Coast, Ghana, Cameroon, Gabon, Congo). Appears to be rare – not simply difficult to locate.

SIMILAR SPECIES No similar small scops owl in the rainforest habitat, but may overlap at forest edge with **African Scops Owl** (p. 306) (greyer overall coloration) or **Eurasian Scops Owl** (p. 310) (grey with black streaks; dark bill; single, short call-note at constant pitch). Small ***Glaucidium* owlets** (pp. 294–300) are dumpier and lack 'ears'.

ALTERNATIVE NAME: Cinnamon Scops Owl

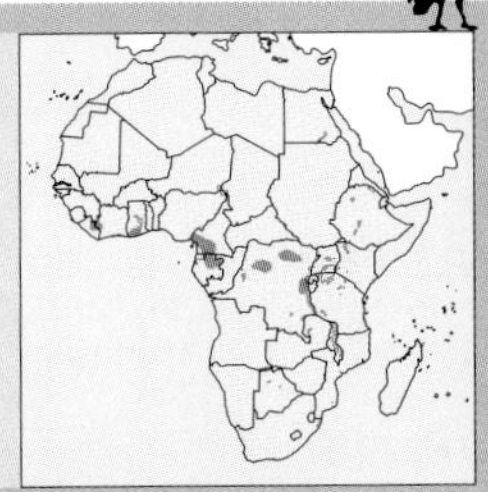

Small, delicate 'eared' owl. Pale rufous, spotted with white. White bar across shoulder. Pale yellow bill and eyes. Call a prolonged, falling, whistling 'twooo' repeated every 3–4 seconds. Very small (about 15 cm tall, 45 cm wingspan). Restricted to equatorial rainforest.

African Scops Owl, Eurasian Scops Owl

Madagascar Scops Owl

Otus rutilus

AT PERCH
'Ears' prominent. Note yellow eyes and pale shoulder-bar. Most appear dark and rather uniformly coloured with only fine darker markings.

IN FLIGHT
Nocturnal, silent, flies by day only if flushed. Flies short distances with rapid wingbeats and direct flight.

DISTINCTIVE BEHAVIOUR
Roosts among thick creepers or in tree holes and crevices, always low down and even on the ground among fallen trees. Same site used for several years. At night, calls frequently with a repetitive series of 3–10 (usually 4) purring notes, 'bruu bruu bruu bruu', fading at end. Mayotte (Maore) birds' call is deeper and slower. Also harsh 'k-r-r-k' call repeated a few times. Uttered with head forward, sometimes several hours before nightfall. Calls high in canopy or lower down in thickets. Rufous form rarer and only common in rainforest habitat. Main diet is insects. Nests in tree holes, less often in old stick-nests.

N. Dennis
Adult rufous form

ADULT Above rufous- or grey-brown, the latter often washed with rufous, and both forms flecked with dark brown. Crown and 'ears' browner, also flecked with dark brown. White ends to scapulars. Face unmarked, but with darker area between bill and eye. Below paler and more streaked with dark brown, least in Mayotte birds. Primaries with pale buff bars. Much individual variation in hue and markings. Bill black, longer in Mayotte birds; eyes yellow; legs feathered and bare toes pinky grey to brown. Sexes apparently similar in plumage and size.
JUVENILE Similar to adult and shows individual colour form from first plumage.
DISTRIBUTION, HABITAT AND STATUS Madagascar and the Comoros island of Mayotte. Occupies a wide range of habitats and altitudes, from urban parks to rainforest, including secondary forest and dry thickets, but avoids exotic plantations. Generally common and widespread on Madagascar and Mayotte. Madagascar and Mayotte populations, with differences in call, may be separate forms or species.
SIMILAR SPECIES Brown form resembles the larger, dumpier **Madagascar Hawk-owl** (p. 288) (prominent pale eyebrows; clearly barred and spotted upperparts; white line across shoulder when perched), buff form similar colour tone to larger **Madagascar Red Owl** (p. 280) ('earless'; orange-brown with fine, dark spots; large-headed; small, dark brown eyes; pale bill). Only small owl on Mayotte, alongside **Common Barn Owl** (p. 282) (much larger; pale golden-brown above spotted with white and mottled with grey; facial disc white and heart-shaped; underparts white or cream with dark spots; pale bill; call a single screech).

ALTERNATIVE NAME: Malagasy Scops Owl

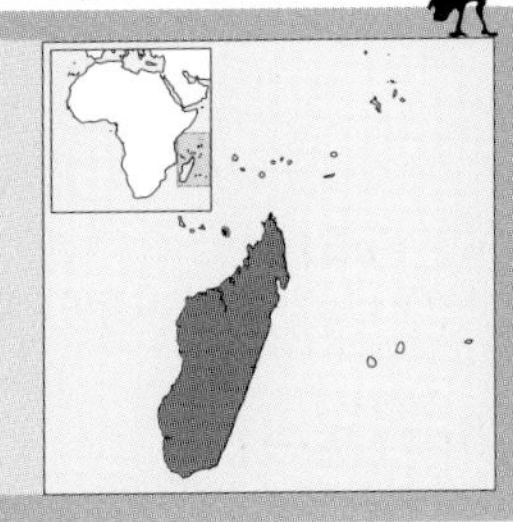

Small 'eared' owl, with 2 forms. Grey (more common) or rufous overall. Above flecked with white. Yellow eyes. Calls with a rapidly repeated series of 4–10 purring 'bruu' notes. Very small (about 15 cm tall, 53 cm wingspan). Restricted to Madagascar and Mayotte (Comoros).

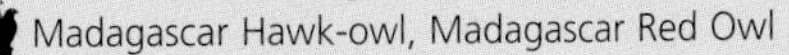

Madagascar Hawk-owl, Madagascar Red Owl

Pemba Scops Owl

Otus pembaensis

AT PERCH

By day, stands upright with 'ears' erect and body feathers sleeked down. Pale scapular bar across shoulder and yellow eyes most obvious features.

IN FLIGHT

Drops low down to fly to next perch, swinging up on landing.

DISTINCTIVE BEHAVIOUR

Main call a 'hu' note, given singly or sometimes repeated at irregular intervals or, at dusk, given as rapid series of 4–6 successive notes, possibly as a pair in a duet. Roosts in dense undergrowth, often rather low down, and sits very tight. Main prey of insects taken from the ground, from foliage or in flight.

ADULT Above plain rufous except for dusky-brown streaks on crown, or brown with dark streaks on head and nape and dark bars on forehead and eyebrow. Scapulars have broad, cream ends. In paler forms vermiculated below with paler rufous, grey and white, with dark streaks on flanks and white bars on abdomen. Flight feathers dark brown with pale spots on inner webs of primaries. Tail plain rufous or brown with dark bars on outer feathers. Bill black with greenish-yellow base and cere; eyes yellow; legs and feet grey. Sexes apparently similar in plumage and size.
JUVENILE Similar to adult of respective rufous or brown colour form. Barred buff juvenile down still on head at fledging. Unstreaked below and on crown, rather finely barred.
DISTRIBUTION, HABITAT AND STATUS Island of Pemba off the northern Tanzanian coast. Occurs in densely wooded areas, including clove and mango plantations. Often considered only a form of Madagascar Scops Owl.
SIMILAR SPECIES The only small owl on Pemba. Only other owl is the much larger **Common Barn Owl** (p. 282) (golden-brown and white or cream; dark eyes; large head; white, heart-shaped facial disc; call a single prolonged screech).

Small 'eared' owl, with 2 forms. Rufous and brown overall, the underparts paler. Call a single 'hu' note given at irregular intervals or in a fast series of 4–6 notes at half-second intervals. Very small (about 15 cm tall, 45 cm wingspan). Restricted to island of Pemba.

Only small owl in range

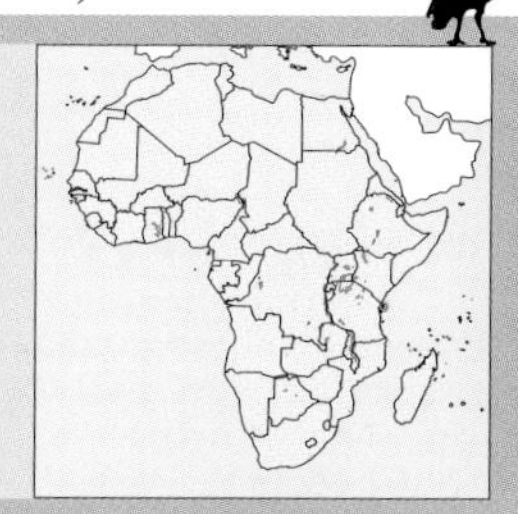

Grande Comore Scops Owl

Otus pauliani

AT PERCH
Appears fluffy and dumpy, with small, indistinct 'ears'. Wings extend to tip of short tail. Body appears finely spotted and barred with white and pale brown. Dark brown or yellow eyes obvious in plain, slightly paler face.

IN FLIGHT
Nocturnal, silent, flies by day only if flushed. At night, fast, shallow, fluttering wingbeats.

DISTINCTIVE BEHAVIOUR
Call a high, whistling 'toot' uttered every second, speeding up to a faster 'cho' every half-second; may continue for up to ten minutes. Very territorial and will approach closely and accelerate its calling even more in response to imitation of its call, sometimes even flailing about in foliage and attacking source. Also a drawn-out whistle 'choeiet'. Diet probably insects.

ADULT Above uniformly grey-brown or brown, with faint, pale barring. Scapulars have dull buff ends. Face grey or brown speckled with white. 'Ears' small. Below rufous-brown with faint, fine, dark streaks, buff bars and white mottling. Underwing and vent white or pale brown with white spots. Flight feathers and tail with paler bases and incomplete buff bars forming spots. Bill black; eyes dark brown or yellow; bare lower legs and small feet dull brown. Sexes similar in plumage and size (single male specimen 70 g).

JUVENILE Undescribed.

DISTRIBUTION, HABITAT AND STATUS Restricted to forested slopes of Mt. Karthala, Grande Comore. Common in forest and forest edge at 1 000–1 800 m, extending into tall heath scrub nearby. Not recorded in exotic plantations.

SIMILAR SPECIES The only small owl on Grande Comore. Only other owl is the much larger **Common Barn Owl** (p. 282) (golden-brown and white or cream; dark eyes; large head; white, heart-shaped facial disc; call a single prolonged screech).

ALTERNATIVE NAMES: Karthala Scops Owl, Ndeu

Small owl with small 'ears'. Dark grey-brown to brown above and rufous-brown below; only faintly spotted and barred. Pale underwing and vent. Buff-spotted flight and tail feathers. Call a high, whistling 'toot' uttered every second, speeding up to every half-second. Very small (about 15 cm tall, 45 cm wingspan). Restricted to Mt. Karthala, Grande Comore.

No other small owls in range

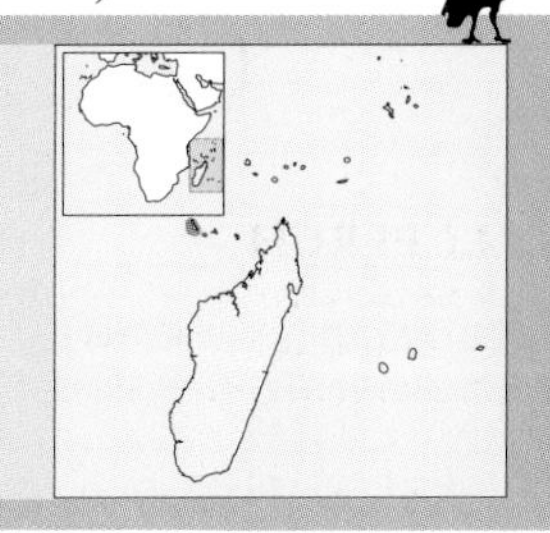

A. Lewis

Anjouan Scops Owl

Otus capnodes

AT PERCH

Roosts in tree holes. When seen at night, 'earless' rounded head obvious and appears dumpy and dark with only fine, paler markings.

IN FLIGHT

Nocturnal, silent, only flies by day if flushed.

DISTINCTIVE BEHAVIOUR

Both colour forms seen together at roosts. Utters a prolonged whistle which is repeated in regular bouts of calling, a drawn-out whistle 'peeoo' or 'peeooee', repeated 3–5 times and said to resemble call of Grey Plover *(Pluvialis squatarola)*. Also a soft, harsh screech. Territorial and responds to imitation of its call. Sometimes calls during daylight. Perches 3–15 m up, usually in dense cover. Roosts in cavities in large trees, sometimes a pair together.

ADULT Above sooty-grey, brown or rufous with fine buff and cream bars. Head and neck dark sooty-grey with fine, buff bars and white spots. Face dark brown with fine streaks. Below brown with fine, dark grey streaks and vermiculations. Flight feathers and tail dark sooty-grey with broad, buff bars. Individual variation in hues and extent of paler markings. Bill and cere dark horn colour; eyes greenish-yellow; feet dull yellowish-green. Sexes presumably similar in plumage and size (single male specimen 119 g).

JUVENILE Undescribed.

DISTRIBUTION, HABITAT AND STATUS Restricted to remaining patches of primary forest on mountain and crater slopes higher than about 800 m a.s.l. on the island of Anjouan (Ndzuani) in the Comoros. Local but in places common in steeply sloping forest and forest edge. Not recorded from exotic plantations in the foothills.

SIMILAR SPECIES The only owl species on the island.

A. Lewis

Adult grey form at night

ALTERNATIVE NAMES: Ndzuani Scops Owl, Badanga

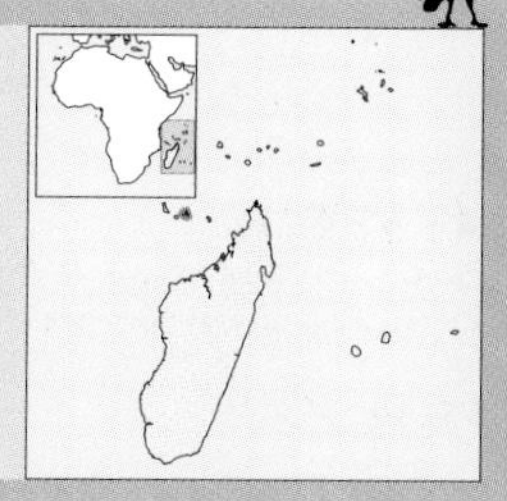

Small 'earless' owl, with 2 forms. Dark grey- or rufous-brown overall. Paler bars on body and wings. Dark, greyer head and neck often obvious. Yellow eyes. Call a drawn-out whistle 'peeooee'. Very small (about 15 cm tall, 45 cm wingspan). Restricted to Anjouan in Comoros.

Only owl on Anjouan

Seychelles Scops Owl

Otus insularis

AT PERCH

Roosts in tall trees and among large rocks. Buff and rufous coloration with pale shoulder-bar. Note yellow eyes, small 'ears' and long, bare legs.

IN FLIGHT

Nocturnal and silent, flies by day only if flushed.

DISTINCTIVE BEHAVIOUR

Difficult to locate at day roost. At night, calls with monotonous, slowly repeated saw-like 'waugh waugh waugh' notes at 3–4 second intervals, or series of clucking, frog-like 'tok tok tok' notes. Only easily located by its calls.

ADULT Buff or rufous-brown overall. Above more rufous-brown, with buff and rufous bars. Crown spotted with dark brown and flecked with white. Pale eyebrow. Face flecked brown; dark rim. Scapulars have white ends, forming shoulder bar. Below lightly streaked with dark brown, especially on chest, and barred with white and buff. Flight feathers and tail dark grey-brown with narrow, buff bars below. Bill black or brown; large eyes yellow; long, bare legs and toes dull brown. Sexes apparently similar in plumage and size.

JUVENILE Similar to adult.

DISTRIBUTION, HABITAT AND STATUS Restricted to upper forest slopes and valleys of the island of Mahé in the Seychelles archipelago. Uncommon and localised in small pockets of forest. Reported previously from Praslin and Félicité.

SIMILAR SPECIES The only small owl on the Seychelles archipelago. Only other owl is the much larger **Common Barn Owl** (p. 282) (golden-brown and white; dark eyes; large head and white, heart-shaped facial disc; call a single prolonged screech).

ALTERNATIVE NAME: Bare-legged Scops Owl

Small, 'eared' buff owl. Note unfeathered legs at close range. Large, yellow eyes. Calls a slowly repeated, harsh 'waugh' note every 3–4 seconds, or a series of clucking 'tok' notes. Very small (about 15 cm tall, 45 cm wingspan). Only small owl on Mahé

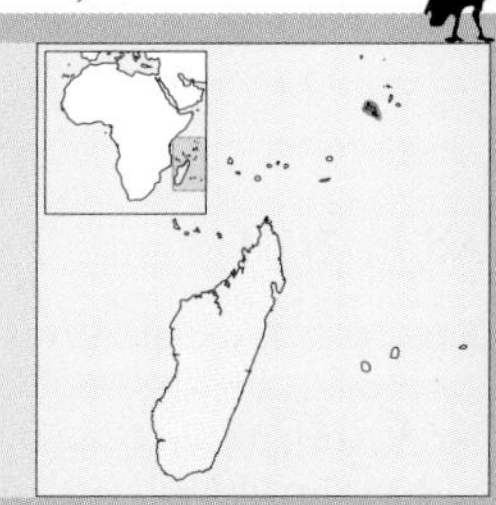

S. Jones & J. Watson/BirdLife International

African Bay Owl

Phodilus prigoginei

DISTINCTIVE BEHAVIOUR

One found roosting on the ground in the grass, another caught while flying low through forest edge. Possible call of this species described as 'wok wok wok'.

ADULT Above deep rufous with fine, black-rimmed, white spots on feather tips. Back, scapulars and upperwing mottled with sooty-grey and fine, white spots. Broad, round facial disc, plain pale rufous with dark brown eye spots and deep rufous rim. Below pale rufous with fine, dark brown spots. Feathered legs plain buff. Bill cream with grey lower mandible; small eyes very dark brown; small, bare feet pale yellow. Only 2 females ever examined (195 g).
JUVENILE Undescribed.
DISTRIBUTION, HABITAT AND STATUS Only 2 specimens handled, and 2 possibly heard or seen. Specimens both from eastern Albertine Rift mountains of extreme eastern Congo, at 1 830 m and 2 430 m. Both associated with mosaic of montane forest, bamboo thicket and grassland. Unconfirmed records, from about 50 km further east on western rift mountains of southwestern Rwanda (2 000 m) and adjacent northwestern Burundi (2 500 m), were for similar habitat.
SIMILAR SPECIES Most resembles in body form the larger **Common Barn Owl** (p. 282) (white facial disc; golden-brown mottled with grey above; white or pale buff below with fine, dark spots; only faint barring on wings; legs only sparsely feathered) and, of similar habitat at lower altitudes, the even larger **African Grass Owl** (p. 284) (dark brown above; buff below with fine, dark spots; buff or white facial disc).

ALTERNATIVE NAMES: Congo Bay Owl, Prigogine's Bay Owl

Shelley's Eagle-owl

Bubo shelleyi

AT PERCH

Appears dark brown overall with white bars below. Sits upright with long, dark 'ears'. Dark eyes, pale cream bill and feet, and plain dark face and 'ears' obvious at close range.

IN FLIGHT

Nocturnal and silent, unlikely to be seen in flight unless flushed.

DISTINCTIVE BEHAVIOUR

Rare and little known. Roosts in dense foliage of tall forest trees. Feeds on medium-sized squirrels, and probably on other mammals and birds. Call described as a loud, long wail 'kooouw', which may be a food solicitation call. Nesting undescribed.

ADULT Dusky-brown above with a few white spots on nape and buff bars at base of covert and back feathers. Facial disc and inside of 'ears' dusky-grey with fine white bars. Underparts white with broad, dusky-brown bars. Flight feathers and tail dusky-brown, with broad, buff to white bars above and paler bars below; tail tipped with white. Bill cream; cere blue-grey; eyes dark brown; heavy, powerful feet pale cream. Sexes similar in plumage and size (single male specimen 1 257 g).
JUVENILE Body and head down has fine, sooty bars all over. Flight feathers, coverts and facial disc dark like adult.
DISTRIBUTION, HABITAT AND STATUS Known only from primary lowland rainforest at scattered localities from Liberia eastwards to northern Democratic Republic of Congo (Zaïre). Apparently very rare and local.
SIMILAR SPECIES Largest, darkest and heaviest of Africa's lowland forest owls; occurs with smaller **Fraser's Eagle-owl** (p. 260) (dark rufous-brown above with broad, buff bars; finely barred below; dark facial rim and long, dark 'ears'; lightly barred legs; dark brown eyes; blue bill and eyelids) and **Akun Eagle-owl** (p. 258) (paler above; wide, brown bars below; forehead and front of 'ears' often white; darkly barred legs; yellow eyes and bill; lightly built).

ALTERNATIVE NAME: Banded Eagle-owl

African Bay Owl

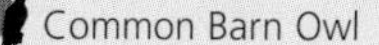

'Earless' and large-headed. Rufous with sooty-grey forearm and shoulder. Body finely spotted with dark brown. Flight feathers and tail barred with dark brown. Broad, buff facial disc; small, dark eyes; cream bill. Small (about 20 cm tall, 63 cm wingspan). Very rare in Congo-Rwanda highlands.

Common Barn Owl

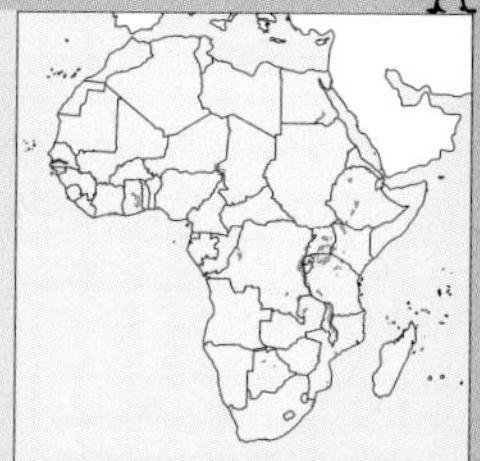

Shelley's Eagle-owl

Very large, heavy, 'eared' owl. Dark brown above; heavily barred below. Dark brown eyes. Cream bill with blue-grey cere. Call a long, loud, wail 'kooouw'. Large (about 50 cm tall, 145 cm wingspan). Very rare in lowland rainforests.

Fraser's Eagle-owl, Akun Eagle-owl

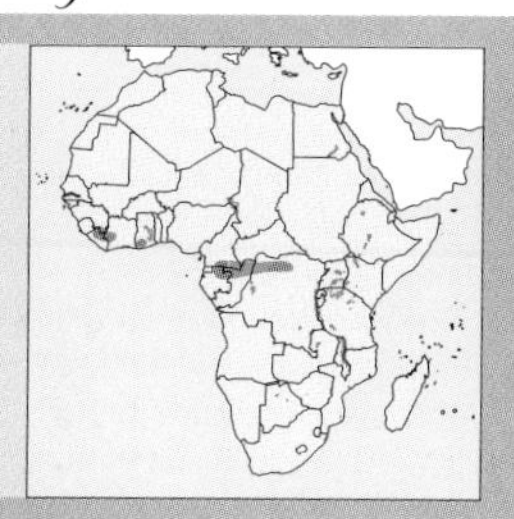

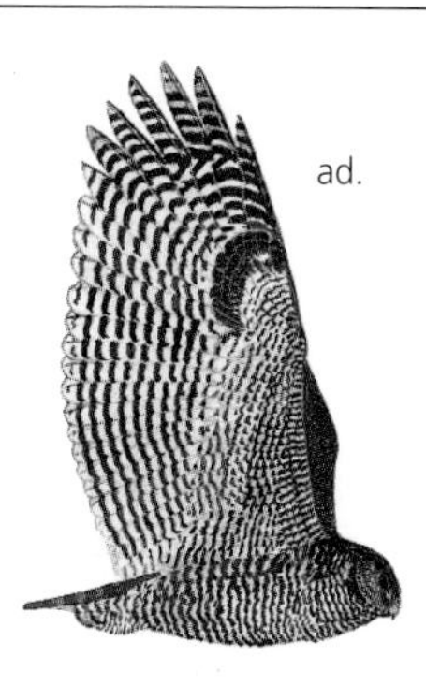

Albertine Owlet

Glaucidium albertinum

DISTINCTIVE BEHAVIOUR

Habits unknown. One bird was collected in dense undergrowth and had insect remains in the stomach.

ADULT Above maroon-brown, head and neck spotted and mantle barred with cream. Large, beige spots on scapulars. Eyebrow and underparts white, face brown streaked with white. Chest barred and breast heavily spotted with maroon-brown. Flight feathers dark brown with broad, pale brown bars. Tail dark brown with 7 narrow, cream bars. Softparts undescribed, bill, cere, eyes and feet probably yellow. Sexes probably similar in plumage and size.
JUVENILE Undescribed.
DISTRIBUTION, HABITAT AND STATUS Restricted to montane forests of the Albertine Rift in eastern Congo and northern Rwanda.
SIMILAR SPECIES May overlap with **African Barred Owlet** (p. 296) (reddish-brown; faintly barred back; white shoulder-bar; barred head, face and chest; narrowly barred flight and tail feathers; call of repeated purring notes). Dumpier than any of the small 'eared' ***Otus* scops owls** (pp. 304–324) (light, small bill and feet, calls widely spaced single notes).

São Tomé Scops Owl

Otus hartlaubi

DISTINCTIVE BEHAVIOUR

Call a low, harsh 'kowe', sometimes ending in a brief rattle. Lives low down in forest, feeding mainly on small lizards and insects. Sometimes descends to the ground.

ADULT Above overall brown with buff vermiculation, or rufous with grey, depending on form. Darkest on wing coverts. Much individual variation. Fine white spots on nape and white ends to dark-tipped scapulars. Throat and facial disc pale grey or rufous with dark rim. Below paler, finely barred with white, brown and rufous, with fine, black shaft streaks. Legs and underwing coverts plain buff. Flight feathers dark sooty-brown, outer primaries mottled with buff and white. Tail brown with incomplete buff bars. Bill and cere yellow; eyes deep yellow; slender unfeathered lower tarsi and feet dusky yellow. Sexes apparently similar in plumage and size.
JUVENILE Similar to adult, but paler and more finely barred with buff and light brown.
DISTRIBUTION, HABITAT AND STATUS Restricted to island of São Tomé in the Gulf of Guinea. Favours secondary forest and plantations but found in any type of forest up to 1 300 m altitude. Unconfirmed report from Príncipe.
SIMILAR SPECIES None in the region.

Albertine Owlet

Small, maroon-brown 'earless' owl. Head and neck spotted with cream. Cream shoulder-bar. Below white, with large, brown spots. Flight feathers with broad, cream bars and tail with narrow, cream bars. Softpart colours and call undescribed. Very small (about 15 cm tall, 35 cm wingspan). Restricted to Albertine Rift of Eastern Congo and northern Rwanda.

African Barred Owlet.

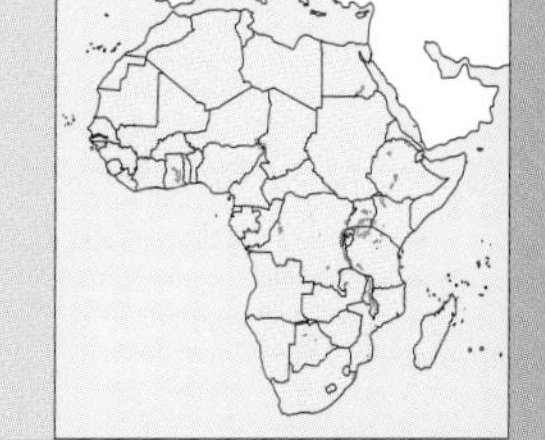

ad.

São Tomé Scops Owl

Small owl with short 'ears', 2 forms. Dark brown or deep rufous overall. Marked with white, dusky and buff spots, bars and streaks, Pale yellow bill and eyes. Call a low, raucous 'kowe'. Very small (about 15 cm tall, 45 cm wingspan).

Only owl species on São Tomé.

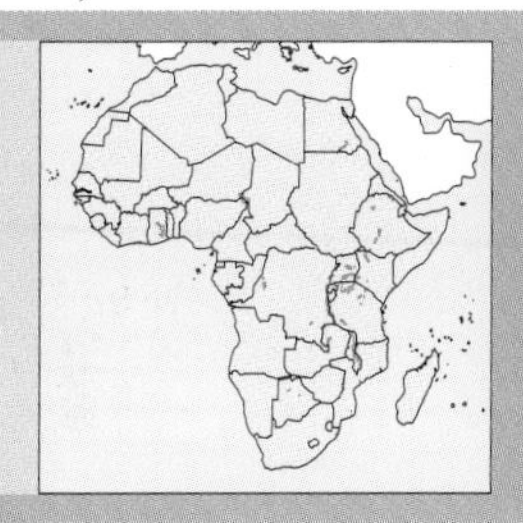

ad. grey form

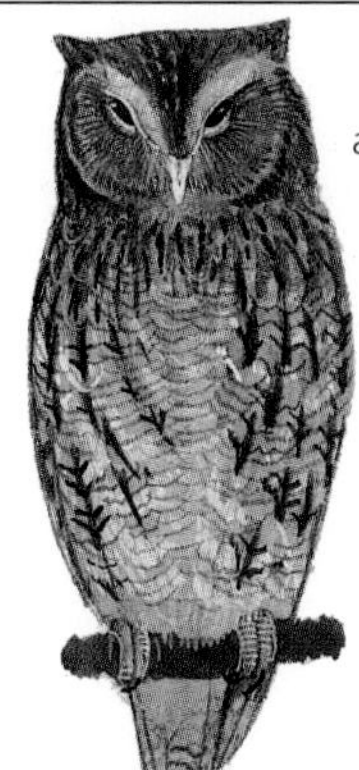

ad. rufous form

FURTHER READING

Allan, D. 1996. *A photographic guide to birds of prey of southern, central and east Africa.* New Holland, Cape Town.

Brown, L. H. & Amadon, D. 1968. *Eagles, hawks and falcons of the world.* Country Life, Feltham.

Brown, L. H., Urban, E. K. & Newman, K. B. (eds). 1982. *The birds of Africa.* Vol. 1. Academic Press, London.

Cramp, S. (ed). 1980. *Birds of the western Palaearctic.* Vol. 2. Oxford University Press, Oxford.

Cramp, S. (ed). 1985. *Birds of the western Palaearctic.* Vol. 4. Oxford University Press, Oxford.

Del Hoyo, J., Elliot, A. & Sargatal, J. 1994. *Handbook of birds of the world.* Vol. 2. Lynx Edicions, Barcelona.

Dowsett, R. J. & Dowsett-Lemaire, F. 1993. A contribution to the distribution and taxonomy of Afrotropical and Malagasy birds. *Tauraco Research Report* 5: 1-389.

Fry, C. H., Keith, S. & Urban, E. K. (eds). 1988. *The birds of Africa.* Vol. 3. Academic Press, London.

Harrison, J. A., Allan, D. G., Underhill, L. G., Herremans, M., Tree, A. J., Parker, V. & Brown, C. J. (eds). 1997. *The atlas of southern African birds. Vol. 1: Non-passerines.* BirdLife South Africa, Johannesburg.

Kemp, A. C. & Calburn, S. 1987. *The owls of southern Africa.* Struik Winchester, Cape Town.

Langrand, O. 1990. *Guide to the birds of Madagascar.* Yale University Press, New Haven.

Maclean, G. L. 1983. *Roberts' birds of southern Africa.* 6th edition. The Trustees of the Voelcker Bird Book Fund, Cape Town.

Mundy, P., Butchart, D., Ledger, J. & Piper, S. 1992. *The vultures of Africa.* Acorn Books and Russell Friedman Books, Randburg and Halfway House.

Newman, K. B. 1995. *Newman's birds of southern Africa.* 5th edition. Southern Book Publishers, Johannesburg.

Porter, R. F., Willis, I., Christensen, S. & Nielsen, B. P. 1986. *Flight identification of European raptors.* T. & A. D. Poyser, Berkhamstead.

Sinclair, J. C. & Hockey, P. 1996. *The larger illustrated guide to birds of southern Africa.* Struik Publishers, Cape Town.

Sinclair, I. & Langrand, O. 1998. *Birds of the Indian Ocean islands.* Struik Publishers, Cape Town.

Steyn, P. 1982. *Birds of prey of southern Africa.* David Philip, Cape Town.

Tarboton, W. & Erasmus, R. 1998. *Sasol owls and owling in southern Africa.* Struik Publishers, Cape Town.

Zimmerman, D. A., Turner, D. A. & Pearson, D. J. 1996. *Birds of Kenya and northern Tanzania.* Russel Friedman Books, Halfway House.

Various articles in the *Bulletin of the African Bird Club, Africa – Birds & Birding* and *Ostrich.*

GLOSSARY OF TERMS

adult sexually mature bird in its final adult plumage and softpart colours after replacement of all immature plumages.

albinism *see* melanistic

albinistic *see* melanistic

alula small bastard wing protruding from wrist joint, actually a finger bone and 3 or 4 associated feathers, used for fine adjustments of airflow over the wing.

arboreal tree-living.

bob to rapidly raise and lower the head.

brood patch patch on belly that loses feathers and becomes suffused with blood capillaries preparatory to incubation of eggs and brooding.

call vocalization of a bird, sometimes separated as single, simple, repeated notes, while song refers to more complex phrases.

colony gathering of birds within a discrete area to nest.

communal describes birds which form a cohesive group to roost or perform some other activity, such as hunting or breeding.

congener species belonging to the same genus.

crepuscular active at dawn and dusk.

crop enlarged part of foregut for storage of food before digestion.

dihedral angle by which the wing is raised above the horizontal.

display standard ritualized form of behaviour, often spectacular and active, combining various body and wing movements and calls, and performed in flight or at a perch.

diurnal active during daylight.

desert dry, barren habit with no permanent vegetation cover.

ear bracket bracket-like marks formed by edges to ear coverts or facial disc.

endemic restricted to a defined geographical area.

flight feathers longest feathers of the wings, formed by the primary (hand) and secondary (forearm) feathers.

flight pattern colour patterns above or below on open wing and tail when bird is in flight.

flush to cause a bird to take to flight.

forest habitat of tall trees with several canopy levels that form a closed canopy, usually with little undergrowth and moist substrate.

fuscous a deep grey-brown colour.

gallery forest strips of tall trees on either side of a watercourse that form a closed-canopy forest over the water and along the banks.

glide to fly without flapping the wings, usually in a straight line with the wings slightly folded, and steadily losing height.

grassland habitat of open country covered in grass with very little or no woody vegetation.

immature a bird not yet sexually mature. In birds of prey, refers to any plumage before full adult colours, despite many medium to large species being able to breed in immature plumage.

juvenile the first distinct plumage on leaving the nest.

kite to hang stationary in the wind like a kite.

m a.s.l. metres above sea level.

melanistic all-black plumage form where melanin is the dominant plumage pigment, as opposed to much rarer cases of albinism with no pigment.

migrant a bird that breeds at one locality and undertakes a regular, usually annual flight to a different (often very distant) non-breeding area.

morph consistent colour variation of the plumage, or some other consistent difference in structure or colour.

moult the replacement of old feathers by new; in smaller species usually repeated during a set period each year; in larger species spread over several years with overlapping waves of moult.

mute droppings or the act of producing droppings, an old falconry term. Technically restricted to falcons (Falconidae) whose droppings just drop, while hawks (Accipitridae) slice or squirt out their droppings.

nocturnal active at night.

nominate subspecies the subspecies from which the original species was named and which shares the same specific name, e.g. *Falco biarmicus biarmicus*, or *F. b. biarmicus*.

Palaearctic a biogeographical region that includes Europe, Asia and northern Africa.

patagial the area of skin which is stretched between the bird's shoulder and wrist (patagium).

pellet undigested food remains, usually of fur, feather, bones or insect exoskeleton, which is regurgitated, usually daily, by a bird of prey.

plumage the feathers of a bird which, since they are dropped or moulted at regular intervals, sometimes refers to the relative age, e.g. first plumage, second plumage, etc.

pump to raise and lower the tail, as opposed to wagging it from side to side.

quelea abundant small African weaver bird; the Red-billed Quelea *Quelea quelea* lives and breeds in huge flocks, earning the name 'locust bird'.

range geographical area in which a bird normally occurs.

ringing pursuit falconry term for a high, circling chase.

riverine associated with a river system.

rouse to raise and ruffle the feathers and wings, an old falconry term.

savanna habitat of mixed grass and woody vegetation, in various proportions between the extremes of grassland and woodland.

scat droppings of small mammal predators, often containing undigested food remains of fur and feathers.

slice *see* mute

soar to fly on extended wings without flapping, usually turning in slow circles within a rising thermal of warm air.

softparts unfeathered fleshy external parts of a bird's anatomy, such as the cere, eye, facial skin, legs and feet.

song *see* call.

steppe habitat type of semi-arid regions with low sparse vegetation cover, often of widely spaced woody shrubs.

stoop to dive at prey from a height, an old falconry term.

subadult any plumage between juvenile and adult, either while in transition from one to the other, or a distinct intermediate plumage grown between the two age extremes.

subspecies geographically separated population of a species with distinct and consistent differences, usually in form, colour or behaviour. Also called form or race, as opposed to morph which is a variation within a population.

taxonomy the science of names, especially scientific names and the classification of organisms. Sometimes includes systematics, the study of relationships between organisms.

terrestrial ground living.

vagrant bird recorded outside its normal geographical range.

vermiculation fine wavy worm-like dark lines within a basal plumage colour.

woodland habitat of large trees whose canopies do not meet entirely and which usually has patches of well-developed shrub and ground cover.

INDEX TO COMMON ENGLISH NAMES

INDEX TO SCIENTIFIC NAMES

BIRD NAMES IN OTHER LANGUAGES

Pg	English	Scientific	French
54	Bateleur	*Terathopius ecaudatus*	Bateleur des savanes
118	Buzzard, Augur	*Buteo augur*	Buse augure
108	Buzzard, Eurasian	*Buteo buteo*	Buse variable
122	Buzzard, Grasshopper	*Butastur rufipennis*	Busautour des saturelles
116	Buzzard, Jackal	*Buteo rufofuscus*	Buse rounoir
114	Buzzard, Long-legged	*Buteo rufinus*	Buse féroce
112	Buzzard, Madagascar	*Buteo brachypterus*	Buse de Madagascar
110	Buzzard, Mountain	*Buteo oreophilus*	Buse montagnarde
120	Buzzard, Red-necked	*Buteo auguralis*	Buse d'Afrique
148	Chanting-goshawk, Dark	*Melierax metabates*	Autour sombre
152	Chanting-goshawk, Eastern	*Melierax poliopterus*	Autour aux ailes grises
150	Chanting-goshawk, Pale	*Melierax canorus*	Autour chanteur
190	Cuckoo-hawk, African	*Aviceda cuculoides*	Baza coucou
192	Cuckoo-hawk, Madagascar	*Aviceda madagascariensis*	Baza malgache
104	Eagle, Booted	*Hieraaetus pennatus*	Aigle botté
90	Eagle, Eastern Imperial	*Aquila heliaca*	Aigle impérial
92	Eagle, Golden	*Aquila chrysaetos*	Aigle royal
86	Eagle, Greater Spotted	*Aquila clanga*	Aigle criard
84	Eagle, Lesser Spotted	*Aquila pomarina*	Aigle pomarin
94	Eagle, Long-crested	*Lophaetus occipitalis*	Aigle huppard
74	Eagle, Martial	*Polemaetus bellicosus*	Aigle martial
88	Eagle, Spanish Imperial	*Aquila adalberti*	Aigle ibérique
82	Eagle, Steppe	*Aquila nipalensis*	Aigle des steppes
80	Eagle, Tawny	*Aquila rapax*	Aigle ravisseur
78	Eagle, Verreaux's	*Aquila verreauxii*	Aigle de Verreaux
106	Eagle, Wahlberg's	*Aquila wahlbergi*	Aigle de Wahlberg
258	Eagle-owl, Akun	*Bubo leucostictus*	Grand-duc tacheté
250	Eagle-owl, Cape	*Bubo capensis*	Grand-duc du Cap
254	Eagle-owl, Eurasian	*Bubo bubo*	Grand-duc d'Europe
260	Eagle-owl, Fraser's	*Bubo poensis*	Grand-duc à aigrettes
248	Eagle-owl, Milky	*Bubo lacteus*	Grand-duc de Verreaux
252	Eagle-owl, Pharaoh	*Bubo ascalaphus*	Grand-duc ascalaphe
326	Eagle-owl, Shelley's	*Bubo shelleyi*	Grand-duc de Shelley
256	Eagle-owl, Spotted	*Bubo africanus*	Grand-duc africain
246	Falcon, African Pygmy	*Polihierax semitorquatus*	Fauconnet d'Afrique
220	Falcon, Amur	*Falco amurensis*	Faucon de l'Amour

German	Spanish	Afrikaans
Gaukler	Aguila Volatinera	Berghaan
Augurbussard	Busardo Augur Oriental	Witborsjakkalsvoël
Mäusebussard	Busardo Ratonero	Bruinjakkalsvoël
Heuschreckenteesa	Busardo Langostero	Sprinkaanjakkalsvoël
Felsenbussard	Busardo Augur Meridional	Rooiborsjakkalsvoël
Adlerbussard	Busardo Moro	Langbeenjakkalsvoël
Madagaskarbussard	Busardo Malgache	Madagaskarjakkalsvoël
Bergbussard	Busardo Montañés	Bergjakkalsvoël
Salvadoribussard	Busardo Cuellirrojo	Rooinekjakkalsvoël
Graubürzel-Singhabicht	Azor-lagartijero Oscuro	Donker Singvalk
Weissbürzel-Singhabicht	Azor-lagartijero Somalí	Oosterse Singvalk
Grosser Singhabicht	Azor-lagartijero Claro	Bleeksingvalk
Kuckucksweihe	Baza Africano	Koekoekvalk
Lemurenweihe	Baza Malgache	Madagaskarkoekoekvalk
Zwergadler	Aguililla Calzada	Dwergarend
Kaiseradler	Aguila Imperial Oriental	Oosterse Keiserarend
Steinadler	Aguila Real	Goue Arend
Schelladler	Aguila Moteada	Groot Gevlekte Arend
Schreiadler	Aguila Pomerana	Klein Gevlekte Arend
Schopfadler	Aguila Crestilarga	Langkuifarend
Kampfadler	Aguila Marcial	Breëkoparend
Spanischer Kaiseradler	Aguila Imperial Ibérica	Spaanse Keiserarend
Steppenadler	Aguila Esteparia	Steppe-arend
Savannenadler	Aguila Rapaz	Roofarend
Kaffernadler	Aguila Cafre	Witkruisarend
Siberadler	Aguila de Wahlberg	Bruinarend
Gelbfussuhu	Búho de Akun	Akun Ooruil
Kapuhu	Búho de El Cabo	Kaapse Ooruil
Uhu	Búho Real	Eurasiese Ooruil
Guineauhu	Búho de Guineea	Fraserse Ooruil
Blassuhu	Búho Lechoso	Reuse Ooruil
Faraouhu	Búho Farao	Farao Ooruil
Bindenuhu	Búho Barrado	Gebande Ooruil
Fleckenuhu	Búho Africano	Gevlekte Ooruil
Halsband-Zwergfalke	Halconcito Africano	Dwergvalk
Amurfalke	Cernícalo del Amur	Oostelike Rooipootvalk

Pg	English	Scientific	French
208	Falcon, Eleonora's	*Falco eleonorae*	Faucon d'Éléonore
200	Falcon, Lanner	*Falco biarmicus*	Faucon lanier
202	Falcon, Peregrine	*Falco peregrinus*	Faucon pèlerin
222	Falcon, Red-footed	*Falco vespertinus*	Faucon kobez
212	Falcon, Red-necked	*Falco chicquera*	Faucon chicquera
206	Falcon, Saker	*Falco cherrug*	Faucon sacre
228	Falcon, Sooty	*Falco concolor*	Faucon concolore
210	Falcon, Taita	*Falco fasciinucha*	Faucon taita
50	Fish Eagle, African	*Haliaeetus vocifer*	Pygargue vocifere
52	Fish Eagle, Madagascar	*Haliaeetus vociferoides*	Pygargue de Madagascar
262	Fishing-owl, Pel's	*Scotopelia peli*	Chouette-pêcheuse de Pel
266	Fishing-owl, Rufous	*Scotopelia ussheri*	Chouette-pêcheuse rousse
264	Fishing-owl, Vermiculated	*Scotopelia bouvieri*	Chouette-pêcheuse de Bouv
172	Goshawk, African	*Accipiter tachiro*	Autour tachiro
156	Goshawk, Gabar	*Micronisus gabar*	Autour gabar
182	Goshawk, Henst's	*Accipiter henstii*	Autour de Henst
184	Goshawk, Northern	*Accipiter gentilis*	Autour des palombes
30	Griffon, African White-backed	*Gyps africanus*	Vautour africain
34	Griffon, Cape	*Gyps coprotheres*	Vautour chassefiente
36	Griffon, Eurasian	*Gyps fulvus*	Vautour fauvre
32	Griffon, Rüppell's	*Gyps rueppellii*	Vautour de Rüppell
132	Harrier, African Marsh	*Circus ranivorus*	Busard grenouillard
138	Harrier, Black	*Circus maurus*	Busard maure
144	Harrier, Hen	*Circus cyaneus*	Busard Saint-Martin
136	Harrier, Madagascar Marsh	*Circus maillardi*	Busard de Maillard
140	Harrier, Montagu's	*Circus pygargus*	Busard cendré
142	Harrier, Pallid	*Circus macrourus*	Busard pâle
134	Harrier, Western Marsh	*Circus aeruginosus*	Busard des roseaux
128	Harrier-hawk, African	*Polyboroides typus*	Gymnogène d'Afrique
130	Harrier-hawk, Madagascar	*Polyboroides radiatus*	Gymnogène de Madagascar
198	Hawk, Bat	*Macheiramphus alcinus*	Milan des chauves-souris
154	Hawk, Lizard	*Kaupifalco monogrammicus*	Buse unibande
146	Hawk, Long-tailed	*Urotriorchis macrourus*	Autour à longue queue
96	Hawk-eagle, African	*Hieraaetus spilogaster*	Aigle fascié
100	Hawk-eagle, Ayres's	*Hieraaetus ayresii*	Aigle d'Ayres
98	Hawk-eagle, Bonelli's	*Hieraaetus fasciatus*	Aigle de Bonelli
102	Hawk-eagle, Cassin's	*Spizaetus africanus*	Aigle de Cassin
76	Hawk-eagle, Crowned	*Stephanoaetus coronatus*	Aigle couronné
288	Hawk-owl, Madagascar	*Ninox superciliaris*	Ninoxe à sourcils blancs

German	Spanish	Afrikaans
Eleonorenfalke	Halcón de Eleonora	Eleonoravalk
Lannerfalke	Halcón Borní	Edelvalk
Wanderfalke	Halcón Peregrino	Swerfvalk
Rotfussfalke	Cernícalo Patirrojo	Westelike Rooipootvalk
Rothalsfalke	Alcotán Turumti	Rooinekvalk
Würgfalke	Halcón Sacre	Sakervalk
Schieferfalke	Halcón Pizarroso	Roetvalk
Taitafalke	Halcón Taita	Teitavalk
Schreiseeadler	Pigargo Vocinglero	Visarend
Madagaskarseeadler	Pigargo Malgache	Madagaskarvisarend
Bindenfischeule	Cárabo Pescador Común	Visuil
Rotrücken-Fischeule	Cárabo Pescador Rojizo	Roesvisuil
Marmorfischeule	Cárabo Pescador Vermiculado	Marmervisuil
Afrikahabicht	Azor Tachiro	Afrikaanse Sperwer
Gabarhabicht	Gavilán Gabar	Kleinsingvalk
Madagaskarhabicht	Azor Malgache	Henstse Sperwer
Habicht	Azor Común	Noordelike Grootsperwer
Weissrückengeier	Buitre Dorsiblanco Africano	Witrugaasvoël
Kapgeier	Buitre de El Cabo	Kransaasvoël
Gänsegeier	Buitre Leonardo	Eurasiese Aasvoël
Sperbergeier	Buitre Moteado	Rüppellse Aasvoël
Froschweihe	Aguilucho Lagunero Etiópico	Afrikaanse Paddavreter
Mohrenweihe	Aguilucho Negro	Witkruispaddavreter
Kornweihe	Aguilucho Pálido	Noordelike Paddavreter
Madagaskarweihe	Aguilucho Lagunero Malgache	Madagaskarpaddavreter
Weisenweihe	Aguilucho Cenizo	Bloupaddavreter
Steppenweihe	Aguilucho Papialbo	Witborspaddavreter
Rohrweihe	Aguilucho Lagunero Occidental	Europese Paddavreter
Höhlenweihe	Aguilucho-caricalvo Común	Kaalwangvalk
Madagaskarhöhlenweihe	Aguilucho-caricalvo Malgache	Madagaskarkaalwangvalk
Fledermausaar	Milano Murielaguero	Vlermuisvalk
Sperberbussard	Busardo Gavilán	Akkedisvalk
Langschwanzhabicht	Azor Rabilargo	Langstertvalk
Afrikanischer Habichtsadler	Aquila-azor Africana	Afrikaanse Jagarend
Fleckenadler	Aguila-azor de Ayres	Kleinjagarend
Habichtsadler	Aguila-azor Perdicera	Bonellise Jagarend
Schwarzachseladler	Aguila-azor Congoleña	Cassinse Jagarend
Kronenadler	Aguila Coronada	Kroonarend
Madagaskarkauz	-	Madagaskar Valkuil

Pg	English	Scientific	French
216	Hobby, African	*Falco cuvierii*	Faucon de Cuvier
218	Hobby, Eurasian	*Falco subbuteo*	Faucon hobereau
126	Honey-buzzard, Crested	*Pernis ptilorhyncus*	Bondrée orientale
124	Honey-buzzard, Western	*Pernis apivorous*	Bondrée apivore
230	Kestrel, Banded	*Falco zoniventris*	Faucon à ventre rayé
232	Kestrel, Common	*Falco tinnunculus*	Faucon crécerelle
226	Kestrel, Dickinson's	*Falco dickinsoni*	Faucon de Dickinson
236	Kestrel, Fox	*Falco alopex*	Crécerelle renard
234	Kestrel, Greater	*Falco rupicoloides*	Crécerelle aux yeux blancs
224	Kestrel, Grey	*Falco ardosiaceus*	Faucon ardoise
238	Kestrel, Lesser	*Falco naumanni*	Faucon crécerellette
240	Kestrel, Madagascar	*Falco newtoni*	Crécerelle malgache
242	Kestrel, Mauritius	*Falco punctatus*	Crécerelle de Maurice
244	Kestrel, Seychelles	*Falco araea*	Crécerelle des Seychelles
196	Kite, African Swallow-tailed	*Chelictinia riocourii*	Elanion naucler
186	Kite, Black	*Milvus migrans*	Milan noir
194	Kite, Black-shouldered	*Elanus caeruleus*	Elanion blanc
188	Kite, Red	*Milvus milvus*	Milan royal
214	Merlin	*Falco columbarius*	Faucon Émerillon
72	Osprey	*Pandion haliaetus*	Balbuzard pêcheur
272	Owl, Abyssinian Long-eared	*Asio abyssinicus*	Hibou d'Abyssinie
326	Owl, African Bay	*Phodilus prigoginei*	Phodile de Prigogine
284	Owl, African Grass	*Tyto capensis*	Effraie du Cap
306	Owl, African Scops	*Otus senegalensis*	Petit-duc africain
278	Owl, African Wood	*Strix woodfordii*	Chouette africaine
322	Owl, Anjouan Scops	*Otus capnodes*	Petit-duc d'Anjouan
282	Owl, Common Barn	*Tyto alba*	Effraie des clochers
310	Owl, Eurasian Scops	*Otus scops*	Petit-duc scops
320	Owl, Grande Comore Scops	*Otus pauliani*	Petit-duc du Karthala
290	Owl, Hume's	*Strix butleri*	Chouette de Butler
302	Owl, Little	*Athene noctua*	Chevêche d'Athéna
276	Owl, Eurasian Long-eared	*Asio otus*	Hibou moyen-duc
274	Owl, Madagascar Long-eared	*Asio madagascariensis*	Hibou de Madagascar
280	Owl, Madagascar Red	*Tyto soumagnei*	Effraie de Soumagne
316	Owl, Madagascar Scops	*Otus rutilus*	Petit-duc de malgache
286	Owl, Maned	*Jubula lettii*	Duc à crinière
268	Owl, Marsh	*Asio capensis*	Hibou du Cap
312	Owl, Pallid Scops	*Otus brucei*	Petit-duc de Bruce
318	Owl, Pemba Scops	*Otus pembaensis*	Petit-duc de Pemba

German	Spanish	Afrikaans
Afrikanischer Baumfalke	Alcotán Africano	Afrikaanse Boomvalk
Baumfalke	Alcotán Europeo	Europese Boomvalk
Schopfwespenbussard	Abejero Oriental	Kuifwespedief
Wespenbussard	Abejero Europeo	Wespedief
Bindenfalke	Cernícalo Malgache	Gebande Rooivalk
Turmfalke	Cernícalo Vulgar	Kransvalk
Schwarzrückenfalke	Cernícalo Dorsinegro	Dickinsonse Valk
Fuchsfalke	Cernícalo Zorruno	Jakkalsvalk
Steppenvalke	Cernícalo Ojiblanco	Grootrooivalk
Graufalke	Cernícalo Pizzaroso	Donkergrysvalk
Rötelfalke	Cernícalo Primilla	Kleinrooivalk
Madagaskarfalke	Cernícalo de Aldabra	Madagaskarrooivalk
Mauritiusfalke	Cernícalo de la Mauricio	Mauritiusrooivalk
Seychellenfalke	Cernícalo de las Seychelles	Seychellesrooivvalk
Schwalbenschwanzaar	Elanio Golondrina	Swaelstertblouvalk
Schwarzmilan	Milano Negro	Swartwou/Geelbekwou
Gleitaar	Elanio Común	Blouvalk
Rotmilan	Milano Real	Rooiwou
Merlin	Esmerejón	Duifvalk
Fischadler	Aguila Pescadora	Visvalk
Abyssinienohreule	Búho Abyssinien	Abissiniese Langooruil
Kongomaskeneule	Lechuza del Congo	Afrikaanse Vosuil
Kapgraseule	Lechuza Malgache	Afrikaanse Grasuil
Afrika-Zwergohreule	Autillo Africano	Afrikaanse Skopsuil
Afrikanischer Waldkauz	Cárabo Africano	Afrikaanse Bosuil
Anjouan-Zwergohreule	Autillo de las Comores	Anjouan Skopsuil
Schleiereule	Lechuza Común	Nonnetjie-uil
Zwergohreule	Autillo Europeo	Eurasiese Skopsuil
Karthala-Zwergohreule	Autillo de Karthala	Grande Comore Skopsuil
Fahlkauz	Cárabo de Hume	Hume se Bosuil
Steinkauz	Mochuelo Europeo	Klein Steenuil
Waldohreule	Búho Chico	Eurasiese Langooruil
Madagaskarohreule	Búho Malgache	Madagaskar Langooruil
Malegasseneule	Lechuza Malgache	Madagaskar Rooi Nonnetjie-uil
Madagaskar-Zwergohreule	Autillo Malgache	Madagaskar Skopsuil
Mähneneule	Cárabo Melena	Maanhaaruil
Kapohreule	Búho Moro	Afrikaanse Vlei-uil
Streifenohreule	Autillo Persa	Bleek Skopsuil
Pemba-Zwergohreule	Autillo de Pemba	Pemba Skopsuil

Pg	English	Scientific	French
314	Owl, Sandy Scops	*Otus icterorhynchus*	Petit-duc à bec jaune
328	Owl, São Tomé Scops	*Otus hartlaubi*	Petit-duc de São Tomé
324	Owl, Seychelles Scops	*Otus insularis*	Petit-duc de scieur
270	Owl, Short-eared	*Asio flammeus*	Hibou des marais
308	Owl, Sokoke Scops	*Otus ireneae*	Petit-duc d'Irène
292	Owl, Tawny	*Strix aluco*	Chouette hulotte
304	Owl, White-faced Scops	*Otus leucotis*	Petit-Duc à face blanche
296	Owlet, African Barred	*Glaucidium capense*	Chevêchette du Cap
328	Owlet, Albertine	*Glaucidium albertinum*	Chevêchette du Graben
300	Owlet, Chestnut-backed	*Glaucidium sjostedti*	Chevêchette à queue barrée
298	Owlet, Pearl-spotted	*Glaucidium perlatum*	Chevêchette perlée
294	Owlet, Red-chested	*Glaucidium tephronotum*	Chevêchette à pieds jaunes
26	Secretarybird	*Sagittarius serpentarius*	Messager serpentaire
68	Serpent-eagle , Congo	*Dryotriorchis spectabilis*	Serpentaire du Congo
70	Serpent-eagle, Madagascar	*Eutriorchis astur*	Serpentaire de Madagascar
166	Shikra	*Accipiter badius*	Epervier shikra
60	Snake-eagle, Beaudouin's	*Circaetus beaudouini*	Circaète de Beaudouin
58	Snake-eagle, Black-chested	*Circaetus pectoralis*	Circaète à poitrine noire
56	Snake-eagle, Brown	*Circaetus cinereus*	Circaète brun
62	Snake-eagle, Short-toed	*Circaetus gallicus*	Circaète Jean-le-Blanc
64	Snake-eagle, Southern Banded	*Circaetus fasciolatus*	Circaète barré
66	Snake-eagle, Western Banded	*Circaetus cinerascens*	Circaète cendré
176	Sparrowhawk, African Little	*Accipiter minullus*	Epervier minule
180	Sparrowhawk, Black	*Accipiter melanoleucus*	Autour noir
164	Sparrowhawk, Eurasian	*Accipiter nisus*	Epervier d'Europe
168	Sparrowhawk, Frances's	*Accipiter francesiae*	Epervier de Frances
170	Sparrowhawk, Levant	*Accipiter brevipes*	Epervier aux pieds courts
162	Sparrowhawk, Madagascar	*Accipiter madagascariensis*	Epervier de Madagascar
158	Sparrowhawk, Ovambo	*Accipiter ovampensis*	Epervier de l'Ovampo
178	Sparrowhawk, Red-thighed	*Accipiter erythropus*	Epervier de Hartlaub
160	Sparrowhawk, Rufous-breasted	*Accipiter rufiventris*	Epervier menu
28	Vulture, Bearded	*Gypaetus barbatus*	Gypaète barbu
46	Vulture, Egyptian	*Neophron percnopterus*	Vautour percnoptère
42	Vulture, Eurasian Black	*Aegypius monachus*	Vautour moine
44	Vulture, Hooded	*Necrosyrtes monachus*	Vautour charognard
40	Vulture, Lappet-faced	*Torgos tracheliotus*	Vautour oricou
48	Vulture, Palm-nut	*Gypohierax angolensis*	Palmiste africain
38	Vulture, White-headed	*Trigonoceps occipitalis*	Vautour à tête blanche

German	Spanish	Afrikaans
Gelbschnabeleule	Autillo Piquigualdo	Kaneel Skopsuil
Hartlaubeule	Autillo de Santo Tomé	Sao Tome Skopsuil
Seychelleneule	Autillo de Seychelles	Seychelles Skopsuil
Sumpfohreule	Búho Campestre	Kortoor Vlei-uil
Sokokeeule	Autillo de Sokoke	Sokoke Skopsuil
Waldkauz	Cárabo Común	Geelbruin Bosuil
Büscheleule	Autillo Cariblanco	Witwanguil
Kapkauz	Mochuelo de El Cabo	Afrikaanse Gebande Uiltjie
Albertseekauz	-	Albertrifuiltjie
Prachtkauz	Mochuelo de Graben	Kastaiingruguiltjie
Perlkauz	Mochuelo Perlado	Witkoluiltjie
Rostbrustkauz	Mochuelo Pechirrojo	Rooiborsuiltjie
Sekretär	Secretario	Sekretarisvoël
Schlangenbussard	Culebrera Congoleña	Kongo Adderarend
Schlangenhabicht	Culebrera Azor	Madagaskar Adderarend
Shikrasperber	Gavilán Chikra	Gebande Sperwer
Beaudouin-Schlangenadler	Culebrera Sudanesa	Sahel Slangarend
Schwanzbrust-Schlangenadler	Culebrera Pechinegra	Swartborsslangarend
Einfarb-Schlangenadler	Culebrera Sombría	Bruinslangarend
Schlangenadler	Culebrera Europa	Korttoonslangarend
Graubrust-Schlangenadler	Culebrera Barreada	Dubbelbandslangarend
Bandschlangenadler	Culebrera Coliblanca	Enkelbandslangarend
Zwergsperber	Gavilancito Chico	Kleinsperwer
Mohrenhabicht	Azor Blanquinegro	Swartsperwer
Sperber	Gavilán Común	Eurasiese Sperwer
Echsenhabicht	Gavilán de Frances	Francesse Sperwer
Kurzfangsperber	Gavilán Griego	Levantse Sperwer
Madagaskarsperber	Gavilán Malgache	Madagaskarsperwer
Ovambosperber	Gavilán del Ovampo	Ovambosperwer
Waldsperber	Gavilancito Muslirrojo	Woud Kleinsperwer
Rotbauchsperber	Gavilán Papirrufo	Rooiborssperwer
Bartgeier	Quebrantahueos	Baardaasvoël
Schmutgeier	Alimoche Común	Egiptiese Aasvoël
Mönchsgeier	Buitre Negro	Eurasiese Swartaasvoël
Kappengeier	Alimoche Sombrío	Monnikaasvoël
Ohrengeier	Buitre Orejudo	Afrikaanse Swartaasvoël
Palmgeier	Buitre Palmero	Witaasvoël
Wöllkopfgeier	Buitre Cabeciblanco	Witkopaasvoël

ILLUSTRATED GLOSSARY

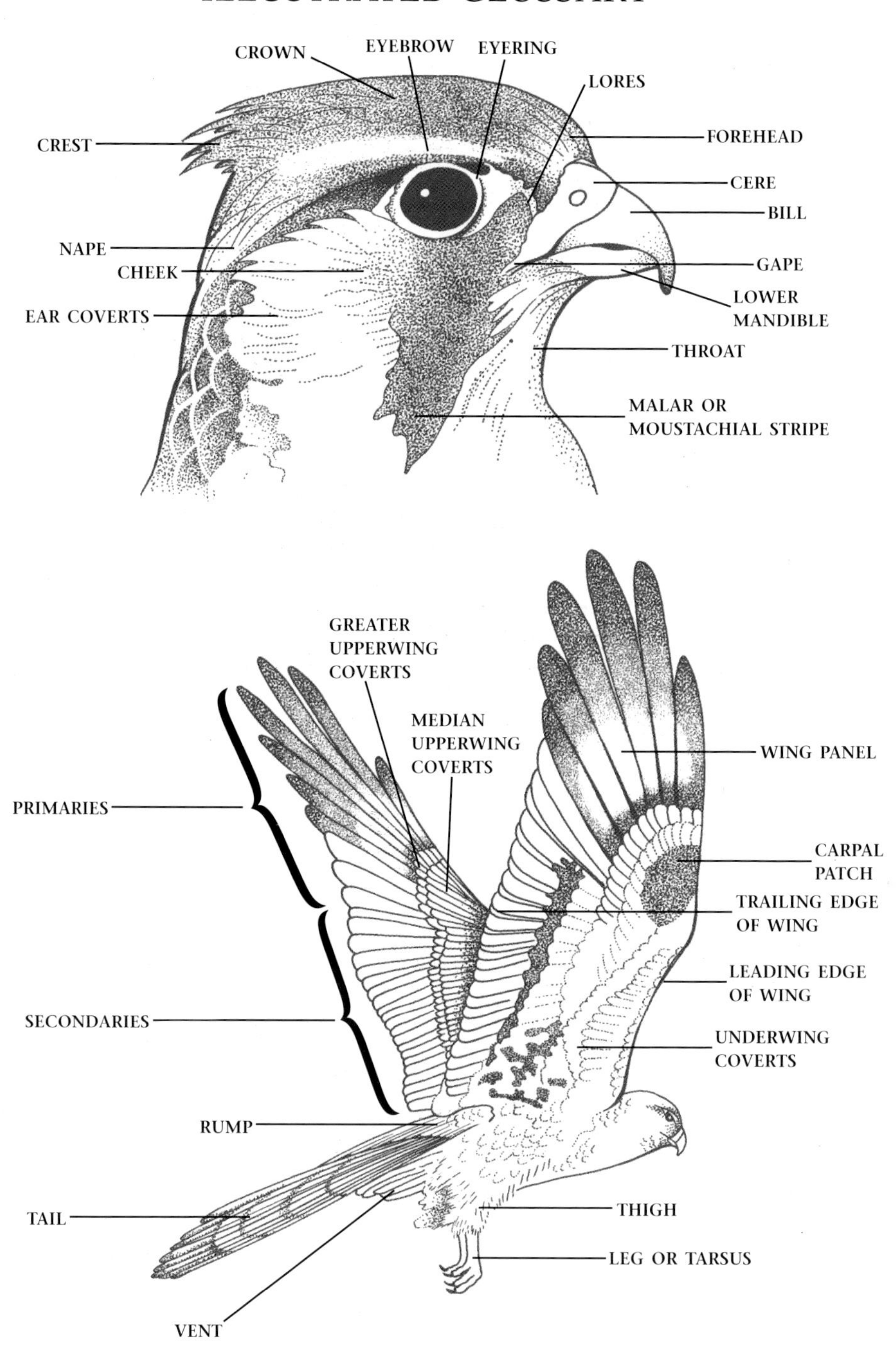